Applied Probability and Statistics

BAILEY · The Elements of Stochastic Processes with Applications to the Natural Sciences

BARTHOLOMEW · Stochastic Models for Social Processes, *Second Edition*

BENNETT and FRANKLIN · Statistical Analysis in Chemistry and the Chemical Industry

BHAT · Elements of Applied Stochastic Processes

BOX and DRAPER · Evolutionary Operation: A Statistical Method for Process Improvement

BROWNLEE · Statistical Theory and Methodology in Science and Engineering, *Second Edition*

CHAKRAVARTI, LAHA, and ROY · Handbook of Methods of Applied Statistics, Vol. II

CHERNOFF and MOSES · Elementary Decision Theory

CHIANG · Introduction to Stochastic Processes in Biostatistics

CLELLAND, deCANI, BROWN, BURSK, and MURRAY · Basic Statistics with Business Applications, *Second Edition*

COCHRAN · Sampling Techniques, *Second Edition*

COCHRAN and COX · Experimental Designs, *Second Edition*

COX · Planning of Experiments

COX and MILLER · The Theory of Stochastic Processes

DANIEL and WOOD · Fitting Equations to Data

DAVID · Order Statistics

DEMING · Sample Design in Business Research

DODGE and ROMIG · Sampling Inspection Tables, *Second Edition*

DRAPER and SMITH · Applied Regression Analysis

DUNN and CLARK · Applied Statistics: Analysis of Variance and Regression

ELANDT-JOHNSON · Probability Models and Statistical Methods in Genetics

FLEISS · Statistical Methods for Rates and Proportions

GOLDBERGER · Econometric Theory

GUTTMAN, WILKS, and HUNTER · Introductory Engineering Statistics, *Second Edition*

HAHN and SHAPIRO · Statistical Models in Engineering

HALD · Statistical Tables and Formulas

HALD · Statistical Theory with Engineering Applications

HOEL · Elementary Statistics, *Third Edition*

HOLLANDER and WOLFE · Nonparametric Statistical Methods

HUANG · Regression and Econometric Methods

JOHNSON and KOTZ · Distributions in Statistics
                                      Discrete Distributions
                                      Continuous Univariate Distributions-1
                                      Continuous Univariate Distributions-2
                                      Continuous Multivariate Distributions

continued on back

Theory of
Probability

Volume 1

Theory of Probability

A critical introductory treatment

Volume 1

BRUNO DE FINETTI

Professor of the Theory of Probability
at the University of Rome

Translated by

ANTONIO MACHÍ

Assistant Professor of Mathematics
at the University of Rome

and

ADRIAN SMITH

Lecturer in Mathematics at the
University of Oxford, and
Fellow of Keble College, Oxford

JOHN WILEY & SONS

London · New York · Sydney · Toronto

First published 1970 © Giulio Einaudi editore s.p.a.,
Torino, under title of *Teoria Delle Probabilità*, by
Bruno de Finetti.

Copyright © 1974, by John Wiley & Sons, Ltd.

Library of Congress Cataloging in Publication Data:

Finetti, Bruno de.
Theory of Probability.

(Wiley series in probability and mathematical statistics)
1. Probabilities. 2. Title.

QA 273.F4913   519.2   73-10744
ISBN 0 471 20141 3

Set on Monophoto Filmsetter and printed in Great Britain
by J. W. Arrowsmith Ltd., Bristol.

This work is dedicated to my colleague Beniamino Segre who about twenty years ago pressed me to write it as a necessary document for clarifying one point of view in its entirety

Foreword

It is an honour to be asked to write a foreword to this book, for I believe that it is a book destined ultimately to be recognized as one of the great books of the world.

The subject of probability is over two hundred years old and for the whole period of its existence there has been dispute about its meaning. At one time these arguments mattered little outside academia, but as the use of probability ideas has spread to so many human activities, and as probabilists have produced more and more sophisticated results, so the arguments have increased in practical importance. Nowhere is this more noticeable than in statistics, where the basic practices of the subject are being revised as a result of disputes about the meaning of probability. When a question has proved to be difficult to answer, one possibility may be that the question itself was wrongly posed and consequently unanswerable. This is de Finetti's way out of the impasse. Probability does not exist.

Does not exist, that is, outside of a person: does not exist, objectively. Probability is a description of your (the reader of these words) uncertainty about the world. So this book is about uncertainty, about a feature of life that is so essential to life that we cannot imagine life without it. This book is about life: about a way of thinking that embraces all human activities.

So, in a sense, this book is for everyone; but necessarily it will be of immediate appeal to restricted classes of readers.

Philosophers have recently increased their interest in probability and will therefore appreciate the challenging ideas that the author puts forward. For example, those of the relationships between possibility and tautology. They will notice the continual concern with reality, with the use of the ideas in practical situations. This is a philosophy intended to be operational and to express the individual's appreciation of the external world.

Psychologists are much concerned with the manner of this appreciation, and experiments have been performed which show that individuals do not reason about uncertainty in the way described in these volumes. The experiments provide a descriptive view of man's attitudes: de Finetti's approach is normative. To spend too much time on description is unwise when a normative approach exists, for it is like asking people's opinion of $2 + 2$, obtaining an average of 4·31 and announcing this to be the sum. It would be better to teach them arithmetic. I hope that this book will divert psychologists' attentions away from descriptions to the important problem, ably discussed in this book, of how to teach people to assess probabilities.

Mathematicians will find much of interest. (Let me hasten to add that some people may approach the book with fear because of the amount of mathematics it contains. They need not worry. Much of the material is accessible with no mathematical skill: yet more needs only a sympathetic appreciation of notation. Even the more mathematical passages use mathematics in a sparse and yet highly efficient way. Mathematics is always the servant—never the master (see Section 1.9.1).) Nevertheless, the mathematician will appreciate the power and elegance of the notation and, in particular, the discussion of finite additivity. He will be challenged by the observation that 'mathematics is an instrument which should conform itself strictly to the exigencies of the field in which it is to be applied'. He will enjoy the new light shed on the calculus of probabilities.

Physicists have long used probabilistic notions in their understanding of the world, especially at the basic, elementary-particle level. Here we have a serious attempt to connect their use of uncertainty with the idea as used outside physics.

Statisticians are the group I can speak about with greatest confidence. They have tended to adopt a view of probability which is based on frequency considerations and is too narrow for many applications. They have therefore been compelled to introduce artificial ideas, like confidence intervals, to describe the uncertainties they need to use. The so-called Bayesian approach has recently made some significant impression, but de Finetti's ideas go further still in replacing frequency concepts entirely—using his notion of exchangeability—and presenting an integrated view of statistics based on a single concept of uncertainty. A consequence of this is that the range of possible applications of statistics is enormously widened so that we can deal with phenomena other than those of a repeatable nature.

There are many other groups of people one would like to see reading these volumes. Operational research workers are continually trying to express ideas to management that involve uncertainty: they should do it using the concepts contained therein. One would like (is it a vain hope?) to see politicians with a sensible approach to uncertainty—what a blessing it would be if they could appreciate the difference between prediction and prevision (p. 98).

The book should therefore be of interest to many people. As the author says (p. 14) 'it is ... an attempt to view, in a unified fashion, a group of topics which are in general considered separately, each by specialists in a single field, paying little or no attention to what is being done in other fields.'

The book is not a text on probability in the ordinary sense and would probably not be useful as a basis for a course of lectures. It would, however, be suitable for a graduate seminar wherein sections of it were discussed and analysed. Which sections were used would depend on the type of graduates, but with the continuing emphasis on unity, it would be valuable in bringing different disciplines together. No university should ignore the book.

It would be presumptious of *me* to say how *you* should read the two volumes but a few words may help your appreciation. Firstly, do not approach it with preconceived ideas about probability. I address this remark particularly to statisticians, who can so easily interpret a formula or a phrase in a way that they have been used to, when de Finetti means something different. Let the author speak for himself. Secondly, the book does not yield to a superficial reading. The author has words of wisdom to say about many things and the wisdom often only appears after reflection. Rather, dip into parts of the book and read those carefully. Hopefully you will be stimulated to read the whole. Thirdly, the style is refreshing—the translators have cleverly used the phrase 'a whimsical fashion' (Section 1.3.3)—so that every now and again delightful ideas spring to view; the idea that we shall all be Bayesian by 2020, or how to play the football pools. But, as I said, this is a book about life.

University College London,
November 1973   D. V. Lindley

Preface

Is it possible that in just a few lines I can achieve what I failed to achieve in my many books and articles? Surely not. Nevertheless, this preface affords me the opportunity, and I shall make the attempt. It may be that misunderstandings which persist in the face of refutations dispersed or scattered over some hundreds of pages can be resolved once and for all if all the arguments are pre-emptively piled up against them.

My thesis, paradoxically, and a little provocatively, but nonetheless genuinely, is simply this:

PROBABILITY DOES NOT EXIST.

The abandonment of superstitious beliefs about the existence of Phlogiston, the Cosmic Ether, Absolute Space and Time, . . . , or Fairies and Witches, was an essential step along the road to scientific thinking. Probability, too, if regarded as something endowed with some kind of objective existence, is no less a misleading misconception, an illusory attempt to exteriorize or materialize our true probabilistic beliefs.

In investigating the reasonableness of our own modes of thought and behaviour under uncertainty, all we require, and all that we are reasonably entitled to, is consistency among these beliefs, and their reasonable relation to any kind of relevant objective data ('relevant' in as much as subjectively deemed to be so). This is Probability Theory. In its mathematical formulation we have the Calculus of Probability, with all its important off-shoots and related theories like Statistics, Decision Theory, Games Theory, Operations Research and so on.

This point of view is not bound up with any particular philosophical position, nor is it incompatible with any such. It is strictly *reductionist* in a methodological sense, in order to avoid becoming embroiled in philosophical controversy.

Probabilistic reasoning—always to be understood as subjective—merely stems from our being uncertain about something. It makes no difference whether the uncertainty relates to an unforseeable future, or to an unnoticed past, or to a past doubtfully reported or forgotten; it may even relate to something more or less knowable (by means of a computation, a logical deduction, etc.) but for which we are not willing or able to make the effort; and so on.

Moreover, probabilistic reasoning is completely unrelated to general philosophical controversies, such as Determinism versus Indeterminism, Realism versus Solipsism—including the question of whether the world 'exists', or is simply the scenery of 'my' solipsistic dream. As far as Determinism and Indeterminism are concerned, we note that, in the context of gas theory or heat diffusion and transmission, whether one interprets the underlying process as being random or strictly deterministic makes no difference to one's probabilistic opinion. A similar situation would arise if one were faced with forecasting the digits in a table of numbers; it makes no difference whether the numbers are random, or are some segment—for example, the 2001st to the 3000th digits—of the decimal expansion of π (which is not 'random' at all, but certain; possibly available in tables and, in principle, computable by you).

The only relevant thing is uncertainty—the extent of our own knowledge and ignorance. The actual fact of whether or not the events considered are in some sense *determined*, or known by other people, and so on, is of no consequence.

The numerous, different, opposed attempts to put forward particular points of view which, in the opinion of their supporters, would endow Probability Theory with a 'nobler' status, or a 'more scientific' character, or 'firmer' philosophical or logical foundations, have only served to generate confusion and obscurity, and to provoke well-known polemics and disagreements—even between supporters of essentially the same framework.

The main points of view that have been put forward are as follows.

The *classical* view, based on physical considerations of symmetry, in which one should be *obliged* to give the same probability to such 'symmetric' cases. But which symmetry? And, in any case, why? The original sentence becomes meaningful if reversed: the symmetry is probabilistically significant, in someone's opinion, if it leads him to assign the same probabilities to such events.

The *logical* view is similar, but much more superficial and irresponsible inasmuch as it is based on similarities or symmetries which no longer derive from the facts and their actual properties, but merely from the sentences which describe them, and from their formal structure or language.

The *frequentist* (or *statistical*) view presupposes that one accepts the classical view, in that it considers an *event* as a class of *individual events*, the latter being 'trials' of the former. The individual events not only have to be 'equally probable', but also 'stochastically independent' ... (these notions when applied to individual events are virtually impossible to define or explain in terms of the frequentist interpretation). In this case, also, it is straightforward, by means of the subjective approach, to obtain, under the appropriate conditions, in a perfectly valid manner, the result aimed at (but unattainable) in the statistical formulation. It suffices to make use of the notion of exchangeability. The result, which acts as a bridge connecting this new approach with the old, has been referred to by the objectivists as 'de Finetti's representation theorem'.

It follows that all the three proposed definitions of 'objective' probability, although useless *per se*, turn out to be useful and good as valid auxiliary devices when included as such in the subjectivistic theory.

The above-mentioned 'representation theorem', together with every other more or less original result in my conception of probability theory, should not be considered as a discovery (in the sense of being the outcome of advanced research). Everything is essentially the fruit of a thorough examination of the subject matter, carried out in an unprejudiced manner, with the aim of rooting out nonsense.

And probably there is nothing new; apart, perhaps, from the systematic and constant concentration on the unity of the whole, avoiding piecemeal tinkering about, which is inconsistent with the whole; this yields, in itself, something new.

Something that may strike the reader as new is the radical nature of certain of my theses, and of the form in which they are presented. This does not stem from any deliberate attempt at radicalism, but is a natural consequence of my abandoning the reverential awe which sometimes survives in people who at one time embraced the objectivistic theories prior to their conversion (which hardly ever leaves them free of some residual).

It would be impossible, even if space permitted, to trace back the possible development of my ideas, and their relationships with more or less similar positions held by other authors, both past and present. A brief survey is better than nothing, however (even though there is an inevitable arbitrariness in the selection of names to be mentioned).

I am convinced that my basic ideas go back to the years of High School as a result of my preference for the British philosophers Locke, Berkeley and, above all, Hume! I do not know to what extent the Italian school textbooks and my own interpretations were valid: I believe that my work based on exchangeability corresponds to Hume's ideas, but some other scholars do not agree. I was also favourably impressed, a few years later, by the ideas of Pragmatism, and the related notions of operational definitions in Physics. I particularly liked the Pragmatism of Giovanni Vailati—who somehow 'Italianized' James and Peirce—and, as for operationalism, I was very much struck by Einstein's relativity of 'simultaneity', and by Mach and (later) Bridgman.

As far as Probability is concerned, the first book I encountered was that of Czuber. (Before 1950—my first visit to the USA—I did not know any English, but only German and French.) For two or three years (before and after the 'Laurea' in Mathematics, and some application of probability to research on Mendelian heredity), I attempted to find valid foundations for all the theories mentioned, and I reached the conclusion that the classical and frequentist theories admitted no sensible foundation, whereas the subjectivistic one was fully justified on a normative–behaviouristic basis. I had some indirect knowledge of De Morgan, and found that some of Keynes' ideas were in partial agreement with mine; some years later I was informed of the similar approach that had been adopted by F. P. Ramsey.

Independent ideas, which were more or less similar, were put forward later by Harold Jeffreys, B. O. Koopman, and I. J. Good (with some beautiful new discussion which illustrated the totally illusory nature of the so-called *objective* definitions of probability). I could add to this list the name of Rudolf Carnap, but this would be not altogether proper in the light of his (to me strange) superposition of the idea of a logical framework onto his own vivid, subjective behaviouristic interpretation. (Richard Jeffreys, in publishing Carnap's posthumous works, seems convinced of his underlying subjectivism.) A singular position is occupied by Robert Schlaifer, who arrived at the subjectivistic approach directly and with impressive freshness and originality, with little knowledge of previous work in the field. A similar thing, although in a different sense, may be said of George Pólya, who discussed *plausible reasoning* in mathematics in the sense of the probability (subjective, of course) of a supposed theorem being true, given the state of mind of the mathematician, and its (Bayesian) modification when new information or ideas appear. The following statement of his is most remarkable: 'It seems to me more philosophical to consider the general idea of *plausible reasoning* instead of its isolated particular cases' like *inductive* (and analogical) *reasoning*. (There have been so many vain attempts to build a theory of induction without beliefs—like a theory of elasticity without matter.)

A very special mention must be reserved, however, for Leonard J. Savage and Dennis V. Lindley, who escaped from the objectivistic school, after having grown up in it, by a gradual discovery of its inconsistencies, and through a comparison of its ambiguities with the clarity of the subjectivistic theory, and the latter's suitability for every kind of practical or theoretical problem. I have often had the opportunity of profitable exchanges of ideas with them, and, in the case of Savage, of actual collaboration, I wrote briefly of Savage's invaluable contributions as a dedication to my book *Probability, Induction and Statistics*, which appeared a few months after his sudden and premature death.

One should note, however, that, even with such close colleagues, agreement ought not to be absolute, on every detail. For example, not all agree with the rejection of countable-additivity.

Finally, having mentioned several of the authors who are more or less connected with the subjectivistic (and Bayesian) point of view, I feel an obligation to recall three great men—the first two, unfortunately, no longer with us—who, although they all shared an opposed view about our common subject, were always willing to discuss, and were extraordinarily friendly and helpful on every occasion. I refer to Guido Castelnuovo, Maurice Fréchet and Jerzy Neyman.

Rome, 16 July 1973   Bruno de Finetti

Translators' Preface

In preparing this English translation, we were concerned to achieve two things: first of all, and most importantly, to translate as accurately as possible the closely argued *content* of the book; secondly, to convey something of the flavour of the author's idiosyncratic *style*; the sense of the painstaking struggle for understanding that runs through the Italian original.

Certain of Professor de Finetti's works have already appeared in English, the principal references being Kyburg and Smokler's *Studies in Subjective Probability* (Wiley, 1964), and the author's *Probability, Induction and Statistics* (Wiley, 1972). For the purpose of comparison—and to avoid any possible confusion—we include the following preliminary notes on the terminological and notational usage that we have adopted.

In common with the above-mentioned translations, we use the word *coherent* when referring to degrees of belief which satisfy certain 'consistency' conditions, *random quantity* in place of the more usual 'random variable', and *exchangeable*, rather than 'equivalent' or 'symmetric'.

We part company with previous translations, however, in our treatment of the concept corresponding to what is usually called 'mathematical expectation'. In Kyburg's translation of de Finetti's monograph 'La Prévision: ses lois logiques, ses sources subjectives' (see Kyburg and Smokler, pp. 93–158), the corresponding word becomes 'foresight'. We shall use the word *prevision*. A discussion of the reasons for this choice is given more fully at the appropriate place in the text (Chapter 1, 10.3) but let us note straightaway that the symbol **P** now very conveniently represents both *probability* and *prevision*, and greatly facilitates their unified treatment as linear operators.

Readers who are familiar with the Italian original will realize that on occasions we have opted for a rather free style of translation; we did so, in fact, whenever we felt this to be the best way of achieving our stated aims. Throughout, however, we have been mindful of the 'misunderstandings' referred to by the author in his Preface, and we can but hope that our translation does nothing to add to these.

Finally, we should like to express our gratitude to Professor de Finetti, who read through our translation and made many helpful suggestions; to the editor at John Wiley & Sons for getting the project under way; and to Mrs Jennifer Etheridge for her care in typing our manuscript.

A. Machí
A. F. M. Smith

Contents

CHAPTER 1

Introduction

1.1 WHY A NEW BOOK ON PROBABILITY?

There exist numerous treatments of this topic, many of which are very good, and others continue to appear. To add one more would certainly be a presumptuous undertaking if I thought in terms of doing something better, and a useless undertaking if I were to content myself with producing something similar to the 'standard' type. Instead, the purpose is a different one: it is that already essentially contained in the dedication to Beniamino Segre

[who about twenty years ago pressed me to write it as a necessary document for clarifying one point of view in its entirety.]

Segre was with me at the International Congress of the Philosophy of Science (Paris 1949), and it was on the occasion of the discussions developed there on the theme of probability that he expressed to me, in persuasive and peremptory terms, a truth, perhaps obvious, but which only since appeared to me as an obligation, difficult but unavoidable.

'Only a complete treatment, inspired by a well-defined point of view and collecting together the different objections and innovations, showing how the whole theory results in coherence in all of its parts, can turn out to be convincing. Only in this way is it possible to avoid the criticisms to which fragmentary expositions easily give rise since, to a person who in looking for a completed theory interprets them within the framework of a different point of view, they can seem to lead unavoidably to contradictions.'

These are Segre's words, or, at least, the gist of them.

It follows that the requirements of the present treatment are twofold: first of all to clarify, exhaustively, the conceptual premises, and then to give an essentially complete exposition of the calculus of probability and its applications in order to establish the adequacy of the interpretations deriving from those premises. In saying 'essentially' complete, I mean that what matters is to develop each topic just as far as is necessary to avoid conceptual misunderstandings. From then on, the reader could follow any other book without finding great difficulty in making those modifications which are needed in order to translate it, if such be desired, according to the point of view which will be taken here. Apart from these conceptual exigencies, each topic will also be developed, in terms of the content, to an extent sufficient for the treatment to turn out to be adequate for the needs of the average reader.

1.2   WHAT ARE THE MATHEMATICAL DIFFERENCES?

1.2.1. If I thought I were writing for readers absolutely innocent of probabilistic–statistical concepts, I could present, with no difficulty, the theory of probability in the way I judge to be meaningful. In such a case it would not even have been necessary to say that the treatment contains something new and, except possibly under the heading of information, that different points of view exist. The actual situation is very different, however, and we cannot expect any sudden change.

> My estimation is that another fifty years will be needed to overcome the present situation, but perhaps even this is too optimistic. It is based on the consideration that about thirty years were required for ideas born in Europe (Ramsey, 1926; de Finetti, 1931) to begin to take root in America (even though B. O. Koopman, 1940, had come to them in a similar form). Supposing that the same amount of time might be required for them to establish themselves there, and then the same amount of time to return, we arrive at the year 2020.

It would obviously be impossible and absurd to discuss in advance concepts and, even worse, differences between concepts to whose clarification we will be devoting all of what follows: however, much less might be useful (and, anyway, will have to suffice for the time being). It will be sufficient to make certain summary remarks which are intended to exemplify, explain and anticipate for the reader certain differences in attitude which could disorientate him, and leave him undecided between continuing without understanding or, on the other hand, stopping reading altogether. It will be necessary to show that the 'wherefore' exists and to give at least an idea of the 'wherefore', and of the 'wherefores', even without anticipating the 'wherefore' of every single case (which can only be seen and gone into in depth at the appropriate time and place).

1.2.2. From a mathematical point of view, it will certainly seem to the reader that either by desire or through ineptitude I complicate simple things; introducing captious objections concerning aspects that modern developments in mathematical analysis have definitively dealt with. Why do I myself not also conform to the introduction of such developments into the calculus of probability? Is it a question of incomprehension? Of misoneism? Of affectation in preferring to use the tools of the craftsman in an era of automation which allows mass production even of brains—both electronic and human?

The 'wherefore', as I see it, is a different one. To me, mathematics is an instrument which should conform itself strictly to the exigencies of the field in which it is to be applied. One cannot impose, for his own convenience, axioms not required for essential reasons, or actually in conflict with them.

I do not think that it is appropriate to speak of 'incomprehension'. I have followed through, and appreciated, the reasons *pro* (which are the ones usually put forward), but I found the reasons *contra* (which are usually neglected) more valid, and even preclusive.

I do not think that one can talk of misoneism. I am, in fact, very much in favour of innovation and against any form of conservatism (but only after due consideration, and not by submission to the tyrannical caprice of fashion). Fashion has its use in that it continuously throws up novelties, guarding against fossilization; in view of such a function it is wise to tolerate with goodwill even those things we do not like. It is not wise, however, to submit to passively adapting our own taste, or accepting its validity beyond the limits which correspond to our own dutiful, critical examination.

I do not think that one can talk of 'affectation' either. If anything, the type of 'affectation' which is congenial to my taste would consist in making everything simple, intuitive and informal. Thus, when I raise 'subtle' questions, it means that, in my opinion, one simply cannot avoid doing so.

1.2.3. The 'wherefore' of the choice of mathematical apparatus, which the reader might find irksome, resides, therefore, in the 'wherefores' related to the specific meaning of probability, and of the theory that makes it an object of study. Such 'wherefores' depend, in part, on the adoption of this or that particular point of view with regard to the concept and meaning of probability, and to the basis from which derives the possibility of reasoning about it, and of translating such reasonings into calculations. Many of the 'wherefores' seem to me, however, also to be valid for all, or many, of the different concepts (perhaps with different force and different explanations). In any case, the critical analysis is more specifically hinged on the conception which we follow here, and which will appear more and more clear (and, hopefully, natural) as the reader proceeds to the end—provided he has the patience to do so.

1.3 WHAT ARE THE CONCEPTUAL DIFFERENCES?

1.3.1. Meanwhile, for those who are not aware of it, it is necessary to mention that in the conception we follow and sustain here only *subjective* probabilities exist—i.e. the *degree of belief* in the occurrence of an event attributed by a given person at a given instant and with a given set of information. This is in contrast to other conceptions which limit themselves to special types of cases in which they attribute meaning to 'objective probabilities' (for instance, cases of symmetry as for dice etc., 'statistical' cases of 'repeatable' events, etc.). This said, it is necessary to add at once that we have no interest, at least for now, either in a discussion, or in taking up a position, about the 'philosophical' aspects of the dispute: in fact, it would be premature and prejudicial because it would entangle the examination of each concrete point in a web of metaphysical misunderstandings.

Instead, we are interested, on the contrary, in clearly understanding what one means *according to one's own conception and in one's own language*, and learning to enter into this conception and language in its motivations and implications (even if provisionally, in order to be able to make pertinent criticism later on). This is, it seems to me, an inviolable methodological need.

1.3.2. There is nothing more disappointing than to hear repeated, presented as 'criticisms', clichés so superficial that it is not possible to infer whether the speaker has even read the arguments developed to confute them and clear them up, or has read them without understanding anything, or else has understood them back to front. The fault could be that of obscure presentation, but a somewhat more meaningful reaction would be required in order to be able to specify accurately, and to correct, those points which lend themselves to misunderstanding.

The fault may be the incompleteness of the preceding, more or less fragmentary, expositions, which, although probably more than complete if taken altogether, are difficult to locate and hold in view simultaneously. If so, the present work should obviate the inconvenience: unfortunately, the fact that it is published is not sufficient; the result depends on the fact that it is read with enough care to enable the reader to make pertinent criticisms.

I would like to add that I understand very well the difficulties that those who have been brought up on the objectivistic conceptions meet in escaping from them. I understand it because I myself was perplexed for quite a while some time ago (even though I was free from the worst impediment, never having had occasion to submit to a ready made and presented point of view, but only coming across a number of them while studying various books and works on my own behalf). It was only after having analysed and mulled over the objectivistic conceptions in all possible ways that I arrived, instead, at the firm conviction that they were all irredeemably illusory. It was only after having gone over the finer details, and developed, to an extent, the subjectivistic conception, assuring myself that it accounted (in fact, in a perfect and more natural way) for everything that is usually accredited, over-hastily, to the fruit of the objectivistic conception, it was only after this difficult and deep work, that I convinced myself, and everything became clear to me. It is certainly possible that these conclusions are wrong: in any case they are undoubtedly open to discussion, and I would appreciate it if they were discussed.

However, a dialogue between the deaf is not a discussion. I think that I am doing my best to understand the arguments of others, and to answer them with care (and even with patience when it is a question of repeating things over and over again to refute trivial misunderstandings). It is seldom that I have the pleasure of forming the impression that other people make a similar effort; but, as the Gospel says, 'And why beholdest thou the mote that is in thy brother's eye, but considerest not the beam that is in thine own eye?': if this has happened to me, or is happening to me, I would appreciate it if someone would enlighten me.

1.3.3. One more word (hopefully unnecessary for those who know me): I find it much more enlightening, persuasive, and in the end more essentially serious, to reason by means of paradoxes; to reduce a thesis to absurdity; to make use of images, even light-hearted ones provided they are relevant, rather than to be limited to lifeless manipulations in technical terms, or to heavy and indigestible technical language. It is for this reason that I very much favour the use of colourful and vivid forms of expression, which, hopefully, may turn out to be effective and a little entertaining, making concrete, in a whimsical fashion, those things which would appear dull, boring or insipid, and therefore inevitably badly understood, if formulated in an abstract way, stiffly or with affected gravity. It is for this reason that I write in such a fashion, and desire to do so; not because of ill-will or lack of respect for other people, or their opinions (even when I judge them wrong). If somebody finds this or that sentence a little too sharp, I beg him to believe in the total absence of intention and animosity, and to accept my apologies as of now.

1.4 PRELIMINARY CLARIFICATIONS

1.4.1. For the purpose of understanding, the important thing is not the difference in philosophical position on the subject of probability between 'objective' and 'subjective', but rather the resulting reversals of the rôles and meanings of many concepts, and, above all, of what is 'rigorous', both logically and mathematically. It might seem paradoxical, but the fact is that the subjectivistic conception distinguishes itself precisely by a more rigorous respect for that which is really objective, and which it calls, therefore, 'objective'.† There are cases in which, in order to define a notion, in formulating

---

† This fact has often been underlined by L. J. Savage (cf. Kyburg and Smokler, p. 178, and elsewhere).

the problem, or in justifying the reasoning, there exists a choice between an unexceptionable, subjectivistic interpretation and a would-be objectivistic interpretation. The former is made in terms of the opinions or attitudes of a given person; the latter derives from a confused transposition from this opinion to the undefinable complex of objective circumstances which might have contributed to its determination: in such cases there is nothing to do but choose the first alternative. The subjective opinion, as something known by the individual under consideration, is, at least in this sense, something objective and can be a reasonable object of a rigorous study. It is certainly not a sign of greater realism, of greater respect for objectivity, to substitute for it a metaphysical chimera, even if with the laudable intention of calling it 'objective' in order to be able to then claim to be concerned only with objective things.

It might be objected that we are in a vicious circle, or engaged in a vacuous discussion, since we have not specified what is to be understood by 'objective'. This objection is readily met, however: statements have *objective* meaning if one can say, on the basis of a well-determined observation (which is at least conceptually possible), whether they are either TRUE or FALSE. Within a greater or lesser range of this delimitation a large margin of variation can be tolerated; with one condition—do not *cheat*. To *cheat* means to leave in the statement sufficient confusion and vagueness to allow ambiguity, second-thoughts and equivocations in the ascertainment of its being TRUE or FALSE. This, instead, must always appear simple, neat and definitive.

1.4.2. Statements of this nature, i.e. the only 'statements' in the true sense of the word, are the object of the *logic of certainty*, i.e. ordinary logic, which could also be in the form of mathematical logic, or of mathematics. They are also the objects *to which* judgements of probability apply (as long as one does not know whether they are true or false) and are called either *propositions*, if one is thinking more in terms of the expressions in which they are formulated, or *events*, if one is thinking more in terms of the situations and circumstances to which their being true or false corresponds.

On the basis of the considerations now developed, one can better understand the statement made previously, according to which the fundamental difference between the subjectivistic conception and the objectivistic ones is not philosophical but *methodological*. It seems to me that no-one could refute the methodological rigour of the subjectivistic conception: not even an objectivist. He himself, in fact, would have unlimited need of it in trying to expose, in a sensible way, the reasons that would lead him to consider 'philosophically correct' this one, or that one, among the infinitely many possible opinions about the evaluations of probability. To argue against this can only mean, even though without realizing it, to perpetuate profitless discussions and to play on the ambiguities which are deeply rooted in the uncertainty.

At this stage, a few simple examples might give some preliminary clarification of the meaning and compass of the claimed 'methodological rigour'; under the condition, however, that one takes into account the necessarily summary character of these preliminary observations. It is necessary to pay attention to this latter remark to avoid both the acceptance of such observations as exhaustive, and the criticism of them that results from assuming that they claim to be exhaustive: one should realize, with good reason, that they are by no means such.

1.5   SOME IMPLICATIONS TO NOTE

1.5.1.   We proceed to give some examples: to save space, let us denote by O statements often made by *objectivists*, and by S those with which a *subjectivist* (or, anyway, *this author*) would reply.

O : Two events of the same type in identical conditions for all the relevant circumstances are 'identical' and, therefore, necessarily have the same probability.†

S : Two distinct events are always different, by virtue of an infinite number of circumstances (otherwise how would it be possible to distinguish them?!). They are equally probable (for an individual) if—and so far as—he judges them as such (possibly by judging the differences to be irrelevant in the sense that they do not influence his judgement).

Remark. An even more fundamental objection should be added: the judgement about the probability of an event depends not only on the event (or on the person), but also on the state of information. This is occasionally recalled, but more often forgotten by many objectivists.

O : Two events are (stochastically) independent‡ if the occurrence of one does not influence the probability of the other.

S : I would say instead: by definition, two events are such (for an individual) if the knowledge of the outcome of one does not make him change the evaluation of probability for the other.

---

† The objectivists often use the word event in a generic sense also, using '*trials*' (or '*repetitions*') *of the same* '*event*' to mean *single events*, '*identical*' *or* '*similar*'. From time to time we will say '*trials*' (or '*repetitions*') *of a phenomenon*, always meaning by event a single event. It is not simply a question of terminology, however: we use 'phenomenon' because we do not give this word any technical meaning; by saying 'trials of a phenomenon' one may allude to some exterior analogy, but one does not mean to assume anything which would imply either equal probability, or independence, or anything else of probabilistic relevance.

‡ Among events, random quantities, or random entities in general, it is possible to have various relations termed 'independence' (linear, logical, stochastic): it is better to be specific if there is any risk of ambiguity.

O: Let us suppose *by hypothesis* that these events are equally probable, e.g. with probability $p = \frac{1}{2}$, and independent, etc.

S: It is meaningless to consider as an 'hypothesis' something which is not an objective statement. A statement about probability (the one given in the example or any other one whatsoever) either *is* the evaluation of probabilities (those of the speaker or of someone else), in which case there is nothing to do but simply register the fact, or it is nothing.

O: These events are independent and all have the same probability which is, however, 'unknown'.

S: This formulation is a nonsense in the same sense as the preceding one but to a greater extent. By interpreting the underlying intention (which, as an intention, is reasonable) one can translate it (cf. Chapter 11) into a completely different formulation, 'exchangeability', in which we do *not* have independence, the probabilities are *known*, and vary, precisely, in depending only on the number of successes and failures of which one has information.

One might continue in this fashion, and it could be said that almost the whole of what follows will be, more or less implicitly, a continuation of this same discussion. Rather let us see, by gathering together the common factors, the essential element in all these contrapositions.

1.5.2. For the subjectivist everything is clear and rigorous when he is expressing something about somebody's evaluation of probabilities; an evaluation which is, simply, what it is. For that somebody, it will have motivations that we might, or might not, know; share, or not share; judge† more or less reasonable, and that might be more or less 'close' to those of a few, or many, or all people. All this can be interesting, but it does not alter anything. To express this in a better way: all these things matter in so far as they determined that unique thing that matters, and that is the evaluation of probability to which, in the end, they have given rise.

From the theoretical, mathematical point of view, even the fact that the evaluation of probability expresses somebody's opinion is then irrelevant. It is purely a question of studying it and saying whether it is coherent or not; i.e. whether it is free of, or affected by, intrinsic contradictions. In the same way, in the logic of certainty one ascertains the correctness of the deductions but not the accuracy of the factual data assumed as premisses.

1.5.3. Instead, the objectivist would like to ignore the evaluations, actual or hypothetical, and go back to the circumstances that might serve as a basis for motivations which would lead to evaluations. Not being able to

---

† With a judgment which is 'subjective squared': our subjective judgment regarding the subjective judgment of others.

invent methods of synthesis comparable in power and insight to those of the human intuition, nor to construct miraculous robots capable of such, he contents himself, willingly, with simplistic schematizations of very simple cases based on neglecting all knowledge except a unique element which lends itself to utilization in the crudest way.

A further consequence is the following. The subjectivist, who knows how much caution is necessary in order to remain within the bounds of realism, will exercise great care in not going far beyond the consideration of cases immediately at hand and directly interesting. The objectivist, who substitutes the abstraction of schematized models for the changing and transient reality, cannot resist the opposite temptation. Instead of engaging himself, even though in a probabilistic sense (the only one which is valid), in saying something about the specific case of interest, he prefers to 'race on ahead', occupying himself with the asymptotic problems of a large number of cases, or even playing around with illusory problems, contemplating infinite cases where he can try, without any risk, to pass off his results as 'certain predictions'.†

1.6 IMPLICATIONS FOR THE MATHEMATICAL FORMULATION

1.6.1. From these conceptual contrapositions there follows, amongst other things, an analogous contraposition in the way in which the mathematical formulation is conceived. The subjectivistic way is the one which it seems appropriate to call 'natural': it is possible to evaluate the probability over any set of events whatsoever; those for which it serves a purpose, or is of interest, to evaluate it; there is nothing further to be said. The objectivistic way (and also the way most congenial to contemporary mathematicians, independently of the conception adopted regarding probability) consists in requiring, as an obligatory starting point, a mathematical structure much more formidable, complete and complicated than necessary (and than it is, in general, reasonable to regard as conceivable).

1.6.2. Concerning a known evaluation of probability, over any set of events whatsoever, and interpretable as the opinion of an individual, real or hypothetical, we can only judge whether, or not, it is *coherent*.‡ If it is not, the evaluator, when made aware of it, should modify it in order to make it coherent. In the same way, if someone claimed to have measured the sides and area of a rectangle and found 3 metres, 5 metres and 12 square-metres,

---

† Concerning the different senses in which we use the terms 'prevision' and 'prediction', cf. Chapter 3 (at the beginning and then in various places, in particular 3.7.3).
‡ Cf. Chapter 3.

we, even without being entitled, or having the inclination, to enter into the merits of the question, or to discuss the individual measurements, would draw his attention to the fact that at least one of them is wrong, since it is not true that $3 \times 5 = 12$.

Such a condition of coherence should, therefore, be *the weakest one* if we want it to be the strongest in terms of absolute validity. In fact, *it must only exclude the absolutely inadmissible evaluations*; i.e. those that one cannot help but judge contradictory (in a sense that we shall see later).

Such a condition, as we shall see, reduces to *finite additivity* (and *non-negativity*). It is not admissible to make it more restrictive (unless it turns out to be necessary if we discover the preceding statement to be wrong); it would make us exclude, erroneously, admissible evaluations.

1.6.3. What the objectivistic, or the purely formalistic, conceptions generally postulate is, instead, that countable additivity holds (as for Borel or Lebesgue measure), and that the field over which the probability is defined be the whole of a Boolean algebra. From the subjectivistic point of view this is both too much and too little: according to what serves the purpose and is of interest, one could limit oneself to much less, or even go further. One could attribute probabilities, finitely but not countably additive, to all, and only, those events which it is convenient to admit into the formulation of a problem and into the arguments required for its solution. One might also go from one extreme to the other: referring to the analogy of events and probability with sets and measure, it might, at times, be convenient to limit oneself to thinking of a measure as defined on certain simple sets (like the intervals), or even on certain sets but not their intersections (for instance, for 'vertical' and 'horizontal' 'stripes' in the (x, y)-plane ($x' \leqslant x < x''$, $y' \leqslant y < y''$), but not on the rectangles); and, at other times, to think of it instead as extended to all the sets that the above-mentioned convention would exclude (like the 'non-Lebesgue-measureable sets').

1.6.4. In a more general sense, it seems that many of the current conceptions consider as a success the introduction of mathematical methods so powerful, or of tricks of formulation so slick, that they permit the derivation of a uniquely determined answer to a problem even when, due to the insufficiency of the data, it is indeterminate. A capable geometer in order to conform to this aspiration would have to invent a formula for calculating the area of a triangle given two sides.

Attempts of this kind are to be found in abundance, mainly in the field of statistical induction (cf. some remarks further on in this Introduction, 1.7.6).

In the present case, the defect is somewhat hidden and consists in the following distinction between the two cases of *measure* and of *probability*.

To extend a *mathematical* notion (measure) from one field (Jordan–Peano) to another (Borel–Lebesgue) is a question of convention. If, however, a notion (like probability) *already has a meaning* (for each event, at least potentially, even if not already evaluated) one cannot give it a value by conventional extension of the probabilities already evaluated *except for the case in which it turns out to be the unique one compatible with them by virtue of the sole conditions of coherence* (conditions pertaining to the meaning of probability, not to motives of a mathematical nature). The same would happen if it were a question of a physical quantity like mass. If one thought of being able to give meaning to the notion of 'mass belonging to any set of points of a body' (for instance those with rational coordinates), in the sense that it were, at least conceptually, possible to isolate such a mass and weigh it, then it would be legitimate, when referring to it, to talk about everything that can be deduced about it by mathematical properties which translate necessary physical properties, and only such things. To say something more (and in particular to give it a unique value when such properties leave the value indeterminate between certain limits), by means of the introduction of arbitrary mathematical conventions, would be unjustified, and therefore inadmissible.

1.7 AN OUTLINE OF THE 'INTRODUCTORY TREATMENT'

1.7.1. The reader must feel as though he has been plunged alternately into baths of hot and cold water: in Section 1.5 he encountered the contraposed examples of the conceptual formulation, presented either as meaningful or as meaningless; in Section 1.6 the mathematical formulations, presented either as suitable or as academic. Following this, a simple and ordered presentation of the topics which will follow may provide a suitable relaxation, and might even induce a return to the preceding 'baths' in order, with a greater knowledge of the motives, to soak up some further meaning.

1.7.2. In Chapter 2 we will *not* talk of probability. Since we wish to make absolutely clear the distinction between the subjective character of the notion of probability and the objective character of the elements (events, or any random entities whatsoever) to which it refers, we will first treat only these entities. In other words, we will deal with the preliminary logic of certainty where there exist only:

TRUE and FALSE as final answers;

CERTAIN and IMPOSSIBLE and POSSIBLE as alternatives, with respect to the present knowledge of each individual.

In this way, the range of uncertainty, i.e. of what is not known, will emerge in outline. This is the framework into which the (subjective) notion of probability will be introduced as an indispensable tool for our orientation and decision-making.

The random events, random quantities, and any other random entities, will already be defined, however, before we enter the domain of probability, and they will simply be events, quantities, entities, well-defined but with no particular features except the fact of not being known by a certain individual.

For any individual who does not know the value of a quantity X, there will be, instead of a unique *certain* value, two, or several, or infinitely many, *possible* values of X. They depend on his degree of ignorance, and are therefore relative to his state of information; nevertheless they are objective because they do not depend on his opinions but only on these objective circumstances.

1.7.3. Up until now the consideration of uncertainty has been limited to the negative aspect of *non-knowledge*. In Chapter 3 we will see how the need arises, as natural and appropriate, to integrate this aspect with the positive aspect (albeit weak and temporary while awaiting the information which would give it certainty) given by the evaluation of probabilities. To any event in which we have an interest, we are accustomed to attributing, perhaps vaguely and unconsciously, a probability: if we are sufficiently interested we may try to evaluate it with some care. This implies introspection in depth by weighing each element of judgment and controlling the coherence by means of other evaluations made with equal accuracy. In this way, each event can be assigned a probability, and each random quantity or entity a distribution of probability, as an expression of the attitude of the individual under consideration.

Let us note at once a few of the points that arise.

Others, in speaking of a random quantity, assume a probability distribution as already attached to it. To adopt a different concept is not only a consequence of the subjectivistic formulation, according to which the distribution can vary from person to person, but also of the unavoidable fact that the distribution varies with the information (a fact which, in any case, makes the usual terminology inappropriate).

Another thing that might usefully be mentioned now is that the conditions of coherence will turn out to be particularly simplified and clarified by means of a simple device for simultaneously handling events and random quantities (or entities of any linear space whatever). Putting the logical values 'True' and 'False' equal to the numbers 1 and 0, an event is a random quantity which can assume these two values: the function $\mathbf{P}(X)$, which for X = event gives its probability, is, for arbitrary X, the 'prevision' of X (i.e. in the usual terminology, the mathematical expectation).

The use of this *arithmetic* interpretation of the events, preferable to, but not excluding, the set-theoretic interpretation, has its utility and motivation, as will be seen. The essential fact is that the *linearity* of the arithmetic interpretation plays a fundamental rôle (which is, in general, kept in the background), whereas the structure of the Boolean algebra enters rather indirectly.

1.7.4. After having extended these considerations, in Chapter 4, to the case of conditional probabilities and previsions (encountering the notions of stochastic independence and correlation), we will, in Chapter 5, dwell upon the evaluation of probabilities. The notions previously established will allow us not only to apply the instruments for this evaluation, but also to relate them to the usual criteria, inspired by partial, objectivistic 'definitions'. We will see that the subjectivistic formulation, far from making the valid elements in the ideas underlying these criteria redundant, allows the best and most complete use of them, checking and adapting, case by case, the importance of each of them. In contrast to the usual, and rather crude, procedure, which consists in the mechanical and one-sided application of this or that criterion, the proposed formulation allows one to behave in conformity with what the miraculous robot, evoked in 1.5.3, would do.

1.7.5. Chapters 6–10 extend to give a panoramic vision of the field of problems with which the calculus of probabilities is concerned. Of course, it is a question of compromising between the desire to present a relatively complete overall view, and the desire to concentrate attention on a small number of concepts, problems and methods, whose rôle is fundamental both in the first group of ideas, to be given straightaway, and, even more, in further developments, which, here, we can at most give a glimpse of.

Also in these chapters, in themselves more concerned with content than critical appraisal, there are aspects, and, here and there, observations and digressions, which are relevant from the conceptual angle, It would be inappropriate to make detailed mention of them, but, as examples, we could quote the more careful analysis of what the knowledge of the distribution function says, or does not say (also in connection with the 'possible' values), and of the meaning of 'stochastic independence' (between random quantities), expressed by means of the distribution function.

1.7.6. The last two chapters, 11 and 12, deal briefly with the problems of induction (or inference) and their applications, which constitute mathematical statistics. Here we encounter anew the conceptual questions connected with the subjective conception, which, of course, bases all inference on the Bayesian procedure (from Thomas Bayes,† 1763). In this way, the theory and the applications come to have a unified and coherent foundation: it is simply a question of starting from the evaluation of the initial probabilities (i.e. before acquiring new information—by observation, experiment, or whatever) and then bringing them up to date on the basis of this new information, thus obtaining the final probabilities (i.e. those on which to base oneself after acquiring such information).

---

† One must be careful not to confuse Bayes' theorem (which is a simple corollary of the theorem of compound probabilities) with Bayes' *postulate* (which assumes the uniform distribution as a representation of 'knowing nothing'). Criticisms of the latter, often mistakenly directed against the former, are not therefore valid as criticisms of the position adopted here.

The objectivistic theories, in seeking to eschew the evaluation and use of 'initial probabilities', lack an indispensable element for proceeding in a sensible way, and appeal to a variety of empirical methods, often invented *ad hoc* for particular cases. We shall use the term '*Adhockeries*', following Good† (1965) who coined this apt expression, for the methods, criteria and procedures, which, instead of following the path of the logical formulation, try to answer particular problems by means of particular tricks (which are sometimes rather contrived).

1.8   A FEW WORDS ABOUT THE 'CRITICAL' APPENDIX

1.8.1. Many of the conceptual questions are, unfortunately, inexhaustible if one wishes to examine them thoroughly; and the worst thing is that, often, they are also rather boring unless one has a special interest in them.

A work which is intended to clarify a particular conceptual point of view cannot do without this kind of analysis in depth, but it certainly seems appropriate to avoid weighing down the text more than is necessary to meet the needs of an ordinary reader who desires to arrive at an overall view. For this reason, the most systematic and detailed critical considerations have been postponed to an Appendix. This is intended as a reassurance that there is no obligation to read it in order to understand what follows, nor to make the conclusions meaningful. This does not mean, however, that it is a question of abstruse and sophisticated matters being set aside for a few specialists and not to be read by others. It is a question of further consideration of different points which might appear interesting and difficult, to a greater or lesser extent, but which might always improve, in a meaningful and useful, though not indispensable, way, the awareness of certain questions and difficulties, and of the motives which inspire different attitudes towards them.

1.8.2. In any case, one should point out that it is a question of an attempt to view, in a unified fashion, a group of topics which are in general considered separately, each by specialists in a single field, paying little or no attention to what is being done in other fields. Notwithstanding the many gaps or uncertainties, and the many imperfections (and maybe precisely also for the attention it may attract to them), I think that such an attempt should turn out to be useful.

---

† Good's position is less radical than I supposed when I interpreted 'Adhockery' as having a derogatory connotation. I gathered this from his talk at the Salzburg Colloquium, and commented to this effect in an Addendum to the paper I delivered there; *Synthese* **20** (1969), 2–16: 'According to it, "adhockeries" ought not to be rejected outright; their use may sometimes be an acceptable substitute for a more systematic approach. I can agree with this only if—and in so far as—such a method is justifiable as an approximate version of the correct (i.e. Bayesian) approach. (Then it is no longer a mere "adhockery".)'

Among other things, we have tried to insert into the framework of the difficulties associated with the 'verifiability' of events in general, the question of 'complementarity' that arose in quantum physics. The answer is the one already indicated, in a summary fashion, elsewhere (de Finetti, 1959), and coinciding with that of B. O. Koopman (1957), but the analysis has been pursued in depth, and related to the points of view of other authors as far as possible (given the margin of uncertainty in the interpretation of the thought of those consulted, and the impossibility of spending more time on this topic in attempting to become familiar with others).

1.8.3. Various other questions, which will be discussed extensively in the Appendix, are currently objects of discussion in various places: for instance, the relationships between possibility and tautology seem to be attracting the attention of philosophers (the intervention of Hacking at a recent meeting, Chicago 1967); while the critical questions about the mathematical axioms of the calculus of probability (in the sense, to be understood, of making it a theory strictly identical to measure theory, or with appropriate variations) are always a subject of debate.

Apart from the points of view on separate questions, the Appendix will also have as a main motive the proposal to model the mathematical formulation on the analysis of the actual needs of the substantive interpretation. Moreover, to do so with the greatest respect for 'realism', which the inevitable degree of idealization must purify just a little, but must never overwhelm or distort, neither for analytical convenience, nor for any other reason.

1.9 OTHER REMARKS

1.9.1. It seems appropriate here to draw attention also to some further aspects, all secondary, even if only to underline the importance that attaches, in my opinion, to 'secondary' things.

One characteristic of the calculus of probability is that mathematical results are often automatically obtained because their probabilistic interpretations are obvious. In all these cases I think it is much more effective and instructive to consider as their proofs these latter expressive interpretations, and as formal verifications their translation into technical details (to be omitted, or left to the reader). This seems to me to be the best way of realizing the ideal expressed in the maxim that Chisini† often repeated: '*mathematics is the art which teaches one how not to make calculations*'.

---

† Oscar Chisini, a distinguished and gifted pupil of Federigo Enriques, was Professor at the University of Milan where the author attended his course on Advanced Geometry. Chisini's generalized definition of the concept of mean (see Chapter 2, Section 2.9) came about as a result of his occasionally being concerned with this notion in connection with secondary-school examinations.

It is incredible how many things are regularly presented in a heavy and obscure fashion, arriving at the result through a labyrinth of calculations which make one lose sight of the meaning, whereas simple, synthetic considerations would be sufficient to reveal that, for those not wishing to behave as if handcuffed or blindfolded, results and meaning are at hand, staring one in the face.

On numerous occasions one sees very long calculations made in order to prove results which are either wrong or obvious. The latter case is the more serious, without any extenuating circumstances, since it implies lack of realization that the conclusion was obvious, even after having seen it. On the other hand, failing to get the result due to a casual mistake merits only half a reproach since the lack of realization only applies before starting the calculations.

Instead, it is often sufficient to remark that two formulae are necessarily identical for the simple reason that they express the same thing in different ways, since they provide the result of the same process starting from different properties which characterize it, or for other similar reasons. Problems which can, more or less 'surprisingly', be reduced to synthetic arguments arise frequently in, amongst other things, questions connected with random processes (ranging from the game of Heads and Tails to cases involving properties of characteristic functions, etc.). Often, on the other hand, it is an appropriate geometric representation which clarifies the situation and also suggests, without calculations and without any doubts, the solution in formulae.

1.9.2. In addition, however, there are even more secondary things which have their importance. These I would like to explain with a few examples so that it does not seem that some small innovation, perhaps in notation or terminology, has been introduced just for the sake of changing things, instead of with reluctance, overcome by the realization that this was the only way of getting rid of many useless complications.

The very simple device, from which most of the others derive, is that mentioned already in 1.7.3. We identify an event E with the random quantity, commonly called the 'indicator of E', which takes values 1 or 0 according as E is true or false. Not only can one operate arithmetically on the events (the arithmetic sum of many events = the number of successes; $E - p$ = the gain from a bet for a person who stakes a sum p in order to receive a sum 1 if E occurs; etc.), but one operates with a unique symbol $\mathbf{P}$ in order to denote both probability and prevision (or 'mathematical expectation') thus avoiding duplication. The 'theorem' $\mathrm{M}(I_E) = \mathbf{P}(E)$, 'the mathematical expectation of the indicator of an event is equal to the probability of the same event' *is rendered superfluous* (it could only be expressed by $\mathbf{P}(E) = \mathbf{P}(E)$!).

1.9.3. The identification TRUE = 1, FALSE = 0 is also very useful as a simple conventional device for denoting, in a straightforward and synthetic way, many mathematical expressions which usually require additional verbal explanation. Applying the same identifications to formulae expressing conditions, for instance, interpreting '$(0 \leqslant x \leqslant 1)$' as a symbol with value 1 for x between 0 and 1, where the inequality is *true*, and value 0 outside, where it is *false*, one can simply write expressions of the type

$$f(x) = g(x) \quad (0 \leqslant x \leqslant 1)$$

(and more complicated forms), which otherwise require verbal explanations, like 'the function $f(x)$ which coincides with $g(x)$ for $0 \leqslant x \leqslant 1$ and is zero elsewhere', or writing in the cumbersome form

$$f(x) = \begin{cases} = 0 & \text{for } x < 0, \\ = g(x) & \text{for } 0 \leqslant x \leqslant 1, \\ = 0 & \text{for } x > 1. \end{cases}$$

It is easy to imagine many cases in which the utility of such a convention is much greater, but I think it is difficult to realize the number and variety of such cases (I am often surprised by new, important applications not previously foreseen).

1.9.4. Other simplifications of this kind, which can sometimes be used in conjunction with the above, result from a parallel (or dual) extension of the Boolean operations to the field of real numbers, coinciding, for the values 0 and 1, with the usual meaning for the events. This natural and meaningful extension will also reveal its utility in many applications† (cf. Chapter 2, Sections 2.5 and 2.11).

1.9.5. A small innovation in notation is that of denoting the three most important types of convergence in the probabilistic field by:

symbol: type of convergence:

$\overset{<}{\rightarrow}$	weak	(in probability)	(in measure)
$\overset{>}{\rightarrow}$	strong	(almost certain)	(almost everywhere)
$\overset{\cdot}{\rightarrow}$	quadratic	(in mean-square)	(in mean (quadratic))

---

† The advantages of these two conventions (0 and 1 for True–False, and $\vee$ and $\wedge$ among numbers) are illustrated, somewhat systematically and with concise examples, in a paper in the volume in honour of O. Onicescu (75th birthday): 'Revue roumaine de mathématiques pures et appliquées', Bucharest (1967), **XII**, 9, 1227–1233. An English translation of this appears in B. de Finetti, *Probability, Induction and Statistics*, Wiley (1972).

(this could also have value in function theory). The innovation seems to me appropriate not only to avoid abbreviations which differ from language to language, but also for greater clarity, avoiding the typographic composition and deciphering of symbols which are either cumbersome or unreadable.

1.9.6. Another device which we will introduce with the intention of simplifying the notation does not have a direct relationship with the calculus of probability. For this reason we were even more hesitant to introduce it, but finally realized that without such a remedy there remained simple and necessary things which could not be expressed in a decently straightforward way.

The most essential is the device of obtaining symbols indicating functions, substituting for the variable (in any expression whatever) a 'placename' symbol: as such $\square$ would seem suitable; it also suggests something which awaits filling in. The scope is the same as obtained by Peano by means of the notation '$|x$', 'varying x', which, applied for instance to the expression $(x \sin x^2 + \sqrt{(3 - x)})/\log(2 + \cos x)$ gives $f = \{[(x \sin x^2 + \sqrt{(3 - x)})/\log (2 + \cos x)]|x\}$, where f is the symbol of the function such that $f(x)$ gives the expression above, and $f(y)$, $f(ax^2 + b)$, $f(e^z),\ldots$ is the same thing in which at each place where an x is found we substitute y, or $ax^2 + b$, or e^z, or whatever. This notation, however, does not lend itself to many cases where it would be required, and where, instead, the notation which puts the 'placename' for the variable, which is left at our disposal,† is very useful. In the preceding example one would write

$$f = \frac{\square \sin \square^2 + \sqrt{(3 - \square)}}{\log(2 + \cos \square)}$$

and to denote $f(x)$, $f(y)$, $f(ax^2 + b)$, $f(e^2)$, it would suffice to write on the right, within parentheses, ( ), the desired variable.

The greatest utility is perhaps obtained in the simplest cases: for instance, in order to denote by $\square$, $\square^2$, $\square^{-1}$ the identity function, $f(x) = x$, or the quadratic, $f(x) = x^2$, or the reciprocal, $f(x) = 1/x$, when the f must be denoted as the argument in a functional. For example, $F(\square)$, $F(\square^2)$, might indicate the first and second moments of a distribution F (according to the conventions of which we shall speak in Chapter 6), and then for any others, $F(\square^n)$, $F(|\square^n|)$, etc.

---

† In the case of many variables (for instance three) one could easily use the same device, putting in their places different 'placenames'; for example, $\square_1, \square_2, \square_3$, with the understanding that $f(x, y, z)$ or $f(5, -\frac{1}{2}, 0)$ or $f(x + y, -\frac{1}{2}x^2, 1 - 2y)$ etc., is what one obtains putting the 1st or 2nd or 3rd elements of the triple in the places indicated by the three 'placenames' with indices 1, 2, 3.

1.9.7. Finally, a secondary device is that of consistently denoting by K any multiplicative constant whatever; and if necessary indicating its expression immediately afterwards, instead of writing it directly, in extensive form, in the formulae. Otherwise, it often happens that a function, of x say, has a rather complicated appearance and each symbol, even those in small print or in the exponents etc., must be deciphered with care in order to see where x appears. Often one subsequently realizes that the function is very simple and that the complexity of the expression derives solely from having expressed the constant in extensive form. We may have a normalizing constant which, at times, could even be ignored because it automatically disappears in the sequel, or can be calculated more easily from the final formula. At times, in fact, it will be left as a 'reminder' of the existence of an omitted multiplicative factor which will always be indicated by K, even if the value might change at each step: the reader should make careful note of this remark.

1.10 SOME REMARKS ON TERMINOLOGY

1.10.1. It is without doubt unreasonable, and rather annoying, to dwell at length on questions of terminology; on the other hand, a dual purpose glossary would be useful and instructive. In the first place, it could improve on a simple alphabetical index in aiding those who forget a definition, or remember it only vaguely; secondly, it could explain the motivation behind the choice, or sometimes the creation, of certain terms, or the fixing of certain conventions for their use.† For those interested, such an explanation would also provide an account of the wherefores of the choices. Such a glossary would, however, be out of place here and, in any case, the unusual terms are few and they will be explained as and when they arise.

1.10.2. More importantly, attention must be drawn to some generic remarks, like paying attention to the nuances of divergencies of interpretation which depend on differences in conception. The main one, that of registering that an *event* is always a single case, has already been underlined (cf. 1.5.1); the same remark holds for a *random quantity* (cf. 1.7.2), and for every kind of 'random entity'. Two clarifications of terminology are appropriate at this juncture: the first to explain why I do not use the term 'variable'; the second to explain the different uses of the terms 'chance', 'random' and 'stochastic'.

To say 'random (or "chance") variable' might suggest that we are thinking of the 'statistical' interpretation in which one thinks of many 'trials' in which the random quantity can *vary*, assuming different values from trial to trial: this is contrary to our way of understanding the problem. Others might think that, even if it is a question of a unique well-determined value, it is 'variable'

---

† A very good example would be that of the *Dictionary* at the end of the 'book' by Bourbaki.

for one who does not know it, in the sense that it may assume any one of the values 'possible' for him. This does not appear, however, to be a happy nomenclature, and, even less, does it appear to be necessary. In addition, if one wanted to adopt it, it would be logical to do so always, by saying: random variable numbers, random variable vectors, random variable points, random variable matrices, random variable distributions, random variable functions, ..., random variable events, and not saying random vector, random point, random matrix, random distribution, random function, random event, and only in the case of numbers not to call it number any more, but variable.

With regard to the three terms—'*chance*', '*random*', '*stochastic*'—there are no real problems: it is simply the convenience of avoiding indiscriminate usage by supporting the consolidation of a tendency which seems to me already present, but not, as far as I know, expressly stated. Specifically, it seems to me preferable to use, systematically:

'*random*' for that which is the *object* of the theory of probability (as in the preceding cases); I will therefore say random process, not stochastic process.

'*stochastic*' for that which is valid 'in the sense of the calculus of probability': for instance; stochastic independence, stochastic convergence, stochastic integral; more generally, stochastic property, stochastic models, stochastic interpretation, stochastic laws; or also, stochastic matrix, stochastic distribution,† etc. As for

'*chance*', it is perhaps better to reserve it for less technical use: in the familiar sense of 'by chance', 'not for a known or imaginable reason', or (but in this case we should give notice of the fact) in the sense of, 'with equal probability' as in 'chance drawings from an urn', 'chance subdivision', and similar examples.

1.10.3. Special mention should be made of what is perhaps the important change in terminology: *prevision* in place of mathematical expectation, or expected value, etc. Firstly, all these other nomenclatures have, taken literally, a rather inappropriate meaning and often, through the word 'expectation', convey something old-fashioned and humorous (particularly in French and Italian, where 'espérance' and 'speranza' primarily mean 'hope'!). In

---

† The case of *matrices* and *distributions* illustrates the difference well. A random matrix is a matrix whose entries are random quantities; a stochastic matrix (in the theory of Markov chains) is the matrix of 'transition probabilities'; i.e. well-determined quantities which define the random process. A random distribution (well-defined but not known) is that of the population in a future census, according to age, or that of the measures which will be obtained in n observations which are to be made; a stochastic distribution would mean distribution of probability (but it is not used, nor would it be useful).

any case, it is inconvenient that the expression of such a fundamental notion, so often repeated, should require two words. Above all, however, there was another reason: to use a term beginning with P, since the symbol **P** (from what we have said and recalled) then serves for that unique notion which in general we call *prevision*† and, in the case of events, also *probability*.‡

1.11 THE TYRANNY OF LANGUAGE

All the devices of notation and terminology and all the clarifications of the interpretations are not sufficient, however, to eliminate the fundamental obstacle to a clear and simple explication, adequate for conceptual needs: they can at most serve as palliatives, or to eliminate blemishes.

That fundamental obstacle is the difficulty of escaping from the tyranny of everyday language, whose viscosity often obliges us to adopt phrases conforming to current usage instead of meditating on more apt, although more difficult, versions. We all continue to say 'the sun rises' and I would not know which phrase to use in order not to seem an anachronistic follower of the Ptolemaic system. Fortunately the suspicion does not even enter one's mind because nobody quibbles about the literal meaning of this phrase.

In the present exposition we shall often, for the sake of brevity, use incorrect language, saying, for example: 'let the probability of E *be* $=\frac{1}{2}$', 'let the events A and B *be* (stochastically) independent', 'let the probability distribution of a random quantity X *be* normal', and so on. This is incorrect, or, more accurately, it is meaningless, unless we mean that it is a question of an abbreviated form to be completed by 'according to the opinion of the individual (for example You) with whom we are concerned and who, we suppose, desires to remain coherent'. The latter should be understood as the constant, though not always explicitly stated, intention and interpretation of the present author.

This is stated, and explicitly repeated, wherever it seems necessary, due to the introduction of new topics, or for the examination of delicate points— perhaps even too insistently, with the risk, and near certainty, of irritating

---

† *Translators' note.* We have used *prevision* rather than *foresight* (as in Kyburg and Smokler, p. 93) precisely for the reasons given in 1.10.3.

‡ In almost all languages other than Italian, the letter E is unobjectionable, and often a single word is sufficient: Expectation (English), Erwartung (German), Espérance mathématique (French), etc. However, the use of E is inconvenient because this is often used to denote an event, and, in any case, it can hardly remain if one seeks to unify it with **P**. It is difficult to foresee whether this unification will command widespread support and lead to a search for terms with initial letter P in other languages (*see footnote above*), or other solutions. We say this to note that the proposed modification causes little difficulty in Italy not only because of the existence and appropriateness of the term 'Previsione', but also because the international symbol E has not been adopted there.

the reader. Even so, notwithstanding the present remark (even imagining that it has been read), I am afraid that the very same reader when confronted with phrases like those we quoted, instead of understanding implicitly those things necessary in order to interpret them correctly, could have the illusion of being in an oasis—in the 'enchanted garden' of the objectivists (as noted at the end of Chapter 7, 7.5.7)—where these phrases could constitute 'statements' or 'hypotheses' in an objective sense.

In our case, in fact, the consequences of the pitfalls of the language are much more serious than they are in relationship to the Copernican system, where, apart from the strong psychological impediments due to man's egocentric geocentrism, it was simply a question of choosing between two objective models, differing only in the reference system. Much more serious is the reluctance to abandon the inveterate tendency of savages to objectivize and mythologize everything;† a tendency that, unfortunately, has been, and is, favoured by many more philosophers than have struggled to free us from it‡. This has been acutely remarked, and precisely with reference to probability, by Harold Jeffreys :§

> 'Realism has the advantage that language has been created by realists, and mostly very naïve ones at that; we have enormous possibilities of describing the inferred properties of objects, but very meagre ones of describing the directly known ones of sensations.'

1.12   REFERENCES

1.12.1. We intend to limit the present references to a bare minimum. The reader who wishes to study the topics on his own can easily discover elsewhere numerous books and references to books. Here the plan is simply to suggest the way which I consider most appropriate for the reader who would like to delve more deeply into certain topics, beyond the level reached

---

† The main responsibility for the objectivizationistic fetters inflicted on thought by everyday language rests with the verb 'to be' or 'to exist', and this is why we drew attention to it in the exemplifying sentences by the use of italics. From it derives the swarm of pseudoproblems from 'to be or not to be', to 'cogito ergo sum', from the existence of the 'cosmic ether' to that of 'philosophical dogmas'.

‡ This is what distinguishes the acute minds, who enlivened thought and stimulated its progress, from the narrow-minded spirits who mortified it and tried to mummify it: those who took every achievement as the starting point to presage further achievement, or those, on the contrary, who had the presumption to use it as a starting point on which to be able to base a definitive systematization.

For the two types, the qualification given by R. von Mises seems appropriate (cf. *Selected Papers*, Vol. II, p. 544): 'great thinkers' (like Socrates and Hume) and 'school philosophers' (like Plato and Kant).

§ Jeffreys, a geophysicist, who as such was led to occupy himself deeply with the foundations of probability, holds a position similar in many aspects to the subjectivistic one. The quotation is taken from H. Jeffreys, *Theory of Probability*, Oxford (1939), p. 394.

here, without the inconvenience of passing from one book to another, with differences in notation, terminology and degree of difficulty.

1.12.2. The most suitable book for consultation according to this plan is, in my opinion, that of Feller:

Willy Feller, *An Introduction to Probability Theory and its Applications*, in two volumes: I (1950) (2nd and 3rd ed., more and more enriched and perfected, in 1956 and 1968); II (1966); Wiley, New York.

The treatment, although being on a high level and as rigorous as is required by the topic, is not difficult to read and consult. This is due to the care taken in abolishing useless complications, in making, as far as possible, the various chapters independent of each other while facilitating the links with cross-references, and in maintaining a constant interplay between theoretical questions and expressive examples. Further discussion may be found in a review of it, by the present author, in *Statistica*, **26**, 2 (1966), 526–528.

The point of view is not subjectivistic, but the mainly mathematical character of the treatment makes differences of conceptual formulation relatively unobtrusive.

1.12.3. For the topics in which such differences are more important, i.e. those of inference and mathematical statistics (Chapter 11 and Chapter 12), there exists another work which is inspired by the concepts we follow here. Such topics are not expressly treated in Feller and thus, with particular reference to these aspects, we recommend the following work, and above all the second volume.

Dennis V. Lindley, *Introduction to Probability and Statistics from a Bayesian viewpoint*, in two volumes: I, *Probability*; II, *Inference*; Cambridge University Press (1965).

Complementing the present work with those of Feller and Lindley would undoubtedly mean to learn much more, and better, than from this work alone, except in one aspect; that is the coherent continuation of the work of conceptual and mathematical revision in conformity with the criteria and needs already summarily presented in this introductory chapter.

The above-mentioned volumes are also rich in interesting examples and exercises, varied in nature and difficulty.

CHAPTER 2

Concerning Certainty and Uncertainty

2.1 CERTAINTY AND UNCERTAINTY

2.1.1. In almost all circumstances, and at all times, we all find ourselves in a state of uncertainty.

Uncertainty in every sense.

Uncertainty about actual situations, past and present (this might stem from either a lack of knowledge and information, or from the incompleteness or unreliability of the information at our disposal; it might also stem from a failure of memory, either ours or someone else's, to provide a convincing recollection of these situations).

Uncertainty in foresight: this would not be eliminated or diminished even if we accepted, in its most absolute form, the principle of determinism; in any case, this is no longer in fashion. In fact, the above-mentioned insufficient knowledge of the initial situation and of the presumed laws would remain. Even if we assume that such insufficiency is eliminated, the practical impossibility of calculating without the aid of Laplace's demon would remain.

Uncertainty in the face of decisions: more than ever in this case, compounded by the fact that decisions have to be based on knowledge of the actual situation, which is itself uncertain, to be guided by the prevision of uncontrollable events, and to aim for certain desirable effects of the decisions themselves, these also being uncertain.

Even in the field of tautology (i.e. of what is true or false by mere definition, independently of any contingent circumstances) we always find ourselves in a state of uncertainty. In fact, even a single verification of a tautological truth (for instance, of what is the seventh, or billionth, decimal place of π, or of what are the necessary or sufficient conditions for a given assertion) can turn out to be, at a given moment, to a greater or lesser extent accessible or affected with error, or to be just a doubtful memory.

2.1.2. It would therefore seem natural that the customary modes of thinking, reasoning and deciding should hinge explicitly and systematically on the factor *uncertainty* as the conceptually pre-eminent and determinative element. The opposite happens, however: there is no lack of expressions referring to uncertainty (like 'I think', 'I suppose', 'perhaps', 'with difficulty', 'I believe', 'I consider it as probable', 'I think of it as likely', 'I would bet', 'I'm almost certain', and so on), but it seems that these expressions, by and large, are no more than verbal padding. The solid, serious, effective and essential part of arguments, on the other hand, would be the nucleus that can be brought within the language of certainty—of what is certainly *true*, or certainly *false*. It is in this ambit that our faculty of reasoning is exercised, habitually, intuitively and often unconsciously.

In reasoning, as in every other activity, it is, of course, easy to fall into error. In order to reduce this risk, at least to some extent, it is useful to support intuition with suitable superstructures: in this case, the superstructure is *logic* (or, to be precise, the *logic of certainty*).

Whether it is a question of traditional verbalistic logic, or of mathematical logic, or of mathematics as a whole, the only difference in this respect is in the degree of extension, effectiveness and elegance. In fact, it is, in any case, a question of ascertaining the *coherence*, the *compatibility*, of stating, believing, or imagining as hypotheses some set of 'truths'. To put it in a different way: thinking of a subset of these 'truths' as *given* (knowing, for instance, that certain facts are true, certain quantities have given values, or values in between given limits, certain shapes, bodies or graphs of given phenomena enjoy given properties, and so on), we will be able to ascertain which conclusions, among those of interest, will turn out to be—on the basis of the data—either *certain* (certainly true), or *impossible* (certainly false), or else *possible*. The qualification 'possible'—which is an intermediate, generic and purely negative qualification—is applied to everything which does not fall into the two extreme limit cases: that is to say, it expresses one's ignorance in the sense that, on the basis of what we know, the given assertion could turn out to be either true or false.

2.1.3. This definition of 'possible' itself reveals an excessive and illusory confidence in 'certainty': in fact, it assumes that logic is always sufficient to separate clearly that which is determined (either true or false), on the basis of given knowledge, from that which is not. On the contrary (even apart from the possibility of deductions which are wrong, or whose correctness is in doubt), to the sphere of the *logically possible* (as defined above) one will always add, in practice, a fringe (not easily definable) of the *personally possible*; i.e. that which must be considered so, since it has not been established either that it is a consequence of one's knowledge, or that it is in conflict with it.

We have already said, in fact, that logic can *reduce the risk of error*, but cannot eliminate it, and that tautological truths are not necessarily accessible.

However, in order not to complicate things more than is required to guard against logical slips, we will always consider the case in which 'possible' can be interpreted as *logically possible*.†

2.2 CONCERNING PROBABILITY

2.2.1. The distinction between that which, at a certain moment, we are ignorant of, and that which, on the other hand, turns out to be certain or impossible, allows us to think about the range of *possibility*; i.e. the range over which our uncertainty extends. However, this is not sufficient as an instrument and guide for orientation, decision or action: to this end—and this is what we are interested in—it will be necessary to base oneself on a further concept; the concept of *probability*.

In this chapter we do not wish to talk about probabilities, however; they will be introduced in Chapter 3. This deferment is undoubtedly awkward: obviously, the awkwardness consists in introducing preliminary notions without, at the same time, exhibiting their use. Didactically this is a bad mistake—one runs the risk of making boring and dull that which otherwise would appear clear and interesting. However, when it is important to emphasize an essential distinction, which otherwise would remain unnoticed and confused, a rigid separation is necessary—even if it seems to be artificial and pedantic. This is precisely the case here.

2.2.2. The study of the range of possibility, to which we shall here limit ourselves, involves learning how to know and recognize all that can be said concerning uncertainty, while remaining in the domain of the logic of certainty; i.e. in the domain of what is *objective*. Probability will be a further notion not belonging to that domain, and therefore a *subjective* notion. Unfortunately, these two adjectives anticipate a question concerning which there could be controversial opinions—their use here is not intended to prejudice the conclusion, however. For the time being, what matters is to make clear a distinction which is methodologically fundamental: afterwards, one can discuss the interpretation of the meaning of the two fields it delineates, the choice of nomenclature, and the points of view corresponding to them. It is precisely in order to be able to discuss them lucidly *afterwards* that it is necessary to avoid an immediate discussion of possibility and probability together; the confusion so formed would be difficult to resolve.

---

† Possibly by *eliminating* some knowledge. For instance, in the case of π it seems reasonable (for the problem under consideration) to imagine that one ignores the properties that permit the calculation of π, and to consider it as an 'experimental constant' whose decimal representation could only be known if somebody had determined it and published the result. I believe that for a mathematician too it would be reasonable to think that everything proceeds as if he were in such a state of ignorance.

Both the distinction and the connection between the two fields are easily clarified: the logic of certainty furnishes us with the range of possibility (and the 'possible' has no gradations); probability is an additional notion that one applies within the range of possibility, thus giving rise to gradations ('more or less probable') that are meaningless in the logic of certainty.

2.2.3. Since it is certain that everyone knows enough about probability to be able to interpret these explanations in a less vague fashion, we can say that 'probability is something that can be distributed over the field of possibility'. Using a visual image, which at a later stage might be taken as an actual representation, we could say that the logic of certainty reveals to us a space in which the range of possibilities is seen in outline, whereas the logic of the probable will fill in this blank outline by considering a mass distributed upon it.

There is no harm in anticipating the developments that the treatment will undergo from the next chapter onwards, provided that, from the fact that they are not talked about here, one understands that they do not belong in the domain that we now consider it important to present as well-delimited and distinct.

2.3 THE RANGE OF POSSIBILITY

2.3.1. *Prologue.* Let us introduce right away the use of 'You', following Good (Savage uses 'Thou'). The characterization of what is *possible* depends on the state of information. The state of information will be that (at a given moment) of a real individual, or it might even be useful to think of a fictitious individual (as an aid to fixing ideas). This individual, real or fictitious, in whose state of information—and, complementarily, of *uncertainty*—we are interested, we will denote by 'You'. We do so in order that You, the reader, can better identify yourself with the rôle of this character. This character—or, better, You—will play a much more important rôle after this chapter, when probabilities will enter the scene. For the moment, You are in the audience, because You have to limit yourself to passively recording what You know for certain, or what You do not know.† All the same, it will be useful for You to at least get used to putting yourself in this character's place, since, even if it is not yet time to speak our lines, we are about to walk onto the stage—i.e. to enter into the range of possible alternatives.

---

† You would have a more personal and autonomous rôle if we took into account the faculty, which You certainly possess, of considering as 'possible' that which You could show to be impossible, but which demands too much deductive effort. However, we have stated, in 2.1.3, that, for the sake of simplicity, we omit consideration of such hypotheses.

With regard to any situation or problem that You have to consider, there will always exist an enormous number of conceivable alternatives. Your information and knowledge will, in general, permit You to exclude some of them as impossible: that is, they will permit You—and this has been said to be the function of science—a 'limitation of expectations'. All the others will remain *possible* for You; neither certainly true, nor certainly false. It will not happen that only one of them will be isolated as *certain*, except in special cases, or unless a rather crude analysis of the situation is given. Obviously, it is always sufficient to take all the possible alternatives and present them as a whole in order to obtain a single alternative which is 'certain'.

The choice of which of the more or less sophisticated, detailed, particularized forms we need, or consider appropriate, in order to distinguish or subdivide such alternatives, according to the problems and the degree of refinement we require in considering them, depends on us, on our judgment. Also, we have available several possible languages in which we can express ourselves in this connection. It is convenient to introduce them straight away, and altogether, in order to show, at the same time, on the one hand their essential equivalence, and, on the other, the differences between them which render their use more or less appropriate in different cases.

2.3.2. *Random events and entities.* Everything can be expressed in terms of *events* (which is the simplest notion); everything can be expressed in terms of *random entities* (which is the most generic and general notion); and so on. One or other of these notions is sufficient as a starting point to obtain all of them. However, it will be instructive to concentrate attention on four notions which will immediately allow us to frame within the general scheme the most significant types of problems, important from both the conceptual and practical points of view.

We will consider:

random events,
random quantities,
random functions,
random entities.

Let us make clear the meaning that we give to 'random': it is simply that of 'not known' (for You), and consequently 'uncertain' (for You), but *well-determined* in itself. Not even the circumstance of 'not known' is to be taken as obligatory; in the same way we could number constants among functions, though we will not call a constant a 'function' if there is no good reason. To say that it is *well-determined* means that it is *unequivocally individuated*. To explain this in a more concrete fashion; it must be specified in such a way that a possible bet (or insurance) based upon it can be decided without question.

2.3.3. First, let us consider *random quantities*: this is an intermediate case from which we can pass more easily to the others, particularizing or generalizing as the case may be. We will denote a number, considered as a random quantity, by a capital letter; e.g. X or Y, etc. It might be an integer, a real number, or even a complex number; but the latter case should be specified explicitly. The true value is unique, but if You call it random (in a non-redundant usage) this means that You do not know the true value. Therefore, You are in doubt between at least two values (possible for You), and, in general, more than two—a finite or infinite number (for instance, all the values of an interval, or all the real numbers). We will denote by $I(X)$ the set of possible values of X, and we will write, in abbreviated form, inf X and sup X for inf $I(X)$ and sup $I(X)$. It is particularly important to distinguish the cases of random quantities which are *bounded* (*from above and below*), i.e. inf X and sup X finite, and those which are only *bounded from above*, or only *bounded from below*, or *unbounded*, i.e. inf $X = -\infty$, or sup $X = +\infty$, or both.

To exemplify what we mean by *well-determined* in the case of random quantities, let us put $X =$ the year of death of Cesare Battisti†. The true value is $X = 1916$. While he was alive this value was not known to anyone and all years from that time on were possible values (for everybody). After the event, it is only random for those who are ignorant of it: for instance, for those who know only that it happened during Italy's participation in the first world war, the possible values are the four years 1915, 1916, 1917, 1918.

Every function of a random quantity, $Y = f(X)$, or of two (or more), $Z = f(X, Y)$, etc., is a random quantity (possibly 'degenerate', i.e. certain, if, for instance, $f(x)$ has the same value for all possible values of X).

2.3.4. An *event* (or *proposition*) admits only two values: TRUE and FALSE. In place of these two terms it is convenient to put the two values 1 and 0 (1 = TRUE, 0 = FALSE); in this way we simply reduce to a special case of the preceding, with an obvious, expressive meaning. Thus, when we wish to interpret the convention in this way, the event is identified with a gain of 1 if the event occurs, and with a gain of 0 if the event does not occur. Moreover, with this convention the logical calculus of the events is simplified.

We continue to denote events with capital letters; in the main, E, H, A, B, ... It is clear, for instance, that $1 - E$ is the *negation of E*, which is false if E is true, and vice-versa (value 0 if $E = 1$, and conversely): it is also clear that AB is the logical product of A and B, i.e. true if both A and B are

---

† Cesare Battisti was deputy for Trento at the Vienna Parliament; he volunteered for the Italian army, was then taken prisoner and hanged by the Austrians in 1916.

(Trento, where the author once lived, is an Italian city which was, in Battisti's time, a part of Austria.)

true, etc. (this is merely an example, the topic will be developed later, in Section 2.5).

An event corresponds to a question which admits only two answers; YES or NO (YES = 1, NO = 0). It is clear that with a certain number of questions of this type we can obtain an answer to a question which involves any number of alternative answers. Given a *partition* into s alternatives (one, and only one, of which is true), we can consider, for instance, the s events (exclusive and exhaustive) which correspond to them. But even less is sufficient: with n events we can imagine 2^n dispositions of YES–NO answers; we therefore have a partition into $s = 2^n$ alternatives if all these answers are possible, or into a smaller number, $s < 2^n$, if some of them are impossible (see Section 2.7 for further details).

Abandoning the restriction to a 'finite number', it is clear that by means of events we can study every case, even those involving an infinite number of possibilities.

2.3.5. By talking about *random entities* in general, we have a means of expressing in a synthetic form the situation presented by any problem whatever. It is a question of referring oneself at all times to the same perspective, the one already implicitly introduced in the case of a random quantity, and which we now wish to make more precise and then to extend.

In the case of a random quantity X we can visualize the situation by considering as the '*space of alternatives*', $\mathscr{S}$, a line, the x-axis,† and on it the set, $\mathscr{Q}$, of the only values (points) *possible* (for You). In this way we consider *en masse*, implicitly, all the events concerning X (that it belongs to a half-line, $X \leqslant x$, or to an interval, $x' \leqslant X \leqslant x''$, or to any arbitrary set, $X \in I$).‡

But now it is obvious that the same representation holds in all cases (in a more intuitive sense, of course, in 3, or fewer, dimensions). If we consider two random quantities, X and Y, we can think of the cartesian plane, with coordinates x and y, as the space $\mathscr{S}$ in which we have a set $\mathscr{Q}$ of *points* (pairs of values for X and Y) *possible* (for You) for a random *point* (X, Y). Every event (proposition, statement) concerning X and Y corresponds to a set I of $\mathscr{S}$: of course, only the intersection with $\mathscr{Q}$ is required, but it is simpler (and innocuous) to think of all sets I. The same could be said in the case of three random quantities X, Y, Z (in this case $\mathscr{S}$ is ordinary space), or for more than three.

---

† We always denote by $x(y,$ etc.) the axis on which $X(Y,$ etc.) is represented.

‡ We omit here critical questions relating to the possibility of giving, or not giving, a meaning to statements of an extremely delicate or sophisticated nature (or at least to the possibility of taking them into consideration). For example, the distinction between $<$ and $\leqslant$, the case of I 'non-measurable' in some sense or other, etc. It will be necessary to say something in Chapter 6; discussion of a critical character will be developed only in the Appendix, apart from brief anticipatory remarks here and there.

Independently of the coordinate system, we could, in this geometric representation, formulate a problem straightaway. It might concern a *random point* on a plane (e.g. that point which would be hit in firing at a target), or in ordinary space (e.g. the position, at a given instant, of a satellite with which we have lost contact). We find an appropriate representation for the situation of a particle (position and velocity) by using 6-dimensional space: the space of dimension $6n$ serves as 'phase space' for the case of n particles.

Independently of the geometrical meaning, or any meaning which suggests (in a natural way) a geometrical representation, we can always imagine, for any *random entity*, an abstract space $\mathscr{S}$ consisting of all possible alternatives (or, if convenient, a larger space of which these form a subset $\mathscr{Q}$). We could consider, for example, *random vectors, random matrices* or *random functions*, and, thus far, the linear structure of the space continues to present itself as natural. But we could also consider *random sets*: for example, *random curves* (the path of a fly, or an aeroplane), random sets on surfaces (that part of the earth's surface in shadow at a given instant, or on which rain fell in the last twenty-four hours); or we could think of random entities inadequate to give any structure to the space.

We can, therefore, accept this representation as the general one, despite some reservations which will follow shortly (the latter are intended not as arguments against the representation, or for its rejection, but rather in favour of its acceptance 'with a pinch of salt').

2.3.6. There is no need to deal with *random functions* separately, by virtue of the particular position they hold with respect to the preceding considerations (just as events and arbitrary entities have extreme positions, and random quantities an intermediate, but instrumentally fundamental, position). It is useful, however, to mention them explicitly for a moment. Firstly, in order to point out an example of applications which become more and more important from now on, and are largely new with respect to the range of problems traditionally recognized. Secondly, because we can allude, in a simple and intuitive way, to certain critical observations of the kind which will be reserved, in general, for the Appendix.

A *random function* is a function whose behaviour is unknown to You: we will denote it by $Y(t)$, assuming for convenience of intuition that the variable t is time.† If the function is known up to certain parameters, for instance $Y(t) = A \cos(Bt + C)$ with A, B, C random (i.e. unknown to You), the whole thing is trivial and reduces to the space of parameters. The case

---

† Our preference for $Y(t)$, rather than the more usual $X(t)$, as a notation for a generic random function, depends mainly on the fact that an X is often used as an 'ingredient' in the construction of $Y(t)$. At other times, x is used as a variable in place of t, and, anyway, in the graphical representation it is always convenient to think of the ordinate as y, and the abscissa as t or x.

which, in general, we have in mind when we speak of a random function—or a *random process*, if we wish to place more emphasis on the phenomenon than on the mathematical translation—is that in which (to use the suggestive, if somewhat vague, phrase of Paul Lévy) the uncertainty exists at every instant (or, in his original expression, 'chance operates instant by instant').

This might mean, for example, that knowing the values of $Y(t)$ at any number of instants, $t = t_1, t_2, \ldots, t_n$, however large the (finite) n, the value at a different instant t will still, in general, be uncertain. Sometimes, either for simplicity or in order to be 'realistic', we imagine that it makes sense to measure Y at a finite (although unrestrictedly large) number of instants, without disposing of other sources of knowledge.† In such cases, the space $\mathscr{S}$ can be thought of as that in which every function is a 'point', but in which the possibility of distinguishing whether or not a function belongs to a set is only possible for those sets defined by a finite number of coordinates: the latter, being observable, are actually events. The simplest form of these events occurs when we ask whether or not the values at given instants fall inside fixed intervals $a_h \leqslant Y(t_h) \leqslant b_h$, $h = 1, 2, \ldots, n$. To give a visual interpretation, we ask whether or not the graph passes through a sequence of n 'doors', like a *slalom*.

2.4   CRITICAL OBSERVATIONS CONCERNING THE 'SPACE OF ALTERNATIVES'

2.4.1.   Having reference to the 'space of alternatives' undoubtedly provides a useful overall visualization of problems. Nevertheless, the systematic and, in a certain sense, indiscriminate use of it, which is fashionable in certain schools of thought, does have its dangers. One should learn to recognize these, and strive to avoid them.

In considering fields of problems of whatever complexity—in which, for instance, random sets, functions, sequences of functions, etc., can occur together—the most general way of interpreting and applying the concepts exhibited in Section 2.3.5 is always the same; i.e. the following.

One goes back to the finest possible partition into 'atomic' events—not themselves subdivisible for the purposes of the problem under consideration —and these are considered as *points* constituting the set $\mathscr{Q}$ of 'possible outcomes'. This abstract space is the 'space of alternatives', or the 'space of outcomes': in certain cases, cf. the examples of Section 2.3, it may be convenient to think of it as embedded in a larger and more 'manageable' space, and to regard this latter as the 'space of alternatives'.

---

† Like, for instance, velocity $Y'(t)$ at an instant, measured with a speedometer; or the maximum or minimum of $Y(t)$ in an interval (t', t''), measured with instruments like a Max–Min thermometer.

In this scheme of representation, each problem (by which we mean problem concerning the alternatives $\mathscr{Q}$) reduces to considering 'the *true* alternative' (or 'the one which will turn out to be verified', or however one wants to express it), as a *random point* in $\mathscr{S}$, or, if we wish to be precise, in $\mathscr{Q}$. Let us call this point Q: it expresses everything there is to be said. Were we to lump together in $\mathscr{S}$ all possible problems, this space would be the space of all possible histories of the universe (explained as far as the most unimaginably minute details), and Q would be that point representing the true history of the universe (explained as far as the most unimaginably minute details).

Each event in this scheme is evidently interpretable as a set of points. E is the set of all points Q for which E is true; for example, it is the set of all individual 'histories of the universe' in which E turns out to be true. With the interpretation $1 = \mathrm{TRUE}$, $0 = \mathrm{FALSE}$, one could also say that E is a function of the point Q with values 1 on points Q of the set E, and 0 elsewhere (the indicator† function of the set E).

Similarly, each random quantity is interpretable as a real-valued function of the points $Q : X = X(Q)$ is the value which X assumes if the *true* point is Q. The preceding case, $E = E(Q)$, is simply the particular case which arises when the function can only take on the values 0 and 1.

The same is true for random entities of any other kind: for example, a random vector is a vector which is a function of the point Q.

2.4.2. That all this can be useful and convenient as a form of representation is beyond question. But things are useful if and only if we retain the freedom to make use of them when, and only when, they are useful, and only up to the point where they continue to be useful. A scheme which is too rigid, too definitely adopted and taken 'too seriously', ends up being employed without checking the extent to which it is useful and sensible, and risks becoming a *Procrustean bed*.

> This is what happens to those who refer themselves too systematically to this scheme. Pushing the subdivision as far as the 'points' perhaps goes too far, but stopping it there creates a false and misleading dichotomy between the problems belonging, and not belonging, to the field under present consideration. The logical inconvenience which this already creates in the range of possibility will become far more dangerous and insidious when probabilities are introduced into such a structure.
>
> An analogy between events and sets exists, but it is nothing more than an analogy. A set is effectively composed of elements (or points), and its subdivision into subsets necessarily stops when subdivision reaches its constituent points. With an event,

---

† In a different terminology, the indicator function is also called the characteristic function: this term has many other meanings, and, in particular, in the calculus of probability it has a different and very important meaning for which it must be reserved (cf. Chapter 6).

however, it is possible, at all times, to pursue the subdivision (although in any application it is convenient to stop as soon as the subdivision is sufficient for the study in progress, otherwise things get unnecessarily complicated). The elements of the 'final subdivision' we have interpreted as 'points', but any idea which does not take into account the relative, arbitrary and provisional nature of such a delimiting of the subdivision, which thinks of it as 'indivisible', or as 'less subdivisible', or in any way different from all other events, is without foundation and misleading. For instance, it would be illusory to wish to distinguish between events corresponding to 'finite' or 'infinite' sets, or belonging to finite, or infinite partitions, as if this had some intrinsic meaning. There is even less justification for retaining, as necessary, topological properties which happen to be meaningful in $\mathscr{S}$. The latter we referred to as 'space', instead of 'set', simply to use a more expressive language, and also because topological structures often exist and have interest in certain spaces by virtue of the nature of the spaces themselves, even when not required for any reason pertaining to the logical or probabilistic meaning.

2.4.3. Other objections, which we will develop a little more in the Appendix, would lead us to impugn even more radically the validity of the above representative scheme (and of many other things which we have hitherto admitted and which, for the moment, we continue to admit). As an example, we note the fact that all sets (or the 'points' of them) must be accepted as having the meaning of events.

In general terms, it will always be a question of examining if, and in which sense, a statement really constitutes an 'event', permitting, in a more or less realistic and acceptable form, and in a unique way, the 'verification' of whether it is 'true' or 'false'.

What should be said concerning statements which are 'verifiable' only by means of an infinite number of observations, or by waiting an infinite length of time, or by attaining an infinite precision? A critical attitude in this respect could lead one not to consider as 'events' the fact that X has exactly the value x, or belongs to a set of measure zero (e.g. is rational), but only the fact that $X \in I$ for a set I 'up to sets of measure zero' (and this, although it eliminates some difficulties, introduces others), or 'up to an error $< \delta$, that can be chosen as small as desired, but non-zero', and so on. Even more radical are the difficulties of 'complementarity' which appeared first in quantum physics, but can be detected on a smaller scale in more everyday examples: A and B are events (observable), but it is not possible to observe both of them, and, therefore, it is not possible to call the product AB an event (observable).

All this, in addition to the specific reasons already given in the main text (and to which we return in the next paragraph), reduces the value of the reduction to 'points'. Indeed, it is symptomatic that, precisely in connection with arguments of this kind, von Neumann developed a 'geometry without points' (in 'Continuous geometries', *Proc. Nat. Acad.*, **22** (1936), 92–100 and exemplified *ibid.*, 101–8) where, as he says: 'The point which we wish to stress is that the investigations described above show an unbroken trend *away from the notion of the point*'. The studies to which he alludes are those of K. Menger and G. Bergmann (on linear spaces), of F. Klein, G. Birkhoff and O. Ore (on lattices), and discussions with J. W. Alexander and H. Veblen.

Even more strictly in accordance with the considerations in the text, appear to be the studies of St. Ulam (in the 'von Neumann lecture', Princeton (1963), still unpublished), since he also refers himself to structures *open* to the adjunction of new entities as new circumstances arise. A 'continuous geometry' of von Neumann, on the other hand, is a closed structure, although very rich, containing linear systems of any dimension c, with c any real number between 0 and 1 (the empty and complete systems, respectively). Ulam says: 'The indications are ... that *there are no atoms of simplicity* and, which is most strange, one would almost be tempted to say that in the physical world the set-theory *axiom of Regularity*—that is to say, that *every set contains a minimal element with respect to the relation of "belonging to a set"*—*does not hold!*'.†

2.5   LOGICAL AND ARITHMETIC OPERATIONS

2.5.1. Having, through the convention 1 = TRUE, 0 = FALSE, given to events an interpretation that makes them particular random quantities, it becomes both possible and useful to take advantage of this unification in order to effect also an appropriate unification of the operations related to them. Usually, and inevitably, prior to such a convention,‡ one considers two distinct series of operations: the (Boolean) *logical operations,*

$$\wedge \quad logical\ product; \qquad \vee \quad logical\ sum; \qquad \sim \quad negation$$

applicable only to *events*; and the *arithmetic operations*

$$\cdot \quad product; \qquad + \quad sum \text{ (and their inverses: } / \text{ and } -)$$

applicable only to *numbers.*

We have already touched upon the utility of certain applications of the arithmetic operations to events, automatically possible by the above convention (cf. Section 2.3.4, and also allusions in Chapter 1). We are now able not only to develop this extension systematically, but also to obtain a complete unification by extending, in the opposite direction, the logical operations into the field of numbers.

In fact, in the field of (real) numbers, we make the definitions:

$$x \wedge y = \min(x, y), \qquad x \vee y = \max(x, y), \qquad \sim x = 1 - x \ (= \tilde{x}).\S$$

It is immediate that the definitions agree with those known in the field of events (that is, of the idempotent numbers 0 and 1), whereas, obviously, the usual properties (which it would be beneficial to interpret and understand through examples in each of the two cases), always hold both for

---

† The italics are present in the original for the last three words only.

‡ Which, as I later discovered, had already been adopted by von Neumann in 1932 in his treatment of quantum mechanics; cf. Appendix, Section 9.

§ As usual, we agree to place the tilde for 'complementary to 1' above, instead of in front, when dealing with a single letter. The same convention—using a bar rather than a tilde—was adopted by L. Dubins and L. J. Savage, *How to Gamble if You Must*, McGraw-Hill (1965), p. 64, and found to be of frequent utility.

numbers and events:

$$\sim(x \wedge y) = \tilde{x} \vee \tilde{y}$$
$$\sim(x \vee y) = \tilde{x} \wedge \tilde{y}$$
$\left. \right\}$ (duality of $\wedge$ and $\vee$ with respect to complements),

$$x \wedge (y \vee z) = (x \wedge y) \vee (x \wedge z)$$
$$x \vee (y \wedge z) = (x \vee y) \wedge (x \vee z)$$
$\left. \right\}$ (distributivity between $\wedge$ and $\vee$),

$$x \wedge x = x$$
$$x \vee x = x$$
$\left. \right\}$ (idempotence for $\wedge$ and $\vee$)

(in addition to the obvious commutative and associative properties of $\wedge$ and $\vee$).

2.5.2. *Operations on events.* By virtue of what has already been said, it is not a question of making new definitions, but only of applying the general definitions to the case of the values 0 and 1; it remains only to establish agreement with the usual meaning.

By the *logical product* of two (or more) events A, B, we mean the event which is true if and only if all the factors are, and therefore false if at least one is false. If the factors can only be 0 and 1, both the arithmetic product and the operation *min* ($\wedge$) obviously enjoy the property that the result is 1 if and only if all the factors are 1. Therefore, *in the field of events*, the two operations of arithmetic product and logical product coincide; thus we could always refer simply to the *product* of two events, without danger of ambiguity, and write $E = AB$. The symbol $\wedge$ might be used for greater clarity only in complicated cases; for instance,

$$E = (X + Y \geqslant 54) \wedge (Z \geqslant Y + 12),$$

where the events are conditions (on random quantities X, Y, Z, etc.), written as parentheses, and the fact that they are events and not numbers could be overlooked.

By the *negation* of an event A, we mean the event which is true if A is false and vice-versa; obviously we have 'not A' $= \sim A = \tilde{A} = 1 - A$, because $\sim 1 = 1 - 1 = 0$, $\sim 0 = 1 - 0 = 1$.

By the *logical sum* of two (or more) events A, B, we mean the event which is true if at least one of the summands is true, and therefore false if and only if they are all false. To this corresponds the operation *max* ($\vee$), which gives 1 if at least one summand is 1, and 0 if all summands are 0. It is also obvious and well-known that, with respect to negation, the operation is dual to that of the product:

$$A \vee B = \sim(\tilde{A} \wedge \tilde{B}).$$

This follows also from the properties stated generally for $\tilde{x} \wedge \tilde{y}$.

This allows us to obtain an arithmetic expression for the logical sum: taking complements and expanding, we obtain

(1) $$A \lor B = 1 - (1 - A)(1 - B) = A + B - AB,$$

and, similarly,

$$A \lor B \lor C = 1 - (1 - A)(1 - B)(1 - C)$$
$$= A + B + C - AB - AC - BC + ABC.$$

In general, for n summands,

(2)
$$E_1 \lor E_2 \lor \ldots \lor E_n = \sum_i E_i - \sum_{ij} E_i E_j$$
$$+ \sum_{ijh} E_i E_j E_h - \ldots \pm E_1 E_2 \ldots E_n,$$

where the sums have to be taken over all the n events E_i, over all the $\binom{n}{2}$ products two at a time, over all the $\binom{n}{3}$ products three at a time, and so on, with alternate signs, up to the last term which is the product of all n events with $+$ if n is odd, $-$ if n is even.

The *arithmetic sum* of two (or more) events A, B, is not, in general, an event, but a random number expressing the *number of successes*. In particular, $A + B$ has either the value 0 (if they are both false), or 1 (if one is true and the other false), or 2 (if they are both true). In general, as in this case, the relation between logical sum and arithmetic sum is the following: *both have the value 0 if every summand happens to be false* (no successes), whereas, otherwise, if true summands (successes) exist and number 1, 2, 3,..., in general m, *the (arithmetic) sum is that number*, whereas the *logical sum* always takes the value *one*; that is, does not take into account multiplicity,

(3) $$(\text{logical sum}) = 1 \land (\text{arithmetic sum})$$

or, explictly,

(3') $$E_1 \lor E_2 \lor \ldots \lor E_n = 1 \land (E_1 + E_2 + \ldots + E_n).$$

The fact of having two distinct notions is not, therefore, inconvenient, but, on the contrary, is an advantage because both have their raison d'être. We are still faced with the problem of eliminating the ambiguity of the terminology—since we do not wish to be obliged to say 'logical sum' or 'arithmetic sum' every time. For this purpose it is sufficient to adopt the natural convention of using *sum* for the arithmetic sum, and *event-sum* for the logical sum (because only this is an event).

2.5.3. We observe that the operations introduced induce, over the field of real numbers, the structure of a *lattice*, with the operation $\sim$ which enjoys many properties of the complement (in the algebraic sense), but is not exactly such, except in the field of events (the numbers 0 and 1). There, in fact, we have $x \vee \tilde{x} = 1$ (because either x or $\tilde{x}$ is 1, and the other 0), in addition to $x + \tilde{x} = 1$, which is also valid for any x.

In addition, we observe that the expressions in arithmetic form for $\sim x, x \wedge y$, $x \vee y$ coincide (in the field of events) with those of Stone, where the sum has to be taken 'mod 2', however, in order to obtain a Boolean ring.

The conventions adopted here do not give rise to algebraic properties of this kind, but seem to be the most suitable for expressing, simply and naturally, many things which are otherwise difficult to express.

We will give examples at the end of this chapter (Section 2.11) in order not to interrupt the flow of the argument, and we will often use similar simplifications. It will be seen that it is not only a question of expressions concerning events or random quantities: for identical reasons, the same conventions meet requirements which also occur in other fields.

2.5.4. We have mentioned, in Section 2.3, the set-theoretic interpretation. It is clear that, by interpreting the events as sets, the operations $\sim$, $\wedge$, $\vee$, which we have introduced, correspond in that context to the set-theoretic operations $\sim$, $\cap$, $\cup$ (*complementation, intersection, union*). For random quantities, understood as functions of the 'point' Q, $Z = X \vee Y$ is the function that, at each point Q, assumes the larger of the two values $X(Q)$ and $Y(Q)$:

(4) $Z(Q) = X(Q) \vee Y(Q)$ (and similarly for $\wedge$).

A geometrical representation (which is formally identical) is especially useful, particularly for didactic purposes, even if a genuine set-theoretic interpretation is lacking: it is that of the so-called 'Venn diagrams'. The events which one wishes to represent are drawn as areas of a rectangle, which itself represents the certain event. The areas are delimited with lines, or, better, distinguished with different types of shading. In this way, one can illustrate visually the relationships which are supposed to exist among the different events: the existence, or not, of a certain intersection— distinguished by the overlapping of different shadings—the inclusion of one event in another; and so on. Of course, it is only in rather simple examples that clear figures, whose areas are not too contorted, are possible.

In Figure 2.1 (a,b) are shown the cases of two and three events, respectively, where all the 4 (or 8) intersections are non-empty, i.e. are possible events; whereas in Figure 2.1(c) two of the pair-wise intersections are not present.

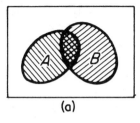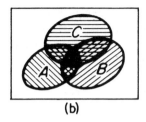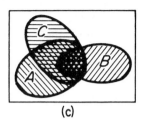

(a)   (b)   (c)

Figure 2.1 Venn diagrams: the representations of events and their logical relationships in the set-theoretic interpretation. (a), (b) The cases of two and, respectively, three events with all (4 and, respectively, 8) constituents possible. (c) An example in which only 6 of the 8 combinations give (possible) constituents

2.6 ASSERTION, IMPLICATION; INCOMPATIBILITY

2.6.1. We began this chapter by saying that, for You, every event, or proposition, can be either certain, or impossible, or possible. We then talked about possibility. The time has now come to translate these premises into a precise argument. We must make a distinction which, in the terminology proposed by B. O. Koopman,† could be called a distinction between *contemplated propositions* and *asserted propositions*. As considered so far, a proposition E is always a contemplated proposition (for which You, or anyone else, could know whether it is true or false). Thus it remains, even if changed into $E = 1$, or $\sim E = 0$, or $(E = 1) = 1$, etc., or, put into words, 'E is true', 'not-E is false', 'it is true that E is true'. Nothing is altered, because these are simply more or less extended ways of saying nothing more and nothing less than E.

To make an *assertion*, we have to step outside of the vicious circle by saying something extra-logical; such as 'I assert that E is true', 'For You, it is certain that E is impossible', 'For me, E is possible': i.e. something expressing not a logical relationship between propositions, but a relationship between the proposition and the *speaker*.

To denote this succinctly, the symbol ⊢ has been introduced. If E is a proposition, an event, then, by using ⊢ as a prefix, ⊢ E becomes the *assertion* that 'E is certain' (for someone). Naturally, ⊢ $\sim E$ is the assertion that 'E is impossible', whereas by $\sim$ ⊢ E we mean to denote the assertion that 'E is possible' (i.e. the non-assertion of both E and of not-E).

2.6.2. We shall not make much use of this symbol, because we think that, in general, the distinction will be clear from the context (for instance, by saying 'certainly'). It is useful, however, to draw attention to the importance of the distinction, and to illustrate the use of the symbol by giving

---

† *The Bases of Probability*, in Kyburg and Smokler, pp. 161–72.

some examples in order to fix all this in the reader's mind. In any case, these observations were necessary at this juncture in order to make it clear that certain expressions, which we will now introduce, *have to be taken as assertions.*

By saying that an event A *implies* the event B, or that A is *contained* in B, we mean to *assert* that A cannot occur unless B also occurs, or that $A\tilde{B}$ is impossible: in symbols $\vdash \sim A\tilde{B}$. Instead of $\sim A\tilde{B}$ one may also write $\tilde{A} \vee B$, or $A\tilde{B} = 0$, or $A \leqslant B$, or $B - A \geqslant 0$ (because the inequality is false only for $1 \leqslant 0$, i.e. for $A = 1$ and $B = 0$). It is always a question of ways of expressing $\sim A\tilde{B}$, independently of the fact that it is certain, or impossible, or possible, and these give assertions, simply by making the assertions. In order to write that 'A implies B', with the meaning, as we have said, of assertion, it will be necessary to write, e.g., $\vdash A \leqslant B$. However, we will introduce some *ad hoc* symbols, to be understood as already having the value of assertions:

$A \subseteq B. = . \vdash A \leqslant B,$ A implies B;

$A \equiv B. = . \vdash A = B,$ A is identical to B (or $A \subseteq B \wedge B \subseteq A$), or, A and B are either both certainly true or both certainly false: (*certain*) *equality of A and B*;

$A \subset B. = . A \subseteq B \wedge \sim A \equiv B,$ A strictly implies B.†

2.6.3. The relationship of implication, which is clearly reflexive and transitive, induces, over any set of events, a partial ordering and in particular a lattice in which the operations $\wedge$ and $\vee$ (in the sense of 'maximal' element contained in those given, and 'minimal' element containing the given ones) coincide with those of logical product and logical sum, already introduced. This is evident above-all under the set-theoretic interpretation: $A \subseteq B$ means that A is *a subset* of B, possibly coincident with B (this being excluded if we write $A \subset B$, affirmed if we write $A \equiv B$); hence the terms 'contains', 'is contained in', have the opposite meaning to 'imply', 'is implied by', instead of being synonymous as they might appear to be if one thought,

---

† The *equality*, $A = B$, is the event that takes place if A and B are both true or both false, and this can happen for any A and B (except in the case of complementary events, $B = \tilde{A}$). However, in order not to make the language unnecessarily heavy, we will continue to say, as usual, 'equal', rather than 'certainly equal', and to write $=$, rather than $\equiv$, except in ambiguous cases.

As regards *strict implication*, observe that it asserts that $A \leqslant B$ with certainty, but that $A = B$ is not certain. In other words, we exclude $A > B$, i.e. A true and B false, but we do not exclude the converse, $A < B$. Nothing is said concerning the possibility or impossibility of A and B being either both true, or both false. Observe that $A \subset B$ means $(A \subseteq B) \wedge \sim (A \equiv B)$ or $(\vdash A \leqslant B) \wedge (\sim \vdash A = B)$, which is very different from $\vdash [(A \leqslant B). \sim (A = B)] = \vdash A < B = \vdash \tilde{A}B$, which denotes the assertion that A is false and B is true.

The meaning of all these relationships is immediately, intuitively obvious under the set-theoretic interpretation.

for both terms, of the interpretation in terms of events.† In other words, in the Venn diagram for two events, which 'in general' (more precisely, for A and B logically independent) has the appearance of Figure 2.1, the part of A not contained in B must be missing (empty); in other words, A must coincide with the doubly shaded area AB (as in Figure 2.2).

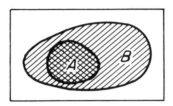

Figure 2.2   Venn diagram: the case of implication (inclusion)

If both regions with single shading are missing we have the case $A \equiv B$, and if the other two regions (double shading and no shading) are missing we have $A \equiv \tilde{B}$. Two other important cases correspond to the absence of the doubly shaded area (case of incompatibility: $AB \equiv 0$), or the absence of the non-shaded area (case of exhaustivity: $\tilde{A}\tilde{B} \equiv 0$).‡

2.6.4. *Incompatibility.* By saying that two events A and B are *incompatible*, we mean to *assert* that it is impossible for them both to occur; i.e. that AB is impossible: in symbols $\vdash \sim AB$. Instead of $\sim AB$ we can write $AB = 0$, or $\tilde{A} \lor \tilde{B}$, or $A + B = A \lor B$, or $A + B \leqslant 1$, or $A \leqslant \tilde{B}$, or $B \leqslant \tilde{A}$, always expressing the event $\sim AB$, independently of the fact that it is certain or impossible or possible. Each of these forms expresses the incompatibility; if it is asserted, we can write, e.g., $\vdash A + B \leq 1$, or $\vdash A \leq \tilde{B}$, which can be expressed, by reduction to the implication, as $A \subseteq \tilde{B}$. By saying that n events $E_1, E_2, \ldots, E_n$ are incompatible, we mean to assert that they are pairwise incompatible ($\vdash E_i E_j = 0$, $i \neq j$); i.e. that at most one of them can occur. As a straightforward extension of $\vdash A + B \leqslant 1$, this can be expressed as $\vdash Y \leqslant 1$, where $Y = E_1 + E_2 + \ldots + E_n$ is the number of 'successes'; i.e. of the events E_i that are true. The same definition also holds for an infinite number of events: in this case, instead of a non-negative integer, Y could also be an infinite cardinal (e.g. that of denumerability, or of the continuum, or any other aleph). We note also that the condition

---

† To avoid possible consequent mnemonic uncertainties about the meaning of $\subseteq$ (and hence the opposite meaning for $\supseteq$), it is sufficient to think of it as corresponding to $\leqslant$ ($\supseteq$ then corresponds to $\geqslant$), whose meaning is clear if we consider operations on the numbers 0 and 1 (events, indicator functions of sets).

‡ The other cases are trivial: A or B or both would be determined, either certain or impossible.

$E_1 + E_2 + \ldots + E_n \equiv E_1 \vee E_2 \vee \ldots \vee E_n$, i.e. the coincidence of the logical and arithmetic sums,† is always characteristic of the case of incompatibility.

In other words, incompatible events are mutually exclusive; in the set-theoretic interpretation it is a question of *disjoint* sets, having an *empty intersection* (pairwise, and hence, *a fortiori*, for three or more).

2.6.5. Exhaustivity. By saying that two events A and B are *exhaustive*, we mean to *assert* that it is impossible for neither of them to occur; i.e. that $\tilde{A}\tilde{B}$ is impossible: in symbols $\vdash \sim \tilde{A}\tilde{B}$. Instead of $\sim \tilde{A}\tilde{B}$ one can (as above) write $\tilde{A}\tilde{B} = 0$, or $A \vee B$, or $\tilde{A} + \tilde{B} = \tilde{A} \vee \tilde{B}$ (i.e. $2 - (A + B) = 1 - AB$, $A + B = 1 + AB$), or $A + B \geqslant 1$, or $\tilde{A} \leqslant B$, or $\tilde{B} \leqslant A$; another form for the exhaustivity is therefore, for instance, $\vdash A + B \geqslant 1$. This lends itself easily to the extension of the definition to the case of n events, or even to an infinite number. By saying that these are exhaustive (or, better, form an exhaustive family—but the phrase is cumbersome), we mean to assert that at least one of them must take place; i.e., in the preceding notation, $\vdash Y \geqslant 1$. This shows the relationship between the two conditions. In the set-theoretic interpretation, it is a question of a family of sets which *covers* the whole set $\mathcal{Q}$ of possible points (of course, there may be some overlapping); i.e. those sets of points for which the complement of the union is empty.‡

2.7 PARTITIONS; CONSTITUENTS; LOGICAL DEPENDENCE AND INDEPENDENCE

2.7.1. Partitions. A *partition* is a family of *incompatible and exhaustive* events—i.e. for which it is *certain* that one and only one event occurs. The coexistence of the conditions $\vdash Y \leqslant 1$ and $\vdash Y \geqslant 1$ means, in fact, $\vdash Y = 1$. A partition can be finite or infinite: partitions (and, for the simplest conclusions, in particular finite partitions) have a fundamental importance in the calculus of probability (which, as already indicated, will consist in distributing a unit 'mass' of probability among the different events of each partition).

It is therefore of importance to see now if, and how, one can reduce the general case, in which one considers any finite number of events $E_1, E_2, \ldots, E_n$, to that of a partition. We observe first of all that if, in particular, the E_i are

---

† For any non-negative (random) numbers the same conclusion is obviously valid: such equality holds if and only if at most one of them can be non-zero.

‡ Suppose that, instead of considering $\mathcal{Q}$—the space of possible points for You, now—one considers a larger space $\mathcal{S}$ which contains, in addition, certain points that are already known to be impossible (for instance, in the light of more recent information). In all the preceding cases, the statement that a set is *empty* must be replaced by *empty of possible points*—i.e. empty of points belonging to $\mathcal{Q}$. In diagrams, one could think of the region $\mathcal{S} \sim \mathcal{Q}$ as drawn in black, and consider it as 'non-existent'.

already incompatible, but not exhaustive, it will be sufficient to add on the extra event

$$E_0 = 1 - (E_1 + E_2 + \ldots + E_n) \quad \text{(i.e., in another form, } E_0 = \tilde{E}_1 \tilde{E}_2 \ldots \tilde{E}_n).$$

In the general case, we must consider the 2^n products $E'_1 E'_2 \ldots E'_n$ where each E'_i is either E_i, or its complement $\tilde{E}_i$; formally, we can obtain them as the individual terms of the expansion $(E_1 + \tilde{E}_1)(E_2 + \tilde{E}_2) \ldots (E_n + \tilde{E}_n)$ which is identically 1, since each factor is 1. Some of the 2^n terms may turn out to be impossible and do not have to be considered: those which remain, and are therefore possible, are called the *constituents* $C_1, C_2, \ldots, C_s$ of the partition determined by $E_1, E_2, \ldots, E_n$, where $s \leqslant 2^n$.

By observing that the given expansion has value 1, we have already established that we are dealing with a partition; on the other hand, the fact is evident *per se* (even more so under the set-theoretic interpretation). A partition is given by a family of disjoint sets which covers the space $\mathscr{Q}$; or, in other words, into which $\mathscr{Q}$ is subdivided—in the same way, for instance, in which Italy is divided into municipalities. If, instead, we perform any other division whatsoever, the partition given by the constituents is that into the 'pieces' resulting from such a subdivision. For instance, Italy east and west of the Monte Mario meridian, north and south of a given parallel, areas of altitude above and below 500 metres, areas more or less than 50 kilometres from the sea, belonging to a province the name of whose capital or main city begins with a vowel or consonant, and so on.†

Sometimes it will also be useful to introduce the (clumsy) notion of a *'multi-event'* for cases in which (provided we do not restrict ourselves to meaning 'event' in a purely technical sense) a partition might correctly be called an 'event with many alternatives'. Such is a game—a football game, for instance—with the three alternatives 'victory', 'draw' and 'defeat' (and possibly a fourth, 'not valid' because of postponement, etc.). The same holds in the case of drawings from an urn containing balls of three or more different colours, e.g. 'white', 'red', 'black'; or throwing a die, or two dice, with possible points in the range 1 to 6, or 2 to 12, respectively. A multi-event with m alternatives—more briefly an *'m-event'*—can always be thought of as a random quantity with m possible values (e.g. $1, 2, \ldots, m$). In the case of a single die, the 'points' are precisely $1, 2, \ldots, 6$, whereas for the two dice it is irrelevant whether we use $2, 3, \ldots, 12$, or $1, 2, \ldots, 11$. The colours, or results of the game, could similarly be coded numerically. In speaking of an m-event we want, essentially, to emphasize the *qualitative* aspects of the

---

† Caution: do not think of separate parts of a unique non-connected 'piece' as 'pieces'—the topology of the representation must be ignored. The 'piece' of Italy north-east of Monte Mario with altitude below 500 metres and more than 50 kilometres from the sea in a province beginning with a vowel is certainly composed of separated parts (for instance in the provinces of Ancona and Udine).

alternatives. It is then appropriate to use the mathematical interpretation of them as unit vectors $(1, 0, \ldots, 0)$, $(0, 1, 0, \ldots, 0)$, $\ldots$, $(0, 0, 0, \ldots, 1)$ in an m-dimensional space. In this way, writing E_h ($h = 1, 2, \ldots, m$) for the events† which consist in the occurrence of the hth alternative, an m-event can be identified with the random vector $(E_1, E_2, \ldots, E_m)$. The (arithmetic) sum of multi-events gives, therefore, the number of occurrences of the single results: for instance, (W, R, B) = the number of drawings of White, Red and Black balls. We observe the analogy with the case of events, which could be handled in this same way, by substituting $(1, 0)$ for 0 and $(0, 1)$ for 1 (if the advantage of the symmetry seemed to compensate for the unnecessary introduction of the doubleton).

2.7.2. *Logical dependence and independence of events.* We define n events (necessarily *possible*) to be *logically independent* when they give rise to 2^n possible constituents. This means that each of these events remains uncertain (possible) even after the outcomes of all the others, whatever they may be, are known: this explains the choice of terminology. In fact, let us suppose that one of the products is impossible, and therefore only a constituent in a formal sense—without loss of generality, take it to be $E_1 E_2 \ldots E_n$. E_1 is possible, $E_1 E_2$ may or may not be, and the same holds for $E_1 E_2 E_3$, $E_1 E_2 E_3 E_4$, etc. If one of these products is impossible obviously all the subsequent ones are; the last one—the product of all the n events—is impossible by hypothesis, and therefore either it or one of the preceding ones must be the first to be impossible: suppose this is $E_1 E_2 E_3 E_4$. This means that it is possible for events E_1, E_2 and E_3 to occur and that, knowing this, we are in a position to exclude the possibility that E_4 can be true. The events are therefore in this case—i.e. if the number of constituents is $s < 2^n$—logically dependent.

Of course, if n events are logically independent the subsets of the n are, *a fortiori*, independent: the converse does not hold. Even if all their proper subsets exhibit logical independence n events can still be logically dependent. As a simple example, let all the constituents in which the number of successes is even be possible, and no others; this imposes no restrictions on the result of any $n - 1$ events whatsoever, but for each event the result is determined once we know the results of the others.

2.7.3. If one wishes to consider more specifically the dependence of a *particular event* E on certain others, $E_1, E_2, \ldots, E_n$, it becomes necessary to consider several cases. It is, in fact, possible that E remains uncertain after we know the results of the E_i, whatever these may be: we then call it *logically independent*. On the other hand, it is possible that it will always be

---

† Necessarily incompatible and exhaustive.

determined (either true or false), in which case we call it *logically dependent*. However, an intermediate case could also arise: the uncertainty or the determination of E might depend on the actual results of the E_i; this we will call *logically semidependence*. We could be more precise and refer to logical semidependence *from below*, or *from above*, or *two-sided*, according to whether there exist outcomes for the E_i which make E *certain*, or *impossible*, or whether there exist outcomes of both types.

In order to characterize the various types of event, with respect to the fixed E_i, it suffices to consider the constituents determined by the E_i. We have $C_1 + C_2 + \ldots + C_s = 1$, and each event E can therefore be decomposed into $E = EC_1 + EC_2 + \ldots + EC_s$. For any one of the summands, say EC_h, there are three possibilities: either $EC_h = C_h$ (if C_h is contained in E), or $EC_h = 0$ (if C_h is contained in $\tilde{E}$), or else $0 \subset EC_h \subset C_h$ (if both EC_h and $\tilde{E}C_h$ are possible). The possible results for the E_i correspond to the occurrence of one of the constituents C_h: according as C_h is of the first, second or third type, E turns out to be *certain, impossible* or remains *uncertain*, respectively.

The conclusions are obvious.

E is *logically dependent* if constituents of the third type do not exist; i.e. if E is a sum of constituents (of the first type). We could also say that E is logically dependent on the E_i if and only if it is expressible as a function of them by means of logical operations: in this case we have dependence by definition. The value (true or false) of such an expression is, in fact, determined by the values of the variables appearing in it; conversely, every such expression reduces to a *canonical form* as a sum of constituents, and, therefore, the condition is also necessary. In this case, constituents of both the first and second types exist; otherwise, E would have been either certain or impossible to begin with, contrary to hypothesis.

E is *logically independent* if all the constituents are of the third type, and *logically semidependent* if some, but not all, are of the third type: in the latter case, we have semidependence *from below* if the others are all of the first type, *from above* if they are all of the second type, *two-sided* if there are some of each type.

If we consider the two events

$E' = $ the sum of all the constituents of the first type, and
$E'' = $ the sum of all the constituents of the first and third types,

it clearly turns out that in each case $E' \subseteq E \subseteq E''$, and E' and E'' are, respectively, the maximal event certainly contained in E, and the minimal event that certainly contains it—i.e. the events giving the best possible bounds.

We can then say that: E is logically dependent if $E' = E''$ (and hence $= E$); logically independent if $E' \equiv 0$ and $E'' = 1$; semidependent from below, from above, or two-sided, if

$$0 \subset E' \subset E'' \equiv 1, \qquad 0 \equiv E' \subset E'' \subset 1, \qquad 0 \subset E' \subset E'' \subset 1,$$

respectively.

2.7.4. These notions of logical dependence and independence are meaningful more generally; they apply not only to the case of events, as considered so far, but also to partitions, or random quantities, or any random entities whatsoever. We will present the development for the case of random quantities, which is the most intuitive; it will then suffice to remark that the concept is always the same.

Random quantities (suppose, to fix ideas, that there are three: X, Y, Z) are said to be *logically independent* if there are no circumstances in which the knowledge of some of them can modify the uncertainty concerning the others. This means that if X, Y, Z have, respectively, r possible values x_i, s possible values y_j, t possible values z_h, then all the rst triples (x_i, y_j, z_h) are possible for (X, Y, Z); i.e. the set $\mathcal{Q}$ of possible points (x, y, z) is the *cartesian product of the sets*, $\mathcal{Q}_x$, $\mathcal{Q}_y$, $\mathcal{Q}_z$ of possible values for X, Y, Z. In this form the definition is general: it is valid not only for n (instead of 3), but also if the random quantities have an infinite number of possible values (for instance, those of an interval), or in the case of random entities of other kinds, or, generically, for partitions.† In other words, the condition means that nothing, no known interdependence, allows any further restriction of the set $\mathcal{Q}$ of possible points over and above that resulting from the fact that the individual random quantities, or entities, must assume values in $\mathcal{Q}_1, \mathcal{Q}_2, \ldots, \mathcal{Q}_n$.

2.7.5. The *logical dependence* of one (random) quantity on others (to fix ideas consider the dependence of Z on X and Y) has exactly the meaning that it has in analysis: Z is a *function* (i.e. a *one-valued* function) of X and Y, $Z = f(X, Y)$, the function $z = f(x, y)$ being defined for all the possible points (x, y) of (X, Y). *Logical independence* means that the set of possible values of Z conditional on the knowledge of the values of X and Y (any pair of possible values (x, y) for (X, Y)) is always the set of $\mathcal{Q}$ of all the (unconditionally) possible values of Z. Intermediate cases, which are not worth listing in further detail, always give *logical semidependence*.

2.7.6. A critical observation is appropriate at this point, both as a refinement of the present argument, and to exemplify various cases in which it

---

† A partition can be reduced to a random quantity by considering as such, for example, the index i of $E_i : X = i$ if E_i is true, provided the partition has at most the cardinality of the continuum, or is denumerable if we require integer values.

is useful to examine whether the logic needs to be taken with a pinch of salt (cf. Appendix).

We will confine ourselves to a single example. Suppose that X and Y have as possible values all the numbers between 0 and 1, with the condition that $X + Y$ is irrational: then $\mathscr{Q}$ is the unit square with an infinite number of 'scratches' removed—parallel to the diagonal and corresponding precisely to the lines $x + y = $ rational. For the partition into *points*, logical independence does not hold; it would hold, however, for every partition into vertical or horizontal stripes, however small.

Is it advantageous to say that we do not have logical independence when its failure is attributable to subtleties of this kind? Clearly, there is no categorical answer. It seems obvious that, depending on the problem and on one's intentions, one decides whether or not to take such subtleties into account (of course, one must be careful to be precise when such is required).

2.7.7. Finally, we make an observation which, strictly speaking, is unnecessary—being implicit in the very definition of 'possibility'—but which it is convenient to make and underline. All the notions we have encountered, or introduced—from incompatibility to logical dependence or independence—are relative to a given 'state of information'. They are valid for You (or for me, or him) according to the knowledge, or the ignorance, determining the uncertainty; i.e. the extent of the range of the *possible*, of the set $\mathscr{Q}$ (yours, mine, his, ...).

It is a question of relative and personal notions, but nonetheless objective, in the sense that they depend on what one knows, or does not know, and not on one's opinion concerning what one does not know, and what is, consequently, uncertain.

In order to avoid ambiguity, we must never forget that we are always speaking about uncertainty in the simple sense of ignorance. In particular, of course, we are dealing with matters traditionally attributed to 'chance'—a trace of this remains in the word 'random',† and in other expressions which we will be using. In general, however, we are concerned with any future matters whatsoever, and also of things in the past concerning which there is no information, or for which no information is available to You, or which You cannot remember exactly: we might even be concerned with tautologies. The various cases differ in one important aspect: that is the existence and degree of possibility and facility of obtaining, in one way or another, further information, should one wish to do so. This fact will, of course, be relevant in determining behaviour in decision problems, where it could be convenient to condition on the acquisition of new information. But apart from this, basically, it is convenient to regard any distinctions of this kind as unimportant. The only essential element, which determines

---

† *Translators' note.* The Italian word here is *'aleatorio'* (cf. French, aléatoire) from the Latin *alea* meaning *die*: 'alea jacta est!'—'the die is cast!'—as Caesar said when crossing the Rubicon.

and characterizes our object of study, is the existence of imperfect information—of whatever kind—and the situations of uncertainty in which, consequently, You might find yourself.

> There is a prejudice that uncertainty and probability can only refer to future matters since these are not 'determined'—in some metaphysical sense attributed to the facts themselves instead of to the ignorance of the person judging them. In this connection, it is useful to recall the following observation of E. Borel: 'One can bet on Heads or Tails while the coin, already tossed, is in the air, and its movement is completely determined; and one can also bet after the coin has fallen, with the sole proviso that one has not seen on which side it has come to rest'.

2.7.8. *Remark.* It might be useful to point out (or, for those who already know it, to recall the fact) that in the theory of probability one often uses the term 'independence' (without further qualification) to denote a different condition, that of *stochastic independence*, which refers to *probability* and will be introduced in Chapter 4.

Be careful not to confuse it with *logical* independence—which we have just discussed—or with *linear* independence, which we will discuss in Section 2.8. Both of these notions have an objective meaning; i.e. independent of the evaluation of the probabilities.

2.8 REPRESENTATIONS IN LINEAR FORM

2.8.1. *Basic notions.* When referring to the set $\mathscr{Q}$ of possible 'points' in the case of two random quantities X and Y, we tacitly interpreted the pair (x, y) as cartesian coordinates in the plane (which it was natural to take as the space of alternatives $\mathscr{S}$). Similarly, for three or more points, we extend to ordinary space, or to spaces of any dimension (always in cartesian coordinates).

This was simply a question of habit and, therefore, of convenience. One could have thought of any coordinate system; of a curved surface instead of a plane, or, in order to say more and in a better way, it is enough to think in terms of a space in a merely abstract sense, for which such distinctions of a geometric nature do not even make sense. With reference to the simplest case, it is sufficient that different pairs (x, y) are made to correspond to distinct 'points'.

For further reasons, which we now wish to take into account—because, as we shall see in Chapter 3, they are essential for the theory of probability—it becomes important instead to think of $\mathscr{S}$ as a *linear (affine) space*. We shall call it the *linear ambit* and denote it by $\mathscr{A}$, because at times it will be convenient to consider as the space $\mathscr{S}$ not the whole of $\mathscr{A}$, but a less extensive manifold which contains $\mathscr{Q}$. For example: if $\mathscr{A}$ is ordinary space, and X, Y, Z are related by the equation $X^2 + Y^2 + Z^2 = R^2$, it might be convenient to think of $\mathscr{S}$ as the spherical surface on which one finds the possible points $\mathscr{Q}$; these may consist of all the points of the surface, or a part of it, or just a few points, depending on other restrictions and circumstances and knowledge.

A representation which is *linear* with respect to certain random quantities (e.g. those considered initially) is such with respect to others which are linear combinations of them (but not with respect to the rest). If we require that linearity holds for the rest too we have to extend the linear ambit $\mathscr{A}$ to new dimensions, as we shall see later.

The random quantities linearly represented in an ambit $\mathscr{A}$ themselves constitute a linear system, which we denote by $\mathscr{L}$, and which is dual to $\mathscr{A}$. One might ask whether it is useful to think of the two dual spaces, $\mathscr{A}$ and $\mathscr{L}$, as superposed. In principle, the answer is no : in fact, only the affine notions have any meaning, and the metric, introduced surreptitiously by means of such a superposition, would be dependent on the arbitrary choice of the coordinate system that has to be superposed onto its dual. In general, for this reason, it is not even practically convenient. A unique exception is perhaps that of the case we considered first, in which we start from *events*, and it is 'natural' to represent them with unit, orthogonal vectors. In any case, whether or not this possibility is useful in a particular case, it is important never to forget that it is only the affine properties which make sense.

These properties also underlie the notions and methods fundamental to the theory of probability. On the other hand, the things in question are very elementary, and are currently applied without first introducing this formulation and terminology—which might well be considered excessively theoretical and, for the purpose in hand, disproportionately so. Nevertheless, if one is prepared to make the small effort necessary to picture the question in terms of the present scheme, many aspects of what follows will appear obvious and well-connected among themselves, instead of, as they might otherwise appear, unrelated and confused. So much so, that the preeminent—one might even say exclusive—rôle of linearity in the theory of probability has always remained very much in the background. This is, in part perhaps, because of the prominence given to the Boolean operations, and because of the non-immediacy of the arithmetic operations on events when the latter are not identified with their 'indicators'. The present treatment is intended to provide the framework within which these observations will find their justification and clarification.

2.8.2. Let us begin by considering events $E_1, E_2, \ldots, E_n$, and, often, in order to be able to think in terms of ordinary space, we will, without essential loss of generality, take n to be 3.

The linear ambit $\mathscr{A}$ is the affine vector space in n dimensions, with coordinate system $x_1, x_2, \ldots, x_n$, in which we will consider the values of the random quantities $X_1, X_2, \ldots, X_n$. In this case, the latter are the events $E_1, E_2, \ldots, E_n$, taking only the values 0 and 1: the set of 'possible' points consists at most, therefore, of the 2^n points (8, if $n = 3$) with coordinates either 0 or 1, and may be a subset of these. One sees immediately—as was inevitable—that the 'possible' points correspond to the s ($s \leqslant 2^n$) constituents.

Given the special rôle of these points, it is convenient to think of the prism, of which they are the vertices, as a cube (or hypercube) and, therefore, to think of the cartesian coordinate system x_i as orthogonal and of unit length—with the reservation that this metric not be taken too 'seriously'.

The linear system $\mathscr{L}$, of linear combinations of $E_1, E_2, \ldots E_n$, consists of random quantities $X = u_1 E_1 + u_2 E_2 + \ldots + u_n E_n$,† interpretable as the *gain* of someone who receives an amount u_1 if E_1 is true, plus an amount u_2 if E_2 is true, and so on (of course, the 'gains' may be positive or negative). The X possess at most as many (distinct) possible values as there are constituents—namely s—and the latter occurs if the corresponding 'possible points' are found on distinct hyperplanes $\sum_i u_i x_i = \text{constant}$.

An important example is that where $Y =$ *the number of successes.* In order to obtain this, it is sufficient to take all the $u_i = 1$—a gain of 1 for each event—obtaining, as we have already shown directly, $Y = E_1 + E_2 + \ldots + E_n$. In this case, it is clearly not true that the possible points occur on distinct hyperplanes; if all the 2^n vertices of the hypercube are possible, they are, in fact, distributed over the $n + 1$ hyperplanes $Y = 0, 1, 2, \ldots, n$ according to the binomial coefficients $(1, n, \frac{1}{2}n(n - 1), \ldots, n, 1)$, $\binom{n}{h}$ being the number of possible ways of obtaining h successes in n events.

For the case $n = 3$, we shall denote the cartesian coordinates of the ambit $\mathscr{A}$ in the usual manner, by x, y, z, and those of the dual system $\mathscr{L}$ by u, v, w. If $X = uE_1 + vE_2 + wE_3$, then $ux + vy + wz$ is the value which X would assume if E_1 takes the value x, E_2 the value y and E_3 the value z. Given the meaning of the E_i, such values can only be either 0 or 1, and the value of the random quantity X (e.g. gain) can only be one of those corresponding to the 8 vertices of the cube (or to a part of it, if not all the vertices are possible).

Here are the coordinates of such vertices Q together with the corresponding values of X:

$Q = (0, 0, 0), (0, 0, 1), (0, 1, 0), (1, 0, 0), (0, 1, 1), (1, 0, 1), (1, 1, 0), \quad (1, 1, 1),$

$X = \quad 0 \qquad w \qquad v \qquad u \qquad v + w \quad u + w \quad u + v \quad u + v + w.$

In particular, for $u = v = w = 1$, we see (as was obvious) that the number of successes is 0 in one case, 1 in three cases, 2 in three cases and 3 in one

---

† In order to simplify this example we omit the constant u_0 (cf. Section 2.8.3).

case. In addition (apart from the combinatorial meaning, $(1 + 1)^3 = 1 + 3 + 3 + 1 = 8$), this shows that, when projected onto a diagonal, the vertices of the cube fall as follows: one at each end, and three each at $\frac{1}{3}$ and $\frac{2}{3}$ of the way along the diagonal.

2.8.3. The sum $\sum_i u_i x_i$ (in particular $ux + vy + wz$) is a linear function both of X (i.e. of its components u_i), and also of $\mathscr{Q}$ (i.e. of its coordinates x_i). We will denote it both by $X(Q)$—thinking of it as 'the value of a given X as Q varies'—and also by $Q(X)$—thinking of it as 'the value assigned to different X by the resultant Q'. The same operation, however, will still turn out to be useful independently of the fact that Q is a possible point (i.e. $Q \in \mathscr{Q}$). That is, by replacing Q by any A in $\mathscr{A}$, writing $X(A)$ or $A(X)$:

$$A(X) = X(A) = \sum_i u_i x_i,$$

where the u_i are the coordinates of the X considered as points of $\mathscr{L}$, or, better, the components of X considered as vectors of $\mathscr{L}$, and similarly the x_i are coordinates (or components) of the A considered as points (or vectors) of $\mathscr{A}$.† The expressions $A(X)$ or $X(A)$ then appear as *products* of vectors, A and X, belonging to the two dual spaces $\mathscr{A}$ and $\mathscr{L}$.‡

What we have said so far in this section (2.8.3) is independent of the assumption that, rather than taking any random quantities whatsoever, we start with events, $X_i = E_i$ (as we did in Section 2.8.2, in order to fix ideas). Since it is convenient to consider not only the homogeneous linear combinations, $X = \sum_i u_i X_i$, as we have up until now, but also complete combinations with an additional constant, say u_0, we will always assume as added to the X_i a fictitious random quantity X_0, taking the single value $X_0 \equiv 1$ *with certainty*. The summand $u_0 X_0$ has precisely the value u_0, with no alteration to the formula; we have only to take into account that there is an additional, fictitious, variable, x_0, and that, for all possible points (and, usually, also for every A to be considered), we will have $x_0 = 1$.

2.8.4. *Linear dependence and independence.* We have considered $X = \sum_i u_i X_i$ ($i = 0, 1, 2, \ldots n$), linear combinations (either homogeneous or complete) of n random quantities X_i ($i = 1, 2, \ldots, n$); X is said to be linearly dependent on the X_i. It may be, however, that the X_i are already linearly

---

† Given that the point O (the origin) has meaning in both $\mathscr{L}$ and $\mathscr{A}$, there is no risk of ambiguity in identifying points and vectors.
‡ If one thinks of the two spaces as superposed—we have already said that, in general, this is not advisable—we would have the scalar product. In any case, one could write AX and XA, instead of $A(X)$ and $X(A)$, thinking in terms of the product rather than writing it as a 'function'. The main application, however, will be when $A = P$ (probability, prevision), and the omission of the parentheses in this case—although used by some authors—seems to give less emphasis to the structure of the formulae, and therefore to the meaning.

dependent themselves; i.e. that one of their linear combinations is identically zero (or constant: due to the inclusion of X_0 the two are essentially identical), in which case at least one of the X_i is a linear combination of the others and can be eliminated (because it already appears as a combination of the others). Geometrically, this means that the set $\mathcal{Q}$ of possible points belongs to a linear subspace $\mathcal{A}'$ of $\mathcal{A}$, and hence it is sufficient to confine attention to $\mathcal{A}'$: the extension from $\mathcal{A}'$ to $\mathcal{A}$ is illusory—one adds only points which are certainly impossible.

We observe that linear dependence is a special case of logical dependence—i.e. that linear dependence is a more restrictive condition. Conversely, it goes without saying that logical independence is more restrictive than linear independence.

We now return, briefly, to the case of events, for even here the distinction between linear dependence and logical dependence is of fundamental importance for the theory of probability. The negation of E depends linearly on E: in fact, $\tilde{E} = 1 - E$. On the other hand, the *logical product* $E = AB$, and the *logical sum* $E = A \vee B$, do not depend linearly on A and B (except when, under the assumption that A and B are incompatible, the logical sum has the form $A \vee B = A + B$). However, the logical sum does depend linearly on the two events and their product: $A \vee B = A + B - AB$. In general, the logical sum of three or more events depends linearly on the events themselves and on their products two at a time, three at a time,..., and finally the product of all of them: cf. Section 2.5.2. Apart from these cases of a general nature, however, it is possible that an event can be a linear combination of others 'by chance' (so to speak): an example can be found in Chapter 3, in connection with a probability problem, where an event E is expressed linearly as a function of others by the following formula

$$E = \tfrac{1}{7}(3 - 2E_1 + E_2 - E_3 + 3E_4 + 5E_5 - 5E_6).$$

How can one tell whether or not such a linear dependence exists? It is sufficient to express all events as sums of constituents and then to see whether the matrix (consisting entirely of zeroes and ones) is zero or not.

2.8.5. The above considerations refer to the system $\mathcal{L}$, but linear dependence is still meaningful and important in the ambit $\mathcal{A}$. The interest there lies in considering the barycentre P of two points Q_1 and Q_2 with 'masses' q_1 and q_2, where $q_1 + q_2 = 1$. By a well-known property in mechanics—which is, on the other hand, an immediate consequence of linearity—each linear function X assumes at P the value $X(P) = q_1 X(Q_1) + q_2 X(Q_2)$, and the same holds for the barycentre of three, or (leaving ordinary space) any number of points whatsoever. The property even holds if some of the masses are negative, but the cases in which we are normally interested are those with non-negative masses (usually, in fact, we will be dealing with probability).

The barycentre can therefore be any point† belonging to the *convex hull* of the points Q_h under consideration. Consideration of the convex hull determined by the 'possible points', $Q \in \mathcal{Q}$, or, in other words, the convex hull of $\mathcal{Q}$, will play a fundamental rôle in the calculus of probability. Dually (and this property too, well-known and intuitive, will turn out to be meaningful in future applications), the convex hull is also the intersection of all the half-spaces containing $\mathcal{Q}$. In other words, if a point P belongs to the convex hull $K(I)$ of a set I, then it is on the same side as I with respect to any hyperplane not cutting the set—i.e. which leaves it all on the same side. On the other hand, if a point does not belong to the convex hull, there exists a hyperplane separating it from I—i.e. which does not cut the latter and leaves it all on the opposite side with respect to the point. Translating all this into an analytic form: every non-negative linear function on I is also such on $K(I)$; conversely, the property does not hold for any point not belonging to $K(I)$.

2.8.6. Returning to the case of the cube (Section 2.8.2), we already have a meaningful example, although a little too simple, of the way in which the convex hull varies as we consider all the 8 vertices or a subset of them (cf. Chapter 3, where the probabilistic meaning will also appear).

With this example in mind, it is now possible to make an observation which, although trivial in this context, is useful for explaining in an intuitive way our immediate intentions (Section 2.8.7) in cases where it could seem less obvious and perhaps strange.

In the space $\mathcal{A}$ we could represent the 8 constituents by the vertices of the cube: we suppose that all 8 actually exist, there is no need to consider other cases here. In the dual space $\mathcal{L}$, however, we could only represent the random quantities depending linearly on E_1, E_2, E_3. The 8 constituents, considered as random quantities, could not be represented, and so neither could the random quantities derived from them linearly—unless these happened to be linearly dependent on the three fundamental events E_i. Does the method create a discrimination between events which have a representation as vectors in $\mathcal{L}$ and those which do not? If so, can we put the situation right?

The answer to the first question is no: the method creates no discrimination. The fact is that it enables us to consider more or fewer dimensions

---

† If the points Q_h are infinite in number, then in order for this to be true we must also allow 'limit cases' of barycentres (which, in other respects, correspond to actual requirements of the calculus of probability, at least according to the version we will follow, in which we do not assume 'countable additivity'). Anyway, apart from questions of interpretation, this simply means that by convex hull we mean the set of barycentres completed by their possible adherent points.

according to what we need. The representation in terms of the cube is sufficient for the separation of the 8 constituents (as points of $\mathscr{A}$), and for the consideration of random quantities linearly dependent on the three E_h. If we wished, we could even reduce to a single dimension by considering only the random quantity $X = 4E_1 + 2E_2 + E_3$: this is sufficient to characterize the 8 constituents since X can assume the values 0, 1, 2, 3, 4, 5, 6, 7. These values, incidentally, are obtained by reading the triple of coordinates as a binary number—e.g. $(1, 0, 1) = 101$ (binary) $= 4 + 0 + 1 = 5$. If we were interested only in such an X (up to linear transformations, $aX + b$), and in distinguishing the constituents, this would be sufficient. Similarly, if in addition to X we are interested, for instance, in the number of successes $Y = E_1 + E_2 + E_3$, and nothing else, we could pass to two dimensions. Suppose, however, that, for reasons which depend on the linearization, we are interested in studying, in $\mathscr{L}$, either one of the constituents, or a linear combination of constituents not reducible to a linear combination of the E_h. In this case, it will be necessary to introduce a third dimension and then, if required, others..., up to seven. In general, if there are s constituents we require $s - 1$ dimensions (s if we include a fictitious one for the constant $X_0 \equiv 1$) *in order that everything geometrically representable in $\mathscr{A}$ is also linearly interpretable in $\mathscr{L}$.*

In fact, if in our case (that of the cube) we consider an eight-dimensional space, whose coordinates x_h give the value of the constituents C_h, the possible points, Q_h, are the points with abscissa 1 on one of the 8 axes (because one, and only one, of the 8 constituents must occur). They are linearly independent in the 7-dimensional space $x_1 + x_2 + \ldots + x_8 = 1$: one of the x_h is superfluous, but it makes no difference whether we leave it, or eliminate it and add a fictitious coordinate $x_0 \equiv 1$. In terms of $\mathscr{L}$, we can therefor obtain all the X either as linear combinations $\sum u_h C_h$, for h from 1 to 8, or for h from 0 to 7 (excluding C_8 but adding the fictitious $C_0 \equiv X_0 \equiv 1$).

Conclusion: everything can be represented linearly provided one takes a sufficient number of dimensions. It is possible, and this provides a simplification, to reduce this number by projecting onto a subspace (although in this way we give up the possibility of distinguishing between those things which have the same projection). Thus, for instance, different possible cases may be confounded into a single one, or even if we take care to avoid it, barycentres arising from different distributions of mass may be confounded. In the case of the cube, for example, each internal point can be obtained as the barycentre of $\infty^{7-3} = \infty^4$ different distributions of mass on the 8 vertices.

2.8.7. In the general case, considering any random quantities whatsoever, the same circumstance arises and has even greater interest. Suppose we consider the ambit $\mathscr{A}$ relative to n random quantities X_i $(i = 1, 2, \ldots, n)$ and, for simplicity, let us assume that all the real values are possible and compatible for the X_i: i.e. that all the points of $\mathscr{A}$ are possible ($\mathscr{A} = \mathscr{Q}$). It follows that every random quantity $Z = f(X_1, X_2, \ldots, X_n)$ is *geometrically* individuated in $\mathscr{A}$ (to each point of $\mathscr{A}$ there corresponds, in a known way, a value of Z), but is not *vectorially* represented in $\mathscr{L}$ unless it is a *linear* function of the X_i. If such a vectorial representation for Z is needed, however, it is sufficient to add on a new dimension for it—i.e. to introduce an extra axis, z, or, if one prefers, x_{n+1}, on which Z can be represented.

To give an intuitive illustration : in the plane (x, y) every function $z = f(x, y)$ already has a geometrical representation (visually through contour lines), but in order for z to appear *linearly* in the representation it is necessary to introduce a new axis, z, and to transfer each contour line to the corresponding height, obtaining the surface $z = f(x, y)$.

As a practical example, in fact one which continuously finds application, an even simpler case will suffice. We have a single random quantity, X: by taking the x-axis as the ambit $\mathscr{A}$, we represent, by means of its points, all the possibilities (values x) which determine, together with x, every function of x, $f(x)$. However, if we are interested in the linear representation of a given $f(x)$ we must introduce a new axis, y, and on it represent $y = f(x)$. The linear ambit $\mathscr{A}$ will be the plane (x, y), but for the space $\mathscr{S}$ we could more meaningfully consider the curve $y = f(x)$, whereas $\mathscr{Q}$ could be a set of points on such a curve (if not all values are possible for X). It will be, so to speak, the set $\mathscr{Q}$, previously thought of on the x-axis, projected onto the curve $y = f(x)$. We note, incidentally, that this illustrates the observation made in Section 2.8.1 regarding the non-identification of $\mathscr{A}$ and $\mathscr{S}$. The criterion which has been followed can be explained in the following way : we delimit $\mathscr{S}$ by taking into account the 'essential' circumstances, considering as such the fact of studying X together with a given $Y = f(X)$, whatever the random quantity X may be; we do not take into account the 'secondary' circumstances, considering as such the particular facts or knowledge which, in certain cases or at certain moments, lead us to exclude the possibility of X attaining certain values.

The most important practical case (which we have already mentioned) is the simplest one : that of X and $Y = f(X) = X^2$. The curve is the parabola $y = x^2$, and the linear system $\mathscr{L}$ consists of all the polynomials of second degree in X; $aX^2 + bX + c$. Suppose that we are interested in barycentres of possible points Q_h with given masses q_h. If the points are taken on the parabola we obtain a point $\bar{x}, \bar{y}$, which is meaningful for both coordinates, whereas if we leave the points on the x-axis the barycentre would give the same $\bar{x}$, but no information about $\bar{y}$.

Obviously, if we were interested in considering $Z = X^3$ also (i.e. extending $\mathscr{L}$ to polynomials of the third degree) it would be necessary to take the space (x, y, z) as the ambit $\mathscr{A}$, the curve $y = x^2$, $z = x^3$ as the space $\mathscr{S}$, and to project onto it the set $\mathscr{D}$ already given; and so on.

2.9 MEANS; ASSOCIATIVE MEANS

2.9.1. Within this representation, we will take the opportunity to present, in an abstract form, a notion which has great practical and conceptual importance in all fields, and which, in what follows, will above all prove useful in connection with probabilistic and statistical interpretations. The notion in question is that of a mean. This is usually defined in terms of mere formal properties of particular cases, but (as Oscar Chisini pointed out) it has a well-defined and important meaning as a useful 'summary' or 'synthetic characteristic' of something more complicated.

A prime example (already considered in the preceding pages) is that of the barycentre, or, arithmetically, that of the arithmetic mean (in general weighted) of the coordinates of the point masses. It is well-known how, in mechanics, for many aspects and consequences, everything proceeds *as if* the whole mass were concentrated at the barycentre. In the language of statistics (which we will encounter mainly in Chapters 11 and 12) one would say that knowledge of the barycentre (and of the mass) constitutes, for certain purposes, a *sufficient statistic* (i.e. an exhaustive summary). For other purposes, in mechanics, it is necessary to know in addition the moments of inertia, and the exhaustive summary is then the collection of these items of information *of first and second orders*. It is convenient to point out in advance that knowledge of the second-order characteristics will also play an important rôle in statistics and in the theory of probability. Above all, it gives a powerful tool for studying problems in a way which is often sufficiently exhaustive, although summary.

2.9.2. Let us now consider the definition of mean according to Chisini, which is based precisely on this concept of an exhaustive summary. In this way we impart to the notion the *relative functional* meaning conveyed by '*tailor-made*' (better the German *Zweckmässig*, whose equivalent is missing in other languages: zweck = purpose, mässig = adequate). According to Chisini,† '*x is said to be the mean of n numbers $x_1, x_2, \ldots, x_n$, with respect to a problem in which a function of them $f(x_1, x_2, \ldots, x_n)$ is of interest, if the function assumes the same value when all the x_h are replaced by the mean value x: $f(x_1, x_2, \ldots, x_n) = f(x, x, \ldots, x)$*'. Here we are considering the simplest case, without *weighting*, but the concept is still the same

---

† O. Chisini, 'Sul concetto di media', in *Periodico di Matematiche* (1929); the topic is taken up again in an article by B. de Finetti in *Giorn. Ist. Ital. Attuari* (1931). The proof of the theorem of Nagumo and Kolmogorov can also be found there.

in the latter case, and in that—as we shall see in Chapter 6—of *distributions*, even continuous ones.

2.9.3. The most important type of mean is the *associative* one. The defining property of associative means is that they are unchanged if some of the quantities are replaced by their mean (in the same way as, in order to find the barycentre, one can concentrate some of the masses at their barycentre). Independently, and almost simultaneously, Nagumo and Kolmogorov proved that the associative means are all, and only, the (increasing) transforms of the arithmetic mean. They are obtained by taking an increasing function $\gamma(x)$, and, given the values x_h with respective weights $p_h(\sum p_h = 1)$, instead of taking the barycentre, $\bar{x} = \sum_h p_h x_h$, one takes the barycentre of the corresponding $y_h = \gamma(x_h)$, $\bar{y} = \sum_h p_h y_h$ and then reverts to the 'scale' x by means of the inverse function $m_y = \gamma^{-1}(\bar{y})$ thus obtaining the γ-mean.

The procedure can be clearly 'seen' in the representation of the preceding paragraph. If we consider the example given there, we have $y = \gamma(x) = x^2$, and, of course, we must limit ourselves to the positive semi-axis in order that γ be increasing.† It is a question of thinking of the masses p_h as placed on the parabola; the barycentre is the point whose coordinates are $\bar{x}$ and $\bar{y}$, whereas $x = m_y$ (obtained as shown in the figure) is the point to which corresponds (on the parabola) the same ordinate of the barycentre, the square root of the mean of the squares.

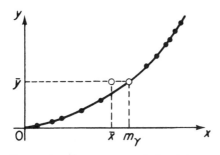

Figure 2.3 Comparison between associative (γ-) means based on comparisons of the convexity of the functions $\gamma(x)$ used to construct them

Considering the other function $z = x^3$ (either by itself in the plane (x, z), or together with $y = x^2$ in the space (x, y, z), as noted in Section 2.8.7), the barycentre would be $\bar{x}, \bar{z}$, respectively, $\bar{x}, \bar{y}, \bar{z}$, where $\bar{z} =$ the mean of the

---

† Or the negative one. In fact, as is easily seen, $\gamma_1(x)$ and $\gamma_2(x)$ are equivalent with respect to the mean if (and only if) $\gamma_1 = a\gamma_2 + b(a \neq 0)$. If we change the sign of a (i.e. change increasing into decreasing) nothing is altered. It is clear from the diagram, in fact, that a change in y, either of scale or sign, or a vertical translation of the curve, makes no difference.

cubes $= \sum_h p_h x_h^3$, and $\sqrt[3]{\bar{z}} =$ the cube root of the mean of the cubes $=$ the cubic mean of the values x_h with weights p_h, and so on. In Chapter 6 we will say something about the most important associative means: these correspond to $\gamma(x) =$ powers (with any positive or negative real exponent whatsoever; if zero we have the limit case of the logarithm), and exponential. At this point, however, it is convenient to consider some general properties related to the notion of convexity of which we have spoken. This will also clarify a few questions which we will meet in Chapter 3.

2.9.4. The barycentre is always in the convex polyhedron (or, in general, the convex hull) determined by the point masses: in our example we can think of it both in the plane and in ordinary space. For the main conclusion of interest to us, the case of the plane is sufficient. If the masses are on a curve whose concavity is always in the same direction, or on a portion of the curve for which this is true, the barycentre is always in the area bounded by the concavity; hence: *the γ-mean is greater than the arithmetic mean if γ (increasing) is concave upwards.* The quadratic mean is therefore greater than the arithmetic mean and so is the cubic mean: the question arises, can these two be compared? Of course; it is sufficient to project the curve $y = x^2$, $z = x^3$ (explicitly, $z = y^{3/2}$) onto the plane (y, z): the concavity is upwards and so the cubic mean is greater.

Even without the graphical comparison, it is sufficient to take into account that 'greater relative concavity' (in the sense that a diagram would display) corresponds, locally, to a greater value of $\gamma''(x)/\gamma'(x)$ (in the interval of interest if the function is not everywhere invertible). In the above example, we have $y''/y' = 2/2x = 1/x$, $z''/z' = 6x/3x^2 = 2/x$ and so, for $x > 0$, z''/z' is always greater. More generally, since for the powers $\gamma(x) = x^c$ one has $\gamma''/\gamma' = c(c-1)x^{c-2}/cx^{c-1} = (c-1)/x$, the mean increases with the exponent; this also holds for $\log x$ (the limit case as $c \to 0$: $\log x \cong (x^c - 1)/c$): in fact, $\gamma''/\gamma' = -x^{-2}/x^{-1} = -1/x = (0 - 1)/x$. This particular choice ($c = 0$, $\gamma = \log$) gives the geometric mean, which, in the case of two, or more generally n, values with equal weights (the 'simple', unweighted case) assumes the more familiar forms: $\sqrt{(x_1 x_2)}$, $\sqrt[n]{(x_1 x_2 \ldots x_n)}$, respectively. For $c = 1$, we have the harmonic mean, the reciprocal of the mean of the reciprocals.

From the fact that $-1 < 0 < 1 < 2 < 3$ it follows that in the above-mentioned cases, for example, we have:

harmonic < geometric < arithmetic < quadratic < cubic.

2.9.5. *Remarks.* Although it may seem strange to do so, we conclude by saying that the following observation is important: the barycentre of points which are on a curve (other than a straight line) is not a point on the curve—unless perhaps 'by chance'. In the same way, the barycentre of points on a surface (not a plane) is not, generally speaking, a point of the surface; and so on, in any dimension. The observation may seem strange because it is so obvious: its obviousness, however, results from the demonstration in terms of the above representation. How many people would recognize the fact before having their attention drawn to it? In facing real problems one often reasons as if what one considers strange, and even absurd, is precisely this fact!

2.10 EXAMPLES AND CLARIFICATIONS

2.10.1. Examples are always useful in order to give a sense of concreteness to concepts introduced in a general and abstract form. In this case, they will serve in addition to underline the meaning and importance of certain refinements, either already mentioned in passing or to be added soon, and also to introduce, before we yet talk about probability, a few of the kinds of situation which we will repeatedly come across in various problems.

Above all, by selecting widely differing examples we intend to remove any possible residual doubts which might lead to restrictive interpretations of the field of uncertainty to which we refer ourselves. The subject matter to which the uncertainty refers is irrelevant: political or economic events, meteorological phenomena, historical or scientific conjectures, judicial investigation, personal or everyday affairs, competition in sport, or any other field in which uncertainty and imperfect knowledge are present. This includes of course—and they in no way differ from the others—the traditional games of chance. This latter is, in fact, the least interesting case, because it leads to a standardized scheme in which all the conceptual and substantial aspects of the problem are made to disappear.

2.10.2. Examples of events. Will a given candidate, on a given occasion, succeed in getting elected (for instance, as a senator, a mayor, a member of a committee, a president of a society or of the university); or in passing (for instance, a student sitting an examination); or in being the winner (in a contest or a lottery, at Bingo, in a sports competition, in a game of cards or chess, or anything else); and so on? Will a vote turn out to be favourable—for instance, for a given law, or for an issue of confidence facing a government, etc.? Is the accused in a given trial really the murderer? And, in any case, will he be convicted as such? Is the approaching tram the one I am waiting for? Will the next child of a given couple be a boy? Will it rain tomorrow at a given place? Will the next attempt at a soft landing on the moon be successful?

In all cases, and in various ways, if we want to be more detailed, or to extend the questions, we often conveniently express ourselves in terms of random quantities. In the examples of elections and voting, we might ask the following sorts of questions. How many votes are favourable? How many against, invalid or abstentions? What is the percentage of those in favour?

In the case of examinations, contests and competitions, what is the mark or position obtained? And when—what year, day or moment—will the event in question occur (moon launch, trial verdict, vote, birth of the particular baby, etc.)? Or, alternatively, how many will succeed—among those participating in an examination, contest, sports event, etc.? Which one among them—identified by entry number, or position in alphabetical order—will attain first place, or second place? Who, within a given age limit, will be best placed? In a competition with several stages, or legs, who among the entrants will maintain, or improve, or worsen, their position with respect to the previous placings?

In other cases one uses different terminology. Random point; for instance, the point of the lunar surface which will next be reached. Random set; the set of those who pass an examination, the set of points on the earth's surface on which rain will fall tomorrow, the set of instants at which the temperature at a given place is below, above, or at, zero. Random function; the temperature at the above-mentioned place, the score during a competition, the number of votes of confidence since a certain date, etc., all considered as functions of time. If one wishes to avoid reference to irrelevant items of information (for instance, by referring to an entry number rather than to the individual concerned) it is preferable to speak of a multi-event, rather than a random quantity, and so on.

2.10.3. It is clear that in all cases it would be possible to go into more and more detail, and if all the cases we have mentioned were considered simultaneously we would arrive at even more minute subdivisions. And to these cases could be added others, ad infinitum. To arrive at a final subdivision into 'points'—not further divisible—would at least imply the construction of all possible 'histories of the universe', distinct in every detail. These would include, for example, the precise specification, instant by instant, of the position of every atom, and of the thoughts and moods of each individual—including, possibly, beings, more or less similar, living on other worlds. Even if we limit ourselves to much more restricted problems, an exhaustive description, though very much reduced in scale, would by no means turn out to be more realistic. Consider a single toss of a coin: unimaginable faculties would be needed if we wished to provide a description, with such absolute precision, of a single one of the possible ways in which a person tosses the coin, the air influences the movement, and every peculiarity of the ground and of the coin at the point and position of the latter's fall gives rise to successive movements, and so on, until the coin comes to rest. But this would still be nothing, because, instead, we must imagine and distinguish the totality of such ways.

We have pushed ourselves to absurd lengths—in a way pointless in itself—but perhaps this will serve to illustrate the thesis that it is inappropriate to distinguish between events represented by 'points', or by 'sets', thinking of it as something systematic, rather than being dependent on momentary conveniences of representation.

2.10.4. This has been said to emphasize the considerations already made (in Section 2.7.7, and elsewhere), but it is even more necessary to underline the sense in which an event (random quantity, etc.) has to be—as we said—something 'well-determined'. This means that the formulation must be unambiguous and complete, in such a way as to rule out any possibility of argument (for instance in the case of a bet which is based on it). To give an example: 'A. N. Other wins the lottery'† is an event only if the person A. N. Other, of whom we are speaking, is perfectly individuated, along with the circumstances that make the statement precise. Examples of the latter might be: win in next week's drawing; or in the first week that he plays; or any week of this year; and so on. It should also be made precise, or understood, whether possible wins in partnership with others are to be included, or not, along with any other possible aspects allowing ambiguity. By changing the individual, or any of the circumstances or provisos, we obtain other events, all different from each other. We say this only to avoid the situation where, being familiar with other terminologies, someone might think that they should be called 'identical events' or, even worse, 'trials' of 'the same event', which consists in 'winning the lottery'.

Conversely, two events expressed in completely different ways are identical —i.e. they are the same event—if we know that the occurrence of either one of them implies the occurrence of the other. Suppose, for instance, that we know for certain that this week A. N. Other is going to play the 'straight' three numbers 21–63–82 on the Roman wheel, and nothing else: in this case, the two events 'A. N. Other is going to win the lottery this week' and 'This week the numbers 21, 63 and 82 will come out on the Roman wheel' are identical. On the other hand, in order to demonstrate that it would be wrong to think in terms of the identification of a 'fact', we note the following: 'A. N. Other is going to win the next time he plays' and 'Next week the youngest person playing is going to win' are two *distinct events* which might, by chance, turn out to be the *same fact* if next week A. N. Other plays and wins and, in addition, happens to be the youngest player. This example also serves the purpose of making clear that there is no need to identify explicitly the person and the drawing (either by the date or, possibly, the wheel) so long as, by some means or other, it turns out that whether we must call the *statement* true or false is well-determined.

---

† *Translators' note.* Every Saturday in Italy, at each of ten cities, a drawing takes place of 5 from 90 possible numbers. To enter the lottery, one places a bet, prior to the drawing, specifying which combination(s) of numbers (up to a maximum of 5) one thinks will be drawn in a chosen city. The device which produces the numbers is known as a 'wheel'.

2.10.5. One could object, with reason, that such a requirement is practically unrealizable, and that, in fact, it is not even realized in the example which we have just given. For instance, how is the statement 'A. N. Other is going to win the lottery the next time he plays' to be evaluated if A. N. Other never plays again for the rest of his life? This should be made clear by means of some arbitrary convention. In most cases of this kind, however, we shall interpret the statement in a sense which falls outside the present concept of an event, but which leads to a generalization (conditional event), which we will consider explicitly later on (in Chapter 4). In addition to being *true* ($=1$) or *false* ($=0$) it could also be *void* ($=\emptyset$). In terms of a bet, this means that it could not only result in gain or loss, but could also, in certain cases, be called off. If these things are not made clear explicitly, in a systematic way, then even the statement 'A. N. Other is going to win the lottery next week' might appear ambiguous, because of the doubt as to whether we mean 'false' or 'void' if A. N. Other does not play: in such a case one implicitly assumes certain refinements, but without justification. We will not labour this point, postponing further discussion until the appropriate place. In the same way, we do not enter into discussion of certain other questions, like perhaps the preceding ones, which may appear sterile but which, if misunderstood, give rise to numerous possible ambiguities and errors. In contrast to the above, these questions can be put off until later. Let us merely remark—in order not to seem mysterious—that, above all, it is a question of discussing the actual possibility of obtaining, within a given time and with greater or lesser certainty and precision, information concerning the events and quantities of interest, about which we are at present uncertain.

2.10.6. We now return to the examples, which we considered before, in order to draw attention to some of the kinds of problems which we will frequently meet in the future, and which will serve, for the time being, to illustrate the notions introduced in the preceding paragraph.

When we ask how many of the participants in an examination will succeed in passing, we have an example of a problem concerning the *number of successes*, $Y = E_1 + E_2 + \ldots + E_n$, where $E_h = $ 'the success of participant h', or, alternatively, one concerning the frequency, or percentage, of successes, Y/n. Other examples, chosen from the infinite number of possibilities, might include the following: the number of 'white balls in n given drawings from an urn'; or of 'males among the first n births registered in Orvieto next year'; or of 'those among the n participants in a competition with many stages who maintain, after a given stage, their previous position'.

Clearly, Y can only assume the values $0, 1, 2, \ldots, n$, and, obviously, these will all be actually possible if the events E_h are logically independent. This means that the set of all those who pass an examination can, in fact, be any one of the 2^n subsets of candidates (including the whole set and the empty set); i.e. for each $h = 0, 1, 2, \ldots, n$, all the $\binom{n}{h}$ subsets of h individuals are subsets for which $Y = h$. In the cases of examinations, drawings from an urn, births, etc., this will be true under most of the usual assumptions (and we shall see what these are shortly, when we turn to counter-examples). For the time being, however, we note that the $n + 1$ values can all be possible even in cases where logical independence does not hold. Suppose, for instance, that E_h means that 'the person placed in the hth position in a competition has reached some minimum prescribed score' (or time in a race, distance with a throw, height with a jump). It is possible that all, or none, or any intermediate number h, will succeed; in the latter case these are obviously the first h and no others. We do not have logical independence since if E_h is true, all the preceding ones are necessarily true, and if false all the following ones are false.

At the other extreme, it is possible that Y is certain. This is the case, for instance, if $E_1, E_2, \ldots, E_n$ represent the drawing of white balls in n successive drawings *without replacement* from n balls, h of which are white; then we certainly have $Y = h$, the number h being known with certainty at the present moment. But in every case (drawings with replacement, examinations, sex of births) we find ourselves in the same situation if we are acquainted with the outcome as a whole, even though ignorant of the results of single drawings, etc. It is important to notice that the E_h are, in this case, not only logically, but also *linearly*, *dependent* $(E_1 + E_2 + \ldots + E_n = Y = h)$. The logical dependence assumes a concrete form in the fact that once all the white balls (or all the others) are out, the result of the subsequent drawings is certain (in any case, this is always so for the last drawing at least).

All intermediate hypotheses can be shown to be possible by the use of examples of a more or less artificial nature. The actual possibility of all $n + 1$ values is also compatible with linear dependence: if $n \geqslant 3$ we could have $E_1 = E_2$ with certainty if one thinks, for example, of the first two balls being drawn from an urn containing balls of the same colour in pairs. Restrictions on Y may exist in the case of competitive examinations with a maximum number of awards available, or in the case of drawings without replacement of n balls from an urn containing N balls of which H are white: in this case $n - (N - H) \leqslant Y \leqslant H$, the restrictions being real if the limits are > 0 and $< n$, respectively.

An important case, of some interest since it is rather less obvious, is that in which all values except $n - 1$ are possible. We meet it in the example of 'maintaining rank in a classification', which, more abstractly, consists in considering the elements that remain fixed under a permutation. One of the many well-known different interpretations is the following: we put, more or less haphazardly, n letters into n envelopes, and we consider the random quantity Y which denotes the number of letters correctly placed. Clearly, all outcomes are possible, except that of making just one error: one letter cannot be misplaced if all the others are in their own envelopes, since only the correct envelope then remains.

2.10.7. In the case of three or more alternatives (for each of n multi-events†) we must consider for each of them the number of successes or realizations: for instance, X, Y, Z, with $X + Y + Z = n$, X, Y, Z being the number of votes for, against or abstentions, out of n votes; or wins, draws and losses out of n games; or of bachelors, married men or widowers out of n males; and so on, and so forth. Similarly for cases involving many alternatives: for example $X_1 + X_2 + \ldots + X_6 = n$ for occurrences of the points $1, 2, \ldots, 6$ when we throw n dice, or a single die n times (or, in the previous example, if we distinguish marital status and sex).

Problems of this kind are called problems of *subdivisions*: here we have been dealing with the subdivisions of the integer n into a given number of (non-negative) integer summands, but more generally we could consider subdivisions of a given quantity q into any kinds of summands whatsoever—non-negative real values $X_1 + X_2 + \ldots + X_m = q$. We often prefer to take $q = 1$, i.e. to reduce to percentages: in the preceding case we could also divide the numbers of occurrences by n, obtaining in this way the frequencies. A classical example is the subdivision of an interval (into m parts with $m - 1$ division points). One could also imagine, however, the masses of the m parts into which an object of mass q breaks on falling; or, alternatively, the masses of m materials from which it is constructed (for example m metals if we are dealing with an alloy). We shall meet these kinds of problems again.

It is of interest to note that in such cases the m random quantities are linearly dependent. Other quantities which have to be considered in connection with questions of this nature are also linearly dependent if they are linear combinations of them. As examples, we note the difference between votes for and against, or the total number of 'points' scored (taking 2 for a win, 1 for a draw). On the other hand, this would not be true, for example, for ratios, such as votes in favour divided by votes against, where one would have logical but not linear dependence.

2.10.8. In the above example of a ratio ($Z = Y/X$), and in others that will follow, the logical dependence will be functional dependence (in the clearest case, with $f(X_1, X_2, \ldots, X_n)$ such that each X_h turns out to be uniquely determined within the permitted field). Naturally, the given definition does not imply anything of this kind. Not only may the uniqueness fail—as when we consider points on the spherical surface $X^2 + Y^2 + Z^2 = 1$ with ad-

---

† Of course, this is also valid in the case of only two alternatives; in this case, however, it is trivial to take into account the number of occurrences of each of them since $Y = n - X$.

missible values not constrained to be non-negative—but one might also consider all points of the sphere as possible (by substituting $\leqslant$ for $=$) without destroying the logical dependence. To see this, note that, given $X = x$ and $Y = y$, the possible values of Z lie in the segment between $\pm\sqrt{(1 - x^2 - y^2)}$, which is a function of x and y. Given that X, Y, Z can all assume values between ± 1, we have logical independence only for the case in which all the points of the cube $-1 \leqslant x, y, z \leqslant +1$ are possible: the exclusion of a single point, for example the origin, is sufficient to give logical dependence (to avoid it, we would have to exclude the points on the co-ordinate planes; i.e. the value 0 for each random quantity separately). One also has logical dependence if one excludes from the cube the points for which, for example, $X + Y + Z$ (or $XYZ, XY/Z$, etc.) is rational, or transcendental, or whatever (to avoid it, one should instead exclude separately, X, for example, being rational, Y being transcendental, Z being zero).

2.10.9. A case of logical dependence, which is of practical importance and frequent occurrence, is the following; given a number of random quantities, say X, Y, Z, we denote, by definition, the smallest of these by X, the middle one by Y, the greatest by Z. In this case, we exclude all those points which are not included in the dihedron $y - x \geqslant 0, z - y \geqslant 0$, even if the coordinates of the points are possible values for X, Y, Z (unless all the possible values for X are less than all the possible values for Y, and these are less than all the possible values for Z, in which case $X \leqslant Y \leqslant Z$ does not constitute a restriction). It is necessary to pay attention to circumstances of this kind since the necessity of establishing and taking appropriate account of them could be overlooked.

If we take the example of a subdivision resulting from the splitting of a fallen object—let us say into three pieces, X, Y, Z—the situation differs according to whether the criterion by which we rank them is the order of magnitude, or something else not depending on it. For instance, we might take the angle formed between the half-line starting from the point of fall and passing through the barycentre of the piece in question and the direction North, the angle being taken in a counter-clockwise direction.

The same thing holds in the example we are about to consider now, where X, Y, Z are the sides of a random prism (rectangle): for example, a block of stone, a building, a suitcase. We may or may not have more or less 'natural' circumstances which lead us to define, in each case, what we mean by 'length' (X), 'breadth' (Y) and 'height' (Z). Without getting bogged down in an analysis which everyone can provide for themselves anyway, the answer seems to be easy for the suitcase, not always such for the building—the distinction between length and breadth may not be clear if there is no recognizable façade— and indeterminate for the block (unless we use conventions based on how it is temporarily situated with respect to North, East and the zenith). If we agree to call the maximum side the length, and the minimum side the height, we are in the other situation.

Given this random prism— and however we think of the problem, with the sides X, Y, Z logically independent or not—let us consider its diagonal U, area V and volume W. In either case, these are random quantities which are logically (and even, in a unique way, functionally) dependent on the preceding ones: $U = \sqrt{(X^2 + Y^2 + Z^2)}, V = 2(XY + XZ + YZ), W = XYZ$. Clearly, however, the dependence is not *linear*; when we return to the question, in Chapter 3, this example will serve to clarify, in an appropriate way, how, and why, certain reasonings about uncertainty, though seemingly obvious, are correct in some cases, but not in others (and this according to whether one has linear dependence or not).

2.11   CONCERNING CERTAIN CONVENTIONS OF NOTATION

2.11.1. As we announced in Section 2.5.3, and briefly mentioned in Chapter 1 (Sections 1.9.3, 1.9.4), we will demonstrate, by means of examples, the utility that can be derived in many cases from the use of conventions introduced in the present chapter for simplifying the notation. To be explicit:

the identification of TRUE and FALSE with 1 and 0;
the 'lattice' operations for numbers.

2.11.2. The convention TRUE $= 1$ and FALSE $= 0$ turns out to be very useful also when applied outside of the field of events, to propositions or any 'conditions' whatsoever.

Examples. $(x \geqslant a)$ is the function which $= 0$ for $x < a$ and $= 1$ for $x \geqslant a$; we could write such a function as $F(x) = (x \geqslant a)$, and, more generally,

$$F(x) = \sum_h p_h(x \geqslant a_h)$$

is the step-function with jumps p_h at the points $x = a_h$; assuming that the a_h are in increasing order of magnitude, this could also be written

$$F(x) = \sum_h c_h(a_h \leqslant x < a_{h+1}),$$

which denotes that in the given interval the value is

$$c_h = \sum_i p_i(i \leqslant h) = \sum_{i=1}^{h} p_i.\dagger$$

---

† Given the purely illustrative purpose of these forms of notation, we omit all the possible refinements which should be added, case by case, in specific applications: for instance, here, hypotheses of convergence if we are dealing with series: in an opposite sense the convention $a_{n+1} = \infty$ if a_n is the last term, etc. The notation $\leqslant$ instead of $<$ etc. will vary from case to case.

In the last example we used the function $(a \leqslant x < b)$, which is $= 1$ in the given interval and $= 0$ outside: more generally, we use $(x \in I)$ to denote the indicator function of the set I (the function which $= 1$ if x is in I, and $= 0$ otherwise).

Using such a function as a multiplier, one obtains immediately the restriction of a function to a given interval or set; for example,

$$x^2(x \geqslant 0) = 0 \quad \text{for} \quad x \leqslant 0 \quad \text{and} \quad = x^2 \quad \text{for} \quad x \geqslant 0\,;$$

$f(x) = x(1 - x)(-1 \leqslant x \leqslant 1) = x(1 - x)(|x| \leqslant 1)$ is equal to $x(1 - x)$ for x in $[-1, 1]$, 0 otherwise; and, more generally, for a function with a different expression in different intervals, for example,

$$f(x) = a(x + 3)^3(-3 \leqslant x < -1) + (b - cx^2)(-1 \leqslant x < 1)$$
$$+ a(3 - x)^3(1 \leqslant x < 3),$$

or even (for a large, or infinite, number of intervals)

$$f(x) = \sum_h f_h(x)(h \leqslant x < h + 1), \quad \text{or} \quad f(x) = \sum_h f_h(x)(a_h \leqslant x < a_{h+1}).$$

Remarks. The examples in which the functions are denoted by $F(x)$ and $f(x)$, respectively, can be interpreted as the *distribution* function (F) and the *density* function $(f = F')$ of a distribution. These notions may already be familiar, but will, in any case, be introduced in Chapter 6.

2.11.3. In the previous cases of summation, we have already seen the expression of the condition functioning as a multiplier in order to define each single sum-function, or (under another equivalent interpretation) specifying, for a given x, which terms had to be summed. The systematic usage of such a convention to this end, even in the absence of a useful interpretation in the first sense, would seem to be very convenient, both for clarity and typographical convenience. It replaces, in an advantageous manner, either explanations in the text, or complicated instructions to be composed under the summation (or integral) sign, etc.

The meaning of the following examples is self-evident:

$$\sum a_h(h \in H), \qquad \sum a_h(b_h \in B), \qquad \sum a_h(h \neq 0), \qquad \sum a_{hk}(h \neq k),$$
$$\sum a_{hk}(h \leqslant k),$$
$$\int f(x)\, dx\, (2n \leqslant x \leqslant 2n + 1), \qquad \int f(x, y)\, dx\, dy\, (x^2 + y^2 \leqslant r^2).$$

2.11.4. *Use of the Boolean operations.* The Boolean operations $\vee$ and $\wedge$ often serve (even better than the system given above) to denote '*truncations*' and similar operations. For instance, the function $F(x) = $ 'x *provided it is not less than zero or greater than one*' could be written in either of the two ways

$$F(x) = x(0 \leqslant x \leqslant 1) + (x > 1) = 0 \vee x \wedge 1,$$

and the second is clearly simpler. In general, the function which $= f(x)$ but is never less than m or greater than M can be written as $m \vee f(x) \wedge M$, and similarly $m(x) \vee f(x) \wedge M(x)$ so long as we always have $m(x) < M(x)$ (otherwise we would not have $(m \vee f) \wedge M = m \vee (f \wedge M)$), and the notation would not be admissible).

This notation, in our context, will serve in particular for random quantities: we present here a few examples in both notations (and the Boolean form seems to be simpler):

$$X(X \geqslant 0) = 0 \vee X, \qquad X(X \leqslant 0) = 0 \wedge X,$$

$$X = X(X \geqslant 0) + X(X \leqslant 0) = 0 \vee X + 0 \wedge X,$$

$$|X| = X(X \geqslant 0) - X(X \leqslant 0) = 0 \vee X - 0 \wedge X;$$

$$\left.\begin{array}{l} X(0 \leqslant X \leqslant K) + K(X > K) = 0 \vee X \wedge K, \\ X(|X| \leqslant K) + K[(X > K) - (X < -K)] = -K \vee X \wedge K, \end{array}\right\} \quad (K > 0)$$

and so on.

CHAPTER 3

Prevision and Probability

3.1 FROM UNCERTAINTY TO PREVISION

3.1.1. So far, even in the way we presented the preceding examples, we have limited ourselves to depicting and representing the situation facing You, when You are interested in distinguishing among a more or less extensive class of alternatives (all those which, in the present state of your information, appear possible to You). This preliminary topic, which we will have to consider more deeply in what follows, is still within the ambit of ordinary logic, the logic of certainty. One should always be careful to distinguish clearly between those things belonging to this domain and those belonging to the *probabilistic* domain—the ambit of the *logic of uncertainty*, the logic of *prevision*—to which we must now turn our attention. It was precisely in order to pin-point this distinction that we decided upon this form of exposition, presenting concepts and related examples which reveal the situation as it is, while leaving undetermined all questions concerning the possible introduction of probability, its conceptual basis and its evaluation. It would certainly be easier, and seemingly more instructive, to go right ahead and take the two steps together, instead of just the one. In other words, we could present right away, fused together in the examples and definitions, both the probability (which answers the need) and the uncertainty (from which the need arises), without first making such a need 'felt', and then pausing to reflect upon it. It is precisely this latter course, however, which must be recommended.

The situation is this: having distinguished the possible cases, and having represented them in the way which seems to You most effective (or in any way convenient to You), if You then wish to restrict yourself to the logic of certainty You have to stop, and consider the question closed. Is this what You *want* to do? And *can* You do it?

For each one of us, it is often the case that we *do not content ourselves* (or are not *able* to content ourselves) with this, and therefore we proceed further. And, strictly speaking, to proceed further means to enter into what we have called the logic of *prevision* (in a sense that we will make clear in

order to draw attention to the distinction between this and other interpretations, whose drawbacks must be pointed out).

3.1.2. *Prevision, not prediction.* In order to use this word, 'prevision',
it will be necessary to give an absolutely precise meaning to it (and to derived
words) and to insist on this meaning and keep it in mind, consistently and
scrupulously, in the sequel. It must be distinguished, and in fact contraposed, to another word, which, in everyday language, is perhaps more
commonly attributed to it, and for which we will reserve the alternative
name, '*prediction*'.

To make a *prediction* would mean (using the term in the sense we propose)
to venture to try to 'guess', among the possible alternatives, the one that will
occur. This is an attempt often made, not only by would-be magicians and
prophets, but also by experts and such like who are inclined to precast the
future in the forge of their fantasies†. To make a 'prediction', therefore,
would not entail leaving the domain of the logic of certainty, but simply
including the statements and data which we assume ourselves capable of
guessing, along with the ascertained truths and the collected data. It is not
enough to tone down the 'prophetic' character of such pronouncements
by taking precautions with feelings ('I think', 'perhaps', etc.) as we have
already mentioned: either these artificial additions remain without any
authentic meaning, or they need to be actually translated into probabilistic
terms, substituting prevision in place of prediction.

> If we remain within the logic of certainty, such additions not only have no authentic
> meaning in themselves, but, in point of fact, they render meaningless the entire
> discussion. If the discussion affirms that something is *true*, and the 'perhaps' means
> that instead of being true it could also be *false*, this is equivalent to retracting the
> preceding statement, declaring it to be invalid and unfounded (cancelling it, dis-
> owning it). If not, then 'perhaps' should be erased as it might give a false impression
> of such a retraction.
>
> Alternatively—and this is the approach indicated below, and which corresponds
> to the subjectivistic conception of probability—the 'perhaps' can be explained as an

---

† Everyone will no doubt have noticed, and had occasion to notice, how often the 'foresights
of experts' turn out to be completely different from the facts, sometimes spectacularly so. In
the main, this is precisely because they are intended as predictions which 'deduce', more or
less logically, a long chain of consequences—still considered necessarily plausible—from the
assumed plausibility of an initial hypothesis. Interesting examples of the lack of connection
between prevision and reality (in the political field) are pointed out and discussed by B. de
Jouvenel in 'Futuribles', *Bulletin Sedeis*, 20 January 1962.

Here also one might note the irrelevant distinction, as far as prevision and prediction are
concerned, between the future and the past: the hypothetical reconstructions of murders or
historical facts made by detectives, scholars or novelists, based on scanty data and meriting
varying degrees of respect, are, in the above sense, 'predictions'.

It is useful to ask oneself, incidentally, whether such 'facile fantasies' are really 'rich fantasies',
or rather 'poor fantasies', in that ineptitude or laziness prevents us from seeing how many other
possibilities there are, besides the first one we happened to think of.

indication, even if crudely qualitative, of a degree of subjective probability which, if we wished, could be made more precise, and even quantified.

All this would be very clear if there did not exist, unfortunately, in the very field of probability and statistics, certain tendencies to avoid the choice, playing precisely on that ambiguity which we drew attention to earlier, and making it worse. In fact, the ambiguity of the 'perhaps' (which could be innocent, due to simple unwariness) is fraudulently concealed beneath a showy exterior. It is translated into technical terms like 'accept' and 'reject', which neither mean YES or NO with certainty, nor are to be interpreted in a probabilistic sense, but simply lay claim to be themselves 'accepted', rather than 'rejected', without giving to those terms any 'acceptable' meaning whatsoever.

3.1.3. *Prevision,* in the sense in which we have said we want to use this word, does not involve guessing anything. It does not assert—as prediction does—something that might turn out to be true or false, by transforming (over-optimistically) the uncertainty into a claimed, but worthless, certainty. It acknowledges (as should be obvious) that what is uncertain is uncertain: in so far as statements are concerned, all that can be said beyond what is said by the logic of certainty is illegitimate. If we think that something might be added, if we think, as we remarked above, that we can *proceed further,* it will necessarily be a question of entering into a completely new field and scheme of things, one which goes beyond the logic of certainty, even if it must be linked to it and superimposed upon it.

When we cease to content ourselves with the logic of certainty, in what sense do we go beyond it? In what sense do You go beyond it? Let us ask ourselves this question. Ask yourself. The thing we are not content with, and neither are You, is the agnostic and undifferentiated attitude towards all those things which, not being known to us with certainty, are uncertain, are possible. There are no degrees† of possibility: it is possible (equally possible) that it snows on a winter or summer day; that a great champion or a novice wins the competition; that every student, whether well-prepared or not, will pass an examination; that next Christmas You will find yourself at any place in the world. However, You do not content yourself with this, and, in fact, it is not your real attitude. Faced with uncertainty, one feels, and You feel too, a more or less strong propensity to expect that certain alternatives rather than others will turn out to be true; to think that the answer to a certain question is YES rather than NO; to estimate that the unknown value of a certain quantity is small rather than large.

These attitudes, of ours and of yours, do not lead us—as in the case of someone who claims to make a spot-on prediction—to assert as certain or

---

† In a certain sense, however, there exists a partial order since one could call, 'not less possible' than another, an event which is a consequence of it (in the same way as one could call, 'not less extended' than another one, a set containing it). In both cases, however, no step forward is made towards a comparability or measurability of the 'possibilities' or of the 'extensions'.

impossible something which, on the basis of the logic of certainty, is possible but uncertain, and which remains such whatever further assertions or thoughts might be added. Uncertain things remain uncertain, but we attribute to the various uncertain events a greater or lesser degree of that new factor which is extralogical, subjective and personal (mine, yours, his, anybody's), and which expresses these attitudes. In everyday language this is called *probability*, a concept that we shall have to clarify and study. *Prevision*, in the sense we give to the term and approve of (judging it to be something serious, well-founded and necessary, in contrast to prediction), consists in considering, after careful reflection, all the possible alternatives, in order to distribute among them, in the way which will appear most appropriate, one's own expectations, one's own sensations of probability.

We all of us enter into this ambit of prevision in a spontaneous fashion; sometimes without a specific need, for the sole reason that one is interested in the object of uncertainty, that there are desires or hopes that certain alternatives occur, anxieties and fears regarding the occurrence of unfavourable alternatives, and that the weighing up of such hopes and fears matters to one. Sometimes, on the other hand, all one's behaviour may necessarily depend on a comparative evaluation, albeit crude and perhaps unconscious, of the various impending risks, and of the various targets that one can set oneself. In this sense, and because of the enormous range of possibilities, one may find oneself *compelled* to weigh up such evaluations, and to express the prevision. In the case of more important and conscious decisions, one might try to *reason about* each choice, and *weigh up the pros and cons* by means of some criterion or other.

3.1.4. *Coherence.* It is precisely in investigating the connection that must hold between evaluations of probability and decision-making ๏nder conditions of uncertainty that one can arrive at criteria for measuring probabilities, for establishing the conditions which they must satisfy, and for understanding the way in which one can, and indeed one must, '*reason about them*'. It turns out, in fact, that there exist simple (and, in the last analysis, obvious) conditions, which we term conditions *of coherence*: any transgression of these results in decisions whose consequences are manifestly undesirable (leading to certain loss).

The 'one must' is to be understood as 'one must if one wishes to avoid these particular objective consequences'. It is not to be taken as an obligation that someone means to impose from the outside, nor as an assertion that our evaluations are always automatically coherent. On the contrary, it is precisely because this is an area where it is particularly easy to slip into incoherence that it is important to learn the *art of prevision* (to adapt the phrase *Ars Conjectandi*, used by James Bernoulli as the title of the first treatise on the calculus of probability).

Given any set of events whatsoever, the conditions of coherence impose no limits on the probabilities that an individual may assign, except that they must not be in contradiction amongst themselves. Without further delay, we will proceed to the construction of the theory of probability, using as a basis the theory of decision-making. For the time being, this will be done in an extremely simplified form, as a preliminary clarification of ideas. In the next paragraph we will discuss certain other aspects of the problem, and then turn to the constructive formulation.

Within this framework, we obtain the greatest insight by considering as a starting point the case of random quantities (especially when we interpret them as random gains). With a more rigorous approach, inspired by decision-theoretic considerations, it is essentially a question of returning to that problem of a *fair evaluation*, or estimation, which, in connection with similar problems of an economic nature, seems to have foreshadowed by centuries the beginnings of the calculus of probability. In this sense, the modern setting of the problem, within decision theory, constitutes, to some extent, a return to its origins.

The definition of the probability of an event will turn out to be contained automatically in that given in the case of random quantities: we simply define events as particular random quantities. From a mathematical viewpoint also, this would appear the appropriate thing to do. In Chapter 2 we saw that in the case of events the most useful arguments, which are very simple if one considers the events as special points in the space of random quantities, are not available if one thinks in terms of the set of events without reference to the space in which this set *is* 'naturally' embedded, and in which it is necessary to *see it* embedded.

Let us proceed then to the matter in hand, starting with the consideration of a *random gain* X : by this we mean a random quantity X having the meaning of gain (the latter intended, of course, in an algebraic sense; a loss is a negative gain). The possible values of X could, therefore, also be negative, either in part, or entirely. We might ask an individual, e.g. You, to specify the *certain gain* which is considered *equivalent* to X. This we might call the *price* (for You) of X (we denote it by $\mathbf{P}(X)$†) in the sense that, on your scale of preference, the random gain X is, or is not, preferred to a certain gain x according as x is less than or greater than $\mathbf{P}(X)$. For every individual, in any given situation, the possibility of inserting the degree of preferability of a random gain into the scale of the certain gains is obviously a prerequisite condition of all decision-making criteria. Among the decisions which lead to different random gains, the choice must be the one that leads to the random gain with the highest price. Moreover, this is not a question of a condition but simply of a definition, since the price is defined only in terms of the very

---

† We could write $\mathbf{P}_i(X)$ to emphasize that we are dealing with the evaluation of a particular individual i. This is an unnecessary precaution, however, since it is understood that we are always referring ourselves to the evaluation of a given individual (real or fictitious).

preference that it means to measure, and which must manifest itself in one way or another.

In general, it is not true that if one is prepared to buy an article A at the price $\mathbf{P}(A)$, and an article B at the price $\mathbf{P}(B)$, one must be prepared to buy both of them together at a price $\mathbf{P}(A) + \mathbf{P}(B)$. It may happen that the purchase of one of them affects, in various ways, the desirability of the other. Similar qualifications hold if instead of two articles A and B we consider two random gains X and Y; this case will be examined in the next paragraph. In both cases, however, additivity is something more than just an interesting simplifying hypothesis, which may be approximately valid. As we shall see later, provided we modify slightly the way in which the notion of price $\mathbf{P}(X)$ is introduced, additivity will turn out to be an exact property, the foundation of the whole treatment.

3.1.5. *Properties of* **P**. If You are indifferent to the exchange of X for $\mathbf{P}(X)$ and of Y for $\mathbf{P}(Y)$, then, if we assume the simplifying hypothesis given above, You are also indifferent to the exchange of $X + Y$ for $\mathbf{P}(X) + \mathbf{P}(Y)$. The value for which You are indifferent to the exchange of $X + Y$ is, however, by definition, $\mathbf{P}(X + Y)$; we therefore conclude that

(a) *the price* **P** *is an additive function*:

(1)
$$\mathbf{P}(X + Y) = \mathbf{P}(X) + \mathbf{P}(Y).$$

A second property, obvious, but equally fundamental, can be derived by noting that $\mathbf{P}(X)$ must not be less than the lower bound of the set of possible values for X, $\inf X$, nor greater than the upper bound, $\sup X$ (otherwise the choice would allow a certain loss). Therefore

(b) *the price* **P** *must satisfy the inequality*:

(2)
$$\inf X \leqslant \mathbf{P}(X) \leqslant \sup X\,;$$

obviously, this condition only imposes a restriction if the random quantity X is *bounded* in at least one direction (either $\inf X > -\infty$ or $\sup X < +\infty$). Generally, but not always, we will restrict our attention to the bounded case (i.e. bounded from above and below).

When we come to formulate and examine this set-up in a more exhaustive fashion, we shall see that the two extremely simple conditions, (a) and (b), are *not only necessary but also sufficient* for coherence—i.e. for avoiding undesirable decisions. This is all that is needed for the foundation of the whole theory of probability: in fact, the definition of probability immediately reduces, as a special case, to that of a price **P**.

We observe, from (a) and (b), that the price $\mathbf{P}$ must also be a *linear* function, in the sense that for every real a we have

$$(3) \qquad\qquad \mathbf{P}(aX) = a\mathbf{P}(X)\dagger,$$

and therefore, more generally,

$$(4) \qquad \mathbf{P}(aX + bY + cZ + \ldots) = a\mathbf{P}(X) + b\mathbf{P}(Y) + c\mathbf{P}(Z) + \ldots$$

for any *finite* number of summands.

Given this property, it is possible to extend the definition of $\mathbf{P}(X)$ to the case in which X is a random quantity (pure number), or a random magnitude not having the meaning of gain (for instance, time, length, etc.). In fact, it suffices to choose a coefficient a whose dimension is such that aX is a monetary value: for instance, in the cases of time and length we could take Lire/sec and \$/cm. We now define $\mathbf{P}(X) = (1/a)\mathbf{P}(aX)$: this is well-defined since the expression is invariant with respect to the choice of a (we can substitute λa in place of a, where λ is a non-zero real number).

In the general case (where we do not have a monetary value), the term 'price' is no longer appropriate: we speak instead of the '*prevision of X*', valid in all cases,‡ and, in particular, of the '*probability of E*' when $X = E$ is an event.

The probability $\mathbf{P}(E)$ that You attribute to an event E is therefore the certain gain p which You judge equivalent to a unit gain conditional on the occurrence of E: in order to express it in a dimensionally correct way, it is preferable to take pS equivalent to S conditional on E, where S is any amount whatsoever, one Lira or one million, \$20 or £75. Since the possible values for a possible event E satisfy $\inf E = 0$ and $\sup E = 1$, for such an event we have $0 \leqslant \mathbf{P}(E) \leqslant 1$, while necessarily $\mathbf{P}(E) = 0$ for the impossible event, and $\mathbf{P}(E) = 1$ for the certain event.§

---

† This is obvious if a is rational, and the extension to every a is straightforward if X is always positive (because then if a lies between a' and a'', we also have aX between $a'X$ and $a''X$). But we can always write $X = Y - Z$, where $Y = X(X \geqslant 0)$ and $Z = -X(X \leqslant 0)$, and these numbers are always non-negative: $Y = X$ if $X > 0$ and zero otherwise, $Z = -X$ if $X < 0$ and zero otherwise. The conclusion is therefore valid for Y and for Z, and hence for $X = Y - Z$.

‡ This corresponds to 'mathematical expectation' in classical terminology, and to 'mean value' in more up-to-date usage. We prefer to reserve the term 'mean value' for *objective* distributions (e.g. statistical distributions).

§ These are the only cases in which the evaluation of the probability is predetermined, rather than permitting the choice of any value in the interval from 0 to 1 (*end-points included*). The predetermination that one meets in these cases arises because there exists no uncertainty and the use of the term probability is redundant. The same thing holds for prevision: $\mathbf{P}(X)$ necessarily has a given value x if and only if X has x as a unique possible value; i.e. if X is not really random. The above is the special case where either $x = 0$ or $x = 1$.

3.2   DIGRESSIONS ON DECISIONS AND UTILITIES

3.2.1. In Section 3.1, we have introduced the notions of *prevision* and *probability* by following the path laid down by certain decision-theoretic criteria of an essentially economic nature: the presentation was, however, in a simplified form.

It follows, therefore, that before going any further we should make some comments and give some further details about the theory of decision-making, and above all about *utility*. The latter, together with probability, is one of the two notions on which the correct criterion of decision-making depends. We warn the reader, however, that this is in the nature of a digression and anyone not interested in the topic can skip it without any great loss: the details (of a non-economic nature) that are given in Section 3.3, and in subsequent sections, will prove quite sufficient.

3.2.2. *Operational definitions.* In order to give an effective meaning to a notion—and not merely an appearance of such in a metaphysical–verbalistic sense—an operational definition is required. By this we mean a definition based on a criterion which allows us to measure it.† We will therefore be concerned with giving an operational definition to the prevision of a random quantity, and hence to the probability of an event.

The criterion, the operative part of the definition which enables us to measure it, consists in this case of testing, through the *decisions* of an individual (which are observable), his *opinions* (previsions, probabilities), which are not directly observable.

Every measurement procedure and device should be used with caution, and its results carefully scrutinized. This is true in physics, despite the degree of perfection attainable, and even more so in a field as delicate as ours, where similar and much more profound difficulties are encountered.

In the first place, if, as is implicit in what we have said so far, we identify, generically, decisions and preferences, then we are ignoring many of the extraneous factors that play a part in decision-making. Nobody accepts all the opportunities or bets that he judges favourable, and perhaps we all sometimes enter into situations that we judge unfavourable. To reduce the influence of such factors it is convenient to effect the observations on the phenomena isolated in their most simple forms: this is in fact what we attempt to do when we construct measuring devices. For the purpose of a formal treatment of the topic, we will present (in the next section) two different procedures by means of which we try to force the individual to make conscious choices, releasing him from inertia, preserving him from whim. Of course, we have to establish that the two procedures are equivalent, and this we shall do.

---

† Cf. P. Bridgman, *The Logic of Modern Physics*, New York (1927).

A doubt might remain, however. Are the conclusions which we draw after observing the actual behaviour of an individual, directly making decisions in which he has a real interest, more reliable than those based on the preferences which he expresses when confronted with a hypothetical situation or decision? Both the direct interest and the lack of it might on the one hand favour, and on the other obstruct, the calmness and accuracy, and hence the reliability, of the evaluations. In any case, it is not really a mathematical question: it is useful to be aware of the problem, but it is mainly up to the psychologists to delve further into the matter. We merely note that between the two extreme hypotheses one could consider an intermediate one which might be of interest; the case of an individual being consulted about a decision in which others are interested. This might well lead to responsibility in the judgment without affecting the calmness of the decision maker. In Chapter 5, 5.5.6, we encounter another example which is similar in spirit to the last one: this is where the accuracy of the evaluation is related to one's self-respect in some competitive situation (with prizes which are materially insignificant, but which are related to the significance of the competition).

At this point, the reader may be wondering on what basis individuals do evaluate their probabilities or previsions: the question is not appropriate, however. Firstly, we must attempt to discover opinions and to establish whether or not they are coherent. Only at the second stage, having acquired the necessary knowledge, could we also apply it to investigate these other aspects, and not until it was very much advanced could this be done in a sufficiently satisfactory way (up to and including the rather complicated justifications for the case of evaluations based on frequencies, a case wrongly considered simple).

3.2.3. *Reservations concerning rigidity.* The main question that we have to face in these 'remarks' is the one already mentioned when we expressed reservations about assuming additivity for the price of a random gain: recall that it is this hypothesis which underlies the definition of prevision, and the special case of probability.

It is well-known, and indeed obvious, that usually this is not realistic because of the phenomenon of risk aversion (or occasionally its opposite, but we shall not bother with such cases). In fact, as we already noted in effect when we introduced it, the hypothesis of additivity expresses an assumption of *rigidity* in the face of risk. Let us now try to make this clear. As a preliminary, it will suffice to restrict ourselves to simple examples that are within our present scope. These will be sufficient to show that in order to obtain a formulation which is completely satisfactory from the economic point of view, it is necessary to eliminate such rigidity by introducing the notion of *utility*. On the other hand, they will also show that one is able to manage without this notion, except when occupied with applications of an expressly economic nature.

Suppose that You are faced with two eventualities which You judge equally probable: taking the standard example, it could be a question of Heads or Tails. Given the hypothesis of rigidity in the face of risk, You should be indifferent between 'receiving with certainty a sum S, or twice the sum if a particular one of the two possible cases occurs': likewise, between 'losing with certainty a sum S, or twice the sum if a particular one of the two possible cases occurs'; and similarly between 'accepting or not accepting a bet which, in the two possible cases, would lead either to a loss, or to a gain, of the same sum S'. This much is obvious, but in any case we shall carry out the calculations as an exercise. Let us denote by A and B the two events: $A + B = 1$ because one and only one of the two occurs. Their probabilities, being supposed equal, must each have the value $\frac{1}{2}$, since $\mathbf{P}(A) = \mathbf{P}(B)$, and $\mathbf{P}(A) + \mathbf{P}(B) = \mathbf{P}(A + B) = \mathbf{P}(1) = 1$. It follows that cases of so-called indifference simply imply the equality of the following: S and $(2S)/2$, $-S$ and $-(2S)/2$, 0 and $\frac{1}{2}S + \frac{1}{2}(-S)$ (since, for instance, the gain $2S$ conditional on the event A is the random quantity $X = 2SA$, $\mathbf{P}(X) = \mathbf{P}(2SA) = 2S\mathbf{P}(A) = 2S \cdot (\frac{1}{2})$.

If instead, as is likely, You are risk averse, then in all cases You will prefer the *certain* alternative to the *uncertain* one (the form and extent of the aversion will depend upon your temperament, or perhaps be influenced by your current mood, or by some other circumstance). To arrive at the actual indifference, You would content yourself with receiving with certainty a sum S' (less than S) in exchange for the hypothetical gain $2S$; You would be disposed to pay with certainty a sum S'' (greater than S) in order to avoid the risk of a hypothetical loss $2S$; You would pay a certain penalty K in order to be released from any bet where the gain and loss are, in monetary terms, symmetric.

This means, however, that by virtue of risk aversion one has symmetry in the scale in which one's judgments of indifference are based: i.e. equal levels in passing from 0 to S' and from S' to $2S$, or in passing from $-2S$ to $-S''$ and from $-S''$ to 0, or in passing from $-S$ to $-K$ and from $-K$ to S. The scale no longer coincides with the monetary one, as in the case of rigidity. In short, as far as we are concerned, things proceed as if successive increments of equal monetary value had for You smaller and smaller subjective value or *utility*. This term—often used in a similar sense, but in a questionable form, in economic science—has been rehabilitated, and adopted with the specific meaning derived from the present considerations about risk.

3.2.4. *The scale of utility.* The above considerations enable us to construct a scale of utility; i.e. a function $U(x)$, the utility of the gain x, whose increments, $U(x_{i+1}) - U(x_i)$, are equal when, and only when, we are indifferent between the corresponding increments of monetary gain, $x_{i+1} - x_i$.

We could proceed, for instance, by dividing an interval into two 'indifferent increments', in the way indicated in the examples above, and in the same way obtain subdivisions into 4, 8, 16, ..., parts. It would be more appropriate, instead of considering the variable x representing the gain, to take $f + x$, where f is the individual's 'fortune' (in order to avoid splitting hairs, inappropriate in this context, one could think of the value of his estate). Anyway, it would be convenient to choose a less arbitrary origin in order to take into account the possibility that judgments may alter because in the meantime variations have occurred in one's fortune, or risks have been taken, and in order not to preclude for oneself the possibility of taking these things into account, should the need arise. Indeed, as a recognition of the fact that the situation will always involve risks, it would be more appropriate to denote the fortune itself by F (considering it as a random quantity), instead of with f (a definite given value).

What we have said concerning the scale of utility makes it intuitively clear—and this is sufficient for the time being—that if, in order to define 'price', we refer to this scale rather than to the monetary scale, then additivity holds. In fact, one might say that such a scale is by definition the monetary scale deformed in such a way as to compensate for the distortions of the case of rigidity which are caused by risk aversion. The formulation put forward in the preceding paragraph could therefore be made watertight, and this we will do shortly by working in terms of the utility instead of with the monetary value. This would undoubtedly be the best course from the theoretical point of view, because one would construct, in an integrated fashion, a theory of decision-making (of the criteria of coherent decisions, under conditions of certainty or uncertainty), whose meaning would be unexceptionable from an economic viewpoint, and which would establish simultaneously and in parallel the properties of *probability* and *utility* on which it depends. The fundamental result lies, in fact, in recognizing that *the criteria of coherent decision-making are precisely those which consist of the choice of any evaluation of the probabilities and any utility function* (with the necessary properties) *and in fixing as one's goal the maximization of the prevision of the utility.* Of course, it is possible to behave coherently with respect to decisions and preferences without knowing anything about probability and utility. The fact is, however, that in this case one must behave *as if* one is acting in the above manner, *as if* obeying an evaluation of probability and a scale of utility underlying one's way of thinking and acting (even if without realizing it). Provided one could succeed in exploring these activities in an appropriate way, it might be possible to trace back and individuate the two components.

This unified approach to an integrated formulation of decision theory in its two components was put forward by F. P. Ramsey (1926) and rigorously developed by L. J. Savage (1954). However, there are also reasons for preferring the opposite approach; the one which we attempt here. This consists in setting aside, until it is expressly required, the notion of utility, in order to develop in a more manageable way the study of probability.

3.2.5. *An alternative approach.* The idea underlying this alternative approach stems from the observation that the hypothesis of rigidity, as considered above, is acceptable in practice—even if we stick to monetary values—provided the amounts in question are not 'too large'. Of course, the proviso has a relative and approximate meaning: relative to You, to your fortune and temperament (in precise terms, to the degree of convexity of your utility function U); approximate because, in effect, we are substituting in place of the segment of the curve U which is of interest the tangent at the starting point. Clearly, the smaller the range considered, the more satisfactory is the approximation. With this in mind, we might consider replacing the previous definition of $\mathbf{P}(X)$—which we temporarily distinguish, denoting it by $\mathbf{P}^*(X)$—with a new one, which we define by means of the relation

$$\mathbf{P}(X) = \lim_{a \to 0} (\tfrac{1}{a})\mathbf{P}^*(aX).$$

Instead, we prefer a less orthodox but more natural and manageable solution, which consists in not changing anything, but merely remarking that in economic examples one must remain within appropriate limits (which, as an aid to understanding, we call 'everyday affairs').

There are several reasons behind this choice (and, more generally, behind rejecting the standard method of considering both probability and utility together, right from the very beginning).

Firstly, on a purely formal level, there is an objection to taking the passage to the limit so seriously as to base a definition on it: in fact, if a becomes too small an evaluation loses, in practice, any reliability. This is the same phenomenon that one encounters when attempting to define density, although the underlying reasons are different. One needs to consider the ratios mass/volume for neighbourhoods small enough to avoid macroscopic inhomogeneities, but not so small as to be affected by discontinuities in the structure of matter. We accept that once we are in the area of mathematical idealization we can leave out of consideration adherence to reality in every tiny detail: on the other hand, it seems rather too unrigorous to act in this way when formulating that very definition which should provide the connection with reality.

This does not mean that it is not useful to accept the form of the passage to the limit (as an innocuous and convenient assumption, although not appropriate to fulfil the function of a definition). In any case, let us suppose that we have introduced the linear prevision $\mathbf{P}(X)$, and that we know the utility function U, which, for the sake of simplicity, we now take to be expressed as a function of the gain x. Then the original $\mathbf{P}(X)$ as it actually turns out to be, assuming that the hypothesis of rigidity is not satisfied (this is denoted above by $\mathbf{P}^*$, but from now on we denote it by $\mathbf{P}_U$), can be expressed immediately as a transform of $\mathbf{P}$ by means of U:

$$(5) \qquad \mathbf{P}_U(X) = U^{-1}\{\mathbf{P}[U(X)]\}.$$

In the standard case, where U is convex, we have $\mathbf{P}_U(X) < \mathbf{P}(X)$, as noted above, and as can be seen from the theory of associative means (Chapter 2, 2.9.3). In order to be able to distinguish between the two concepts, when referring to them, we will say that:

a transaction is according as	$\mathbf{P}_U$	*indifferent,* remains constant,	*advantageous,* or increases,	*disadvantageous* or decreases;
a transaction is according as	$\mathbf{P}$	*fair,* remains constant,	*favourable,* or increases,	*unfavourable* or decreases.

A fair transaction is such for everyone agreeing on the same evaluation of probabilities, even for the other contracting party $(\mathbf{P}(-X) = -\mathbf{P}(X))$; an indifferent transaction is not such as U varies, and cannot be such for both contracting parties if they both have convex utility functions (in this case $\mathbf{P}_U(-X) < \mathbf{P}(-X) = -\mathbf{P}(X) < -\mathbf{P}_U(X)$).

3.2.6. *Some further remarks.* Finally, let us turn to the other reasons for preferring this approach: these are essentially concerned with simplicity. The separation of probability from utility, of that which is independent of risk aversion from that which is not, has first of all the same kind of advantages as result from treating geometry apart from mechanics, and the mechanics of so-called rigid bodies without taking elasticity into account (instead of starting with a unified system).

The main motivation lies in being able to refer in a natural way to combinations of bets, or any other economic transactions, understood in terms of monetary value (which is invariant). If we referred ourselves to the scale of utility, a transaction leading to a gain of amount S if the event E occurs would instead appear as a variety of different transactions, depending on the outcome of other random transactions. These, in fact, cause variations in one's fortune, and therefore in the increment of utility resulting from the possible additional gain S: conversely, suppose that in order to avoid this one tried to consider bets, or economic transactions, expressed, let us say, in 'utiles' (units of utility, definable as the increment between two fixed situations). In this case, it would be practically impossible to proceed with transactions, because the real magnitudes in which they have to be expressed (monetary sums or quantity of goods, etc.) would have to be adjusted to the continuous and complex variations in a unit of measure that nobody would be able to observe.

Essentially, our assumption amounts to accepting as practically valid the hypothesis of rigidity with respect to risk: in other words, the identity of monetary value and utility† within the limits of 'everyday affairs'. One should be concerned, however, to check whether this assumption is sufficiently realistic within a wide enough range: actually, it seems safe to say that under the heading of 'everyday affairs' one can consider all those transactions whose outcome has no relevant effect on the fortune of an individual (or firm, etc.), in the sense that it does not give rise to substantial improvements in the situation, nor to losses of a serious nature.

There is no point in prolonging this discussion, but it seems appropriate to mention an analogy from economics, and one from insurance: these—in the same spirit as the preceding geometrical–mechanical analogy—are sufficient to clarify the question, both from a conceptual and practical point of view. Using the prices $P(X)$ as they appear in our hypothesis of rigidity is to do the same thing as one does in economics when one considers the total price of a set of goods, of given amounts, on the basis of the unit-prices in force at the time, without taking into account the variation that a possible transaction would cause by changing the supply and demand situation. On the other hand, these variations are only noticeable if the quantities under consideration are sufficiently large. Even more apposite is the example of actuarial mathematics: indeed, the latter is nothing other than a special case of the theory we are discussing. In the main, it is traditionally concerned with the terms of an insurance under *fair* conditions ('pure' is the usual terminology: pure premium, etc.), and only in special cases—for instance, the theories dealing with the risk of the insurer, or with the advantage for those exposed to risk in insuring themselves—does one speak in terms of utility (or something equivalent, if such a notion is not introduced explicitly). Notwithstanding the fact that this stems less from deliberate choice than from a tradition which lacks an awareness of the questions involved, the 'rigid' approximation has turned out to be satisfactory for the greater part of this most classical field of application of the calculus of probability to economic questions. We intend to use only the simplified version; the above considerations suggest that this is a reasonable thing to do.‡

On the other hand, we shall see (in Chapter 5) how, although starting from the hypothesis of rigidity, one can arrive at the evaluation of probabilities by means of criteria which are neutral with respect to it. The method, which takes as its basis the most meaningful concept, and then clarifies it by means of this simplifying hypothesis, therefore achieves its objective without prejudicing the conclusions.

---

† Except for (obviously inessential) changes of origin and unit of measurement.
‡ For all these topics cf. de Finetti–Emanuelli (1967), Part I.

3.3 BASIC DEFINITIONS AND CRITERIA

3.3.1. We must now translate into actual definitions and proofs those things that we have hitherto put forward in an introductory form, bringing in any necessary refinements, and beginning the developments.

We have given some idea of what a *prevision-function* **P** is, and what conditions it must satisfy in order to be *coherent*. The function **P** represents the opinion of an individual who is faced with a situation of uncertainty. To each random magnitude X, there corresponds the individual's evaluation **P**(X), the *prevision* of X, whose meaning, operationally, reduces, in terms of gain, to that of the (fair) *price* of X. This includes, in particular, the special case of *probability* (which is the more specific name given to prevision when X is an event). A prevision-function **P** is *coherent* if its use cannot lead to an inadmissible decision (i.e. such that a different possible decision would have certainly led to better results, whatever happened). We have remarked already that coherence reduces to linearity and convexity.

3.3.2. In order to fix the formulation in a precise way, we will now put forward two *criteria* (in the sense of *devices* or *instruments* for obtaining a measurement). Each one furnishes an *operational definition* of probability or prevision **P**, and, together with the corresponding *conditions of coherence*, can be taken as a foundation for the entire theory of probability.

Let us recall that the term 'operational' applies to those definitions which allow us to reduce a concept not merely to sentences, which might have only an apparent meaning, but to actual experiences, which are at least conceptually possible. Think of Einstein's definition of 'simultaneity' by means of signals: until that time no-one had even doubted that the term lacked an absolute meaning. That definitions should be operational is one of the fundamental needs of science, which has to work with notions of ascertained validity, in a pragmatic sense, and which must not run the risk of taking as concepts illusory combinations of words of a metaphysical character.

In our case, for the definition of **P**(X), it is a question of stating exactly what 'the rules of the game' are. To state, in other words, what, in the application of a given criterion, are the practical consequences that You know You must accept, and which You do accept, when You enunciate your evaluation of **P**(X) (whose meaning as 'price' is already essentially given). From a conceptual point of view, in the case of coherence too the pointers given in 3.1.5 are sufficient in themselves. To make them explicit in a compact form for specific criteria provides, however, a more incisive schematization of the theory by reducing it to a really small nucleus of initial assumptions.

3.3.3. As far as the extension of the domain of definition of a function of prevision **P** is concerned, we assume that in principle **P** could be evaluated (by You, by anybody) for *every* event E or random quantity X: this is in contrast to what is assumed in other theories and so it is appropriate to point it out explicitly. It will be sufficient for You to place yourself under the restriction of a certain criterion, which we shall soon make explicit, and being forced to answer—i.e. to make a choice among the alternatives at your disposal—to reveal your evaluation of **P**(X) (or, in particular, of **P**(E)). This is valid, as we have said, *in principle*: in other words, we intend not to acknowledge any distinction according to which it would make sense to speak of probability for some events, but not for others.† On the other hand, however, we certainly do not pretend that **P** could actually be imagined as determined, by any individual, for *all* events (among which those mentioned or thought of during the whole existence of the human race only constitute an infinitesimal fraction, even though an immense number). On the contrary, we can at each moment, and in every case, assume or suppose **P** as defined or known for all (and only) the random quantities (or, in particular, events) belonging‡ to some completely arbitrary set $\mathscr{X}$: for instance, those for which we know the evaluation explicitly expressed by the individual under consideration.

Without leaving this set, whatever it may be, we can recognize whether or not **P** includes any incoherence: if so, the individual, when made aware of this fact, should eliminate it, modifying his evaluations after reconsideration. The evaluation is then coherent, and can be extended to any larger set whatsoever: the extension will be uniquely determined up to the point that the coherence demands, and is, to a large extent, more or less arbitrary outside that range. One can only proceed, therefore, by interrogating the individual, and alerting him if he violates coherence with respect to the preceding evaluations.

Among the answers that do not make sense, and cannot be admitted, are the following: 'I do not know', 'I am ignorant of what the probability is', 'in my opinion the probability does not exist'. Probability (or prevision) is not something which in itself can be known or not known: it exists in that it serves to express, in a precise fashion, for each individual, his choice in his given state of ignorance. To imagine a greater degree of ignorance which would justify the refusal to answer would be rather like thinking that in a statistical survey it makes sense to indicate, in addition to those whose sex is unknown, those for whom one does not even know 'whether the sex is unknown or not'.

---

† For instance, the two following distinctions are quite common: yes for 'repeatable' events, no for 'single' instances; yes if X belongs to a measurable set I, no otherwise. Cf. Appendix.
‡ And not, necessarily, belonging to something reducible to a ring (or to a σ-ring) of events. Again, cf. Appendix.

Other considerations and restrictions may enter in if we consider functions of probability defined other than as an expression of the opinion of a given individual. If, for instance, after having considered and interrogated many individuals, we want to study 'their common opinion', **P**, this will only exist in the domain of those X for which all the $\mathbf{P}_i(X)$ coincide (in this way defining $\mathbf{P}(X)$), and will not exist elsewhere. We can also confine ourselves (there is nothing to prevent us) to evaluations which conform to more restrictive criteria to which one would prefer to limit the investigation, excluding in this way events for which one would like to say that the probability 'does not exist' or 'is not known', knowing all along that such motivations remain, nonetheless, meaningless within the present formulation. I may please a friend of mine by not inviting along with him a person whom he judges 'a jinx', without myself believing that such things exist, nor understanding how others can believe in it.

As far as *coherence* is concerned, we will again underline here in what sense the notion is and must be *objective*. The conditions of coherence must exclude the possibility of certain consequences whose unacceptability appears expressible and recognizable to everyone, independently of any opinions or judgments they may have regarding greater or lesser 'reasonabler ↞s' in the opinions of others. Let this be said in order to make clear that such conditions, although *normative*, are not (as some critics seem to think) unjustified impositions of a criterion which their promoters consider 'reasonable': they merely assert that 'you must avoid this if you do not want . . . ' (and there follows the specification of something which is obviously undesirable). We will see this immediately—note it well!—in the two criteria we are about to put forward.

3.3.4. *Criteria for the evaluations.* We now present the details of the two criteria mentioned above; each will consist of the following:

a *scheme of decisions* to which an individual (it could be You) can subject himself in order to reveal—in an operational manner—that value which, *by definition*, will be called his *prevision* of X, or in particular his *probability* of E,

and a *condition of coherence* which enables one to distinguish (so that the distinction has an objective meaning) whether an individual's set of previsions is coherent, and therefore acceptable, or, conversely, intrinsically contradictory.

In both cases, the prevision of X will be a value $\bar{x}$ which can be chosen at will as an 'estimate' of X: along with such a choice goes the necessity of making precise, according to which scheme is used, the otherwise completely indeterminate† meaning of the word 'estimate'. To anticipate the outcome in words; both criteria start by considering the random magnitude given by the difference, or deviation, $X - \bar{x}$, between the actual value X and that chosen by You. Both lead to the same $\bar{x}$ if applied coherently.

The first criterion stipulates that You must accept a bet proportional to $X - \bar{x}$, in whatever sense chosen by your opponent (i.e. positively proportional either to $X - \bar{x}$ or to $\bar{x} - X$). This means that there is no advantage to You in deviating, one way or the other, from the value which makes the two bets indifferent for You; otherwise, one or other would be unfavourable to You, and the opponent could profit from this by an appropriate choice.‡

The second criterion stipulates that You will suffer a penalty (positively) proportional to the square of the deviation $(X - \bar{x})^2$, increasing as one deviates in either direction from the actual value.

> This is evident if one recalls the properties of the barycentre (stable equilibrium, minimum of the moment of inertia) which give an analogy and, in fact, a perfect interpretation. Those who already know something about probability or statistics will be well acquainted with the fact that these properties characterize $\mathbf{P}(X)$. The latter is usually called '(mathematical) expectation', or 'mean value', and is denoted by $\mathbf{E}(X)$ or $\mathbf{M}(X)$: the only novelty lies in making use of it as an operational and direct definition of $\mathbf{P}(X)$, and in particular of probability. Given the probabilities of all possible values (if they are finite in number), it is clear how $\mathbf{P}(X)$ can be expressed as a function of them: the extension of this result to the general case will be immediate when we introduce the notion of a 'probability distribution' (cf. Chapter 6). In the latter approach, however, one introduces the simpler notion (that of 'prevision') by means of the more complicated one (that of 'distribution'), which itself becomes a prerequisite, and forces us to use more advanced mathematical tools (Stieltjes integrals) than necessary. The same thing happens in the case of a solid body: the barycentre is easily determined and, it might be said, is always useful; the exact distribution of mass can never be determined in practice, and is of relatively little interest.

Two further remarks. Firstly, let us recall 'the hypothesis of rigidity with respect to risk', which we continue to assume in what follows (not without noting where appropriate, under the heading of 'Remarks', any implications of this hypothesis at those points where it merits attention). In order to fulfil more easily the resulting requirement of considering only 'moderate amounts', and to omit certain delicate points which are better reconsidered later on (in Section 3.11, etc.), we restrict ourselves for the time being to *bounded* random magnitudes (i.e. those whose possible values are contained in some interval; in other words, $-\infty < \inf X, \sup X < +\infty$).

Concerning preferences for one or other of the two criteria of definition, it is merely a question of individual taste, since—as we have stated, and will

---

† Or, even worse, open to being interpreted as 'prediction'!

‡ This is the same criterion as 'divide the cake into two parts and I will choose the larger', which ensures that the person dividing it does so into parts he judges to be equal.

later show—the two definitions (together with their respective conditions of coherence) are equivalent. The first has a meaning which is slightly more immediately intuitive, but, as far as actual deductions are concerned, the second is more meaningful and fits better into a decision-theoretic framework. A third criterion, which has useful applications, will be derived in Chapter 5, but does not lend itself to an autonomous presentation.

3.3.5. *The first criterion.* Given a random quantity (or random magnitude) X, You are obliged to choose a value $\bar{x}$, on the understanding that, after making this choice, You are committed to accepting any bet whatsoever with gain $c(X - \bar{x})$, where c is arbitrary (positive or negative) and at the choice of an opponent.

Definition. $\mathbf{P}(X)$, the prevision of X according to your opinion, is by definition the value $\bar{x}$ which You would choose for this purpose.

Coherence. It is assumed that You do not wish to lay down bets which will with *certainty* result in a loss for You.† A set of your previsions is therefore said to be *coherent* if among the combinations of bets which You have committed yourself to accepting there are none for which the gains are *all uniformly negative*.‡

Analytic conditions. Expressed mathematically, this means that we must choose the values $\bar{x}_i = P(X_i)$ such that there is no linear combination

$$Y = c_1(X_1 - \bar{x}_1) + c_2(X_2 - \bar{x}_2) + \ldots + c_n(X_n - \bar{x}_n)$$

with sup Y negative (conversely, inf Y cannot be positive, because then $\sup(-Y) = -\inf Y$ would be negative).

Remark. Observe the objective character of these conditions, revealed by the fact that only 'possible values' are referred to.

3.3.6. *The second criterion.* You suffer a penalty L§ proportional to the square of the difference (or deviation) between X and a value $\bar{x}$, which You are free to choose for this purpose as you please:

$$L = \left(\frac{X - \bar{x}}{k}\right)^2$$

---

† Giving rise to what is sometimes called a 'Dutch Book'.
‡ The reason why we cannot simply say 'all negative' (i.e. < 0), but must add 'uniformly' (i.e. $< -\varepsilon$ with ε positive) will be given later (for the time being we do not worry about the finer points). By 'combinations' we always mean linear combinations *of a finite number* of the bets (even if there are infinitely many of them).
§ From *Loss*, the terminology introduced by A. Wald.

(where k, arbitrary, is fixed in advance, possibly differing from case to case).†

Definition. $\mathbf{P}(X)$, the prevision of X according to your opinion, is the value $\bar{x}$ which You would choose for this purpose.

Coherence. It is assumed that You do not have a preference for a given penalty if You have the option of another one which is *certainly* smaller. Your set of previsions is therefore said to be *coherent* if there is no other possible choice which would certainly lead to a uniform reduction in your penalty.

Analytic conditions. The definition of coherence implies that there exist no values x_i^* which, when substituted for the chosen $\bar{x}_i = \mathbf{P}(X_i)$, lead to the penalty

$$L^* = \sum_i \left(\frac{X_i - x_i^*}{k_i} \right)^2$$

being uniformly less than

$$L = \sum_i \left(\frac{X_i - \bar{x}_i}{k_i} \right)^2,$$

for *any possible points* $(X_i, X_2, \ldots, X_n)$; i.e. belonging to the set $\mathcal{Q}$.

Remark. As for the first criterion.

3.3.7. *The equivalence of the two criteria.* The identity of the previsions given by the two criteria can be verified immediately.

Let $\bar{x}$ be the prevision of X based on the first criterion, and $\bar{\bar{x}}$ that based on the second; this implies, respectively, that:

(1) in the first case, the random gain X is judged equivalent to the certain gain $\bar{x}$ (hence: preferable to each $x < \bar{x}$, but not to any $x > \bar{x}$);

(2) in the second case, the gain $-(X - \bar{\bar{x}})^2$—negative, since a penalty!— is judged preferable to any other $-(X - x)^2$ with $x \neq \bar{\bar{x}}$; in other words, the gain

$$G = (X - x)^2 - (X - \bar{\bar{x}})^2$$

is preferred to 0 (for all $x \neq \bar{\bar{x}}$).

---

† It is convenient to think of k as being homogeneous with X, so that the expression turns out to be a pure number; with the further understanding that we multiply by a monetary unit u, L has the dimension of a monetary value. This avoids the complication of writing it, or assuming it as included in k by conjuring up a strange factor of dimension $u^{1/2}$.

More generally, let us compare preferences between the penalties corresponding to any two values of x, say $x = a$ and $x = b$, and let us denote by $c = \frac{1}{2}(a + b)$ the mid-point of the interval $[a, b]$.

The choice of a is preferred to that of b, if the gain $G = (X - b)^2 - (X - a)^2$ if preferred to 0; in other words, expanding, if

$$G = (X^2 - 2bX + b^2) - (X^2 - 2aX + a^2) = 2(a - b)X - (a^2 - b^2)$$

$$= 2(a - b)(X - c)$$

is preferred to 0. Preferring G to 0 means that $\mathbf{P}(G) > 0$; on the basis of the first criterion it turns out that $\mathbf{P}(G) = 2(a - b)(\bar{x} - c)$, an expression which is positive if $a > b$ and $\bar{x} > c$, or, conversely, if $a < b$ and $\bar{x} < c$. In other words, in either case, if $\bar{x}$ lies in the subinterval between c and a; i.e. if $\bar{x}$ is *closer* to a than it is to b.

Our assertion is an obvious corollary of this result (which it seemed useful to put forward in this more general form): the optimal choice, $x = \bar{\bar{x}}$, is given by $\bar{\bar{x}} = \bar{x}$.

The equivalence of the conditions of coherence can also be verified by expansions of this sort (and we shall do so, writing them out in full, for those who would like to check them and apply them directly). Conceptually, however, we can make everything incomparably easier, and intuitively meaningful, by presenting an obvious geometrical interpretation.

3.4 A GEOMETRIC INTERPRETATION: THE SET $\mathscr{P}$ OF COHERENT PREVISIONS

3.4.1. Any prevision in the linear ambit $\mathscr{A}$ of the n random quantities $X_1, X_2, \ldots, X_n$ consists in fixing, in the n-dimensional space with co-ordinates $x_1, x_2, \ldots, x_n$ (the linear ambit $\mathscr{A}$), the n values $\bar{x}_1, \bar{x}_2, \ldots, \bar{x}_n$, where $\bar{x}_i = \mathbf{P}(X_i)$, and hence corresponds to a point in the said space. The conditions of coherence state—as we shall immediately verify—that *the set $\mathscr{P}$ of coherent previsions is the closed convex hull of the set $\mathscr{Q}$ of possible points.*

For the first criterion: in a form which is more directly suited to the purpose in hand, the necessary and sufficient condition for coherence can be expressed by saying that *every linear relation (or inequality) between the random quantities X_i*

$$c_1 X_1 + c_2 X_2 + \ldots + c_n X_n = c \text{ (or } \geqslant c)$$

must be satisfied by the corresponding previsions $\mathbf{P}(X_i)$:

$$c_1 \mathbf{P}(X_1) + c_2 \mathbf{P}(X_2) + \ldots + c_n \mathbf{P}(X_n) = c \text{ (or } \geqslant c).$$

Geometrically, a point P represents a coherent prevision if and only if *there exists no hyperplane separating it from the set $\mathcal{Q}$ of possible points;* this characterizes the points of the convex hull.

For the second criterion: here one introduces into the (*affine!*) linear ambit $\mathcal{A}$ a *metric* of the form $\rho^2 = \sum_i (x_i/k_i)^2$, setting

'penalty' $= L = (P - Q)^2$

$\quad\quad\quad = $ 'the square of the distance between the prevision-point P and the outcome-point Q, according to the given metric'.

The necessary and sufficient condition for coherence requires, in geometrical terms, that *P cannot be moved in such a way as to reduce the distance from all points Q;* this is another characterization of the convex hull.†
Further explanations, and diagrams in the simple cases, are given in Chapter 5, Section 5.4.

3.4.2. *Other interpretations.* Every prevision-point P, which is admissible in terms of coherence, is a barycentre of possible points Q_j with suitable weights (or is a limit case‡). On the other hand, the possible points are themselves particular cases of previsions; degenerate cases, in that the probability is concentrated at a unique point Q_j. In words, one could say, according to this interpretation, that *a prevision turns out to be a mixture of possibilities.*

Of course, one can also form linear combinations of different coherent previsions (with non-negative weights, summing to 1) again obtaining

---

† If we move the point P to another position P^*, its distance from a generic point A increases or decreases according as A is on the same side as P or P^*, with respect to the hyperplane which bisects the segment PP^* orthogonally.

If P is not in the convex hull of $\mathcal{Q}$ there exists a hyperplane separating it from $\mathcal{Q}$. Moving P to P^*, its orthogonal projection onto such a hyperplane, diminishes its distance from all points Q in $\mathcal{Q}$ (which are on the opposite side). More precisely, the diminution of the penalty, $L - L^*$, i.e. the square of the distance, is alway at least $(P - P^*)^2$: in fact, $(Q - P)^2 - (Q - P^*)^2 = [(Q - P^*) - (P - P^*)]^2 - (Q - P^*)^2 = (P - P^*)^2 - 2(Q - P^*) \times (P - P^*)$, and this scalar product is negative, since the component of the first vector parallel to the second is in the opposite direction.

Suppose instead that P belongs to the convex hull of $\mathcal{Q}$. Then to whatever point P^* we move P, it always follows that for some point Q the distance increases: if, with respect to the bisecting hyperplane, they were all on the same side as P^*, the point P would be separated from the convex hull of $\mathcal{Q}$, contrary to the hypothesis.

‡ To be precise: either they can be obtained as barycentres of at most $n + 1$ points Q_j (in the n-dimensional space), or they are *adherent* points of $\mathcal{Q}$ (but not belonging to $\mathcal{Q}$). For instance, we could think of $\mathcal{Q}$ as the set of points on the circumference of a circle, having rational angular distance from one of its points (with respect to the complete angle). We are in the plane, $n = 2$, and each point inside the circle is inside triangles with vertices in $\mathcal{Q}$: hence it is the barycentre of $3 = n + 1$ points (two would suffice if it were on chords connecting rational points, and only one if it coincided with such a point). The points which are on the circumference, but not rational, are required in order to complete the closed convex hull: they are adherent points of $\mathcal{Q}$ (i.e. there are points of $\mathcal{Q}$ in each of their neighbourhoods).

coherent previsions. More generally, if $\mathscr{P}_0$ is any set of coherent previsions, then its closed convex hull is also a set of coherent previsions, the *mixtures* of those in $\mathscr{P}_0$. Let us denote it by $\mathscr{P}_1$.

3.5 EXTENSIONS OF NOTATION

It is convenient, in addition to being natural (and also useful for compactness of notation), to exploit the linear structure of **P** in order to extend the range of applicability of this symbol to any random elements whatsoever belonging to a linear space (vectors, matrices, n-tuples of numbers or magnitudes, functions, etc.), or even just to a linear manifold (a linear subspace which also contains the zero: for example, the points of a space in which the differences between points $\mathbf{u} = A - B$, constitute a linear space of vectors).

As a formal definition, it is sufficient to state that **P** is always intended to be *linear*, so that if f is any linear function—i.e. $f(A)$ is a scalar linear function of the points or elements A of our linear space or manifold—we have $f(\mathbf{P}(A)) = \mathbf{P}(f(A))$. For practical purposes, it is enough to note that **P** operates on the components or coordinates, so that, if

$$A = 0 + X\mathbf{i} + Y\mathbf{j} + Z\mathbf{k} \quad \text{(or, in conventional notation, } A = (X, Y, Z)\text{)},$$

we could write

$$\mathbf{P}(A) = \mathbf{P}(X)\mathbf{i} + \mathbf{P}(Y)\mathbf{j} + \mathbf{P}(Z)\mathbf{k} = \bar{x}\mathbf{i} + \bar{y}\mathbf{j} + \bar{z}\mathbf{k}$$

(in other words, $\mathbf{P}(X, Y, Z) = (\mathbf{P}(X), \mathbf{P}(Y), \mathbf{P}(Z))$).

A case of particular importance is the following: if Z is a *complex* random quantity, and we denote by X and Y, respectively, the real and imaginary components, its prevision will be

$$\mathbf{P}(Z) = \mathbf{P}(X + iY) = \mathbf{P}(X) + i\mathbf{P}(Y) = \bar{x} + i\bar{y}.$$

As a practical rule, it is sufficient to replace the random component X by the corresponding prevision $\bar{x}$; for example:

$$\mathbf{P}(X_1, X_2, \ldots, X_n) = (\bar{x}_1, \bar{x}_2, \ldots, \bar{x}_n), \qquad \mathbf{P}(\|X_{rs}\|) = \|\bar{x}_{rs}\|, \quad \text{etc.}$$

In the case of a random function $X(t)$ (where t is the independent variable; for example, time) it would appear to be an unnecessary subtlety (but it is not) to say that one could write $f = \mathbf{P}(X)$ to mean that f is the function which for each t gives $f(t) = \mathbf{P}(X(t))$. It would be a little more explicit to write $f(\cdot) = \mathbf{P}(X(\cdot))$ in order to indicate that it is a question of operating on a variable whose position is denoted by the point. Here, however (at least if one does not want to be limited to considering not more than a finite number of t_h at once), one would step outside of the ambit in which, for the time being, we have expressed our intention of remaining.

3.6 REMARKS AND EXAMPLES

3.6.1. The properties which we have established in Section 3.4 could be said to contain the whole calculus of probability, even though we have not as yet mentioned probability, except to point out that it is a special case of prevision. Sections 3.8 and 3.9 will be devoted to this special case, giving it the attention it merits. However, from our point of view it turns out to be better formulated and much clearer if embedded in the general case, where the basic properties present themselves as simple, clear and 'practical'. It is precisely for this reason (and certainly not because of any dubious motive of wishing to start, come what may, by showing off, with no justification, the greatest generality and abstraction) that we did not begin the discussion with the case of events, and have still not stopped to consider it. Otherwise, we would have found ourselves, at this moment, having defined $\mathbf{P}(E)$ and not $\mathbf{P}(X)$, in more difficulty than if we had defined a unique concept, and with the unavoidable problem of producing $\mathbf{P}(X)$ as something not equally immediate, but as the combination of the $\mathbf{P}(E)$ and who knows what mathematical definition of integral.

3.6.2. *Some remarks concerning the two criteria.* Every operational definition, if one wants to take it too seriously as an actual method of measurement, carries with it the difficulty that the discussions of principle become mixed up with the doubts deriving from the practical imperfections inherent in any tool or procedure (these, however, often arise for reasons which may be important). Let us accept that this difficulty is unavoidable, but that it is by no means a tragedy since the definition deals with an idealized case, or limit case, of conceptually possible experiments. Having said this, and having repeated that it is always infinitely better than any attempt at a mere verbal definition, emptily 'philosophical', there remains, nonetheless, the necessity of making oneself aware of the weak points in order to keep in mind the appropriate precautions.

We have already discussed, in Section 3.2, 'rigidity in the face of risk'; in other words, the temporary identification of utility and monetary value. At this juncture, a brief mention, with specific reference to the two criteria which we have put forward, will suffice. They both assume, implicitly, the hypothesis of rigidity. In the first place, they take the different bets, which are used as 'tests', to be summable; to be rigorous, the stipulation of any one of them should modify slightly the conditions for the stipulations of the others: secondly, by virtue of having a homogeneous character, in the sense that the procedure itself presupposes that $\mathbf{P}(aX) = a\mathbf{P}(X)$ (this adds no further restrictions, it is the same rigidity). This is useful if one attempts to limit the bets to be of moderate size; it is dangerous to allow it to be used indiscriminately. In the first criterion one has to think that in practice the opponent cannot impose excessively large bets (although the explicit inclusion of this kind of regulation in the definition would lead to a hybrid and tediously wordy exposition).

3.6.3. A defect of the first criterion is, in any case, the intervention of an 'opponent': this can make it difficult to avoid the risk, or at least the suspicion, of other factors intruding (such as the possibility of taking advantage of differences in information, competence, or shrewdness). By and large, such possibilities are in 'his' favour (he being the one who decides how much to bet, and in which direction; especially if he is the same person who has chosen the events for which the evaluation of probability is required). Sometimes, however, they can be in your favour (for instance, if, imagining the opponent to have a very distorted opinion, You enunciate an evaluation which induces him to stipulate a bet in a way that You judge favourable†).

3.6.4. Under the second criterion these negative features are not present (apart from that inherent in thinking of the various bets as summable). However, given that by choice of the coefficients k_i one can arrange the sizes of the penalties in whatever way one considers most appropriate, even this consequence of 'rigidity' becomes practically negligible. Observe, on the other hand, that the (arbitrary) choice of such coefficients—i.e. of the *metric*—has no influence at all on the implications of the criterion, since these are always based on merely *affine* notions: the notion of barycentre, and therefore its property of yielding a minimum for the moment, remains invariant under whatever metric one introduces for other purposes, and which occurs in the definition of the moment.

Another doubt arises: one might ask whether there is any good reason for considering the minimization of a penalty L, rather than the maximization of a prize $K - L$. Formally, there is no difference, but if one wants to fix K greater than any possible value of L (in order that the 'prize' always turns out to be positive) one is faced with an annoying limitation (which is impossible anyway if X is not bounded). There is, moreover, an historical reason: when introducing a similar theory in statistical applications, Wald found it natural to posit a Loss in the case of 'wrong decisions' (zero for

---

† On the eve of a certain football match, You attribute a probability of 40% to the victory of team A, but You think that your opponent, being a supporter of team A, evaluates it at 70%. You can then enunciate a probability of 65%, confident that he will hasten to pay 65 for that which to You is worth 40, but to him 70. But be careful! If he evaluates the probability at 50% instead, and decides to bet in the opposite way, he will pay You 35 for that which is worth 60 to You, and 50 to him—not 30 as You thought.

correct decisions). Finally, one might exercise more care in attempting to prudently minimize a loss (which, in any case, involves uneasiness and disappointment), than in assuring oneself, in a reasonable manner, of the highest level of gain (in this context, the temptation to take a chance is often irresistible; naturally enough, since one cannot lose whatever happens).

3.6.5. One further remark, which is so deeply rooted in what we have said over and over again about the subjective meaning of probability that it is perhaps unnecessary. The two criteria are operational in the sense that they provide a means by which the opinions that an individual carries within himself, whatever they may be, turn out to be observable from the outside. There is no connection with questions like 'what is the *true* value of the probability?'—a question whose meaning finds no place within the present formulation, and whose meaning I was unable to discover in the attempts made by other theories to provide one—and not even with questions like 'how well-founded, or how reasonable, are certain evaluations and their associated motivations?'.

This last question can, in a certain sense, be dealt with by reflecting on various problems which will present themselves from time to time as we proceed to study probability, and to examine various attitudes to both concrete applications and conceptual questions. However, it is mainly a question of arguments (of a rather psychological nature) concerning the choice of a single prevision, **P**, from among the infinite set of coherent previsions, $\mathscr{P}$ (which are equally acceptable from the mathematical point of view). The question does not concern mathematics, except in that it may give a still more enriched description of the various aspects of each choice, so that such a choice, always made absolutely freely, can be made by each individual after an accurate and straightforward examination of everything that in his personal judgment appears relevant for his decision.

3.7 PREVISION IN THE CASE OF LINEAR AND NON-LINEAR DEPENDENCE

3.7.1. Let us return to the two examples already introduced (Chapter 2, Section 10) under the guise of the logic of certainty. They lend themselves not only to illustrating in practice the application of the two criteria and the consequences of the properties we have established, but above all to developing necessary and instructive insights of a general character. First of all, one notes the essential connection between *linearity* and *prevision* (and the way in which this makes inapplicable to prevision certain arguments which would be valid for prediction). In this connection, it will become clearer how and why it is appropriate to extend the linear ambit $\mathscr{A}$ in relation to the questions to be examined (cf. the brief explanations given in Chapter 2, Section 2.8).

In the case of a ballot with n voters, we denoted by X, Y, Z, the number of votes cast in favour, against or abstaining, and considered in addition the difference and the ratio of votes for and against, which we denote by $U = X - Y$, $V = X/Y$.

If invited to make a prevision of the outcomes—on the basis of the first or second of the criteria put forward—you provide values $\bar{x}$, $\bar{y}$, $\bar{z}$, $\bar{u}$, $\bar{v}$. These are based on your knowledge, information, impressions or conjectures, about the inclinations or moods of the voters. If your values constituted a prediction, or if You intended to put them forward as *sure*, they would have to be chosen as *integers*, satisfying $\bar{x} + \bar{y} + \bar{z} = n$, $\bar{u} = \bar{x} - \bar{y}$, $\bar{v} = \bar{x}/\bar{y}$. In a prevision they might not even be integers: would the three relations hold? Is it valid to argue that they necessarily hold because they must hold for the true values? In fact, it might seem completely obvious that if the votes for and against *are* x and y (in reality, or in a prediction, or an estimation, or any prevision whatever) then their difference *is* $x - y$, and their ratio *is* x/y, whereas the number abstaining *is* $n - x - y$. For the two linear relations, given the linearity of **P**, this is certainly true, and, considering the ambit $\mathscr{A}$ of (x, y, u)—respectively, of (x, y, z)—is easily interpreted as follows: $\mathscr{Q}$ is the set of points with non-negative integer coordinates x and y with sum $\leqslant n$ in the plane $u = x - y$—respectively, in the plane $z = n - x - y$—and a barycentre of such points (with arbitrary weights) can only be some point in the given plane, in the triangle defined by $x \geqslant 0$, $y \geqslant 0$, $x + y \leqslant n$. If, however, we consider the $\mathscr{A}$ of (x, y, v), the points $\mathscr{Q}$ are those having the same x, y, but now on the surface (hyperbolic paraboloid) $v = x/y$, and the conclusion is no longer valid.†

In the case of a random prism with sides X, Y, Z, we denoted the diagonal by U, the area by V, and the volume by W. Since these are not linear functions of the sides, one would not expect that, having evaluated the previsions of the three sides as $\bar{x}$, $\bar{y}$, $\bar{z}$, those of the other elements, say $\bar{u}$, $\bar{v}$, $\bar{w}$, would satisfy the same relations as those holding between the true magnitudes. In other words, in the (six-dimensional) linear ambit $\mathscr{A}$ of (x, y, z, u, v, w), in which $\mathscr{Q}$ is the three-dimensional manifold with equations $u^2 = x^2 + y^2 + z^2$, $v = 2(xy + yz + zx)$, $w = xyz$ (with x, y, z, u positive), one would not necessarily expect a barycentre of points of $\mathscr{Q}$ to lie in this manifold. It could be any point **P** whatsoever of $\mathscr{P}$, the convex hull of the given $\mathscr{Q}$: once we have evaluated $\bar{x}$, $\bar{y}$, $\bar{z}$, the previsions $\bar{u}$, $\bar{v}$, $\bar{w}$ can only turn out to be some point of the intersection of $\mathscr{P}$ with $x = \bar{x}$, $y = \bar{y}$, $z = \bar{z}$.

---

† In this example, provided we do not evaluate the probability that $Y = 0$ as *zero*, we actually have $\bar{v} = \mathbf{P}(V) = +\infty$. This is of more use in showing how one can encounter, in a natural way, cases where the hypothesis of boundedness does not hold, than in illustrating the proposition, which will be clearer after the following example.

If one were interested in a complete solution to the problem, it would be necessary to determine $\mathscr{P}$, or this intersection with it. More generally, one should consider the same problem with certain restrictions on $\mathscr{Q}$: for instance, we might we aware of restrictions like $a' \leqslant X \leqslant a'', b' \leqslant Y \leqslant b'', c' \leqslant Z \leqslant c''$, or $d' \leqslant X + Y + Z \leqslant d'', X \leqslant Y \leqslant 2X, Y \leqslant Z \leqslant 2Y$, or that only integer values are admissible for X, Y, Z, or that there are only a finite number of values (x_i, y_i, z_i), and so on. For the purpose of illustration, a simpler version will do : let us suppose that Z is known, $Z = a$ say, and let us consider the restrictions that, given $\bar{x}$ and $\bar{y}$, result for $\bar{u}, \bar{v}$ and $\bar{w}$, *separately*, instead of jointly. In this way, everything is represented each time in a three-dimensional ambit $\mathscr{A}$, which is directly 'visible'.

In the ambit of (X, Y, U) the possible points $\mathscr{Q}$ lie on the circular hyperboloid $u^2 = a^2 + x^2 + y^2$; in fact, if there are no further restrictions, they are all the points on the 'quarter' $x \geqslant 0, y \geqslant 0$ of the sheet $u \geqslant 0$; otherwise, they are a subset of these. The barycentre of masses placed on this surface necessarily falls in the convex region that it encompasses (except in the trivial case where the mass is concentrated at a single point, a case where nothing is really random). In a coherent prevision, the diagonal must therefore *necessarily* be estimated longer than it would be if the lengths of the sides coincided exactly with their respective previsions. In the absence of other constraints, given $\bar{x}$ and $\bar{y}$, all the values lying between that minimum and $a + \bar{x} + \bar{y}$ are in fact admissible for $\bar{u}$: one approaches $a + \bar{x} + \bar{y}$ asymptotically by placing two small masses, $\bar{x}/k$ and $\bar{y}/k$, at the points $(k, 0)$ and $(0, k)$, with the rest at the origin, and then letting k become arbitrarily large.

In the ambit of (X, Y, W) the possible points $\mathscr{Q}$ lie instead on the hyperbolic paraboloid $w = axy$ (on the 'quarter' $x \geqslant 0, y \geqslant 0$). In the absence of other restrictions, the convex hull $\mathscr{P}$ is the entire positive orthant, since the barycentre can lie anywhere in this region. In other words, given $\bar{x}$ and $\bar{y}$, $\bar{w}$ can either coincide with $w = a\bar{x}\bar{y}$, or can be less (but bounded below by zero), or can be greater (with no constraint). The two limit cases can be approached by simply placing all the mass at the origin, with the exception, in the first case, of two masses $\bar{x}/k$ and $\bar{y}/k$ at the points $(k, 0)$ and $(0, k)$, respectively, and, in the second case, a single mass $1/k$ at the point $(k\bar{x}, k\bar{y})$. The case $\bar{w} = a\bar{x}\bar{y}$ occurs, for example, under a very important assumption— that of stochastic independence—which we are not yet in a position to discuss. The case of V reduces immediately (having set $Z = a$) to that of W ; in fact, $2(XY + aX + aY) = 2W/a + 2a(X + Y)$.

The different behaviour in the two cases we looked at is due to that fact that the points of the first surface were all elliptic, always presenting the concavity in the same direction (delimiting in this way its convex hull), whereas for the second surface, whose points were all hyperbolic, the convex hull of each of its parts is necessarily formed of two parts, adhering to the two faces.

3.7.2. *Functional dependence and linear dependence.* In this context, the representation we have already introduced (Chapter 2, Section 2.8) by means of the dual spaces, $\mathscr{A}$ and $\mathscr{L}$, is appropriate, and provides some insight. $A(X)$ denotes the value which $X = f(X_1, X_2, \ldots, X_n)$ would assume if the X_h (belonging to $\mathscr{L}$) assumed the values x_h (the coordinates of the point A of $\mathscr{A}$): in other words, $A(X) = f(x_1, x_2, \ldots, x_n)$. This is only meaningful if A is one of the possible points Q of $\mathscr{Q}$: i.e. the values x_h of the X_h are not incompatible. However, we now know that the other points in $\mathscr{A}$—i.e. the points P of $\mathscr{P}$ the convex hull of $\mathscr{Q}$—can be interpreted as previsions,† and one might ask whether $P(X)$ (understood as above, with $P = A$) is actually the prevision of X. It is clear that this only holds if X belongs to $\mathscr{L}$; in other words, if it is a linear function in the ambit $\mathscr{A}$, or, alternatively, if X is given not just by any function f of the X_h, but in fact by a linear function $X = \sum u_h X_h$ $(h = 0, 1, \ldots, n)$. The extension is only valid in this case, and that is why we always confine ourselves to using the notation $A(X)$. The above considerations give us another way of exhibiting the importance and the compass of the linearity. A point P (an admissible prevision) can be either a Q (that is a possible point) in the linear ambit $\mathscr{A}$, or a barycentre of possible points. The knowledge of the barycentre is sufficient, however, to determine only those things which remain invariant under any choice of the points Q and distribution of mass over them so long as one keeps the barycentre fixed.

In other words; one has always to recall that a **P**, defined on any set of random quantities $\mathscr{X}$ whatsoever (or, in particular, on any set of events $\mathscr{E}$), is uniquely extendible—and therefore defined—only on the linear space $\mathscr{L}$ of the (finite) linear combinations of $\mathscr{X}$, or, dually, in the corresponding linear ambit $\mathscr{A}$. If $\mathscr{L}$ (or $\mathscr{A}$) is enlarged, one can determine **P** more precisely by more or less arbitrary extensions. So long as we remain in a given ambit $\mathscr{A}$, each point **P** represents, in a manner of speaking, all the **P*** in some larger ambit which have **P** as their projection onto $\mathscr{A}$. This also holds in the infinite-dimensional case, but we postpone any explicit discussion until the Appendix. In order to clear up the simplest cases—one- or two-dimensions— it is sufficient to recall the examples already given in 3.7.1 and to examine these further aspects in that context.

3.7.3. *Conclusion.* We conclude, therefore, that whereas it is well-known that coherent previsions *preserve* linear dependence, this *only* happens, in fact, *in this case.* In any other case it does not (unless by chance, or under suitable additional hypotheses) because the barycentre of masses lying in a

---

† Moreover, it is possible to see that it can even be meaningful to consider an $A(X)$ where A does not belong to $\mathscr{Q}$, and not even to $\mathscr{P}$. For example, one might be interested in the difference between two **P**, $A(X) = \mathbf{P}_1(X) - \mathbf{P}_2(X)$, and $A = \mathbf{P}_1 - \mathbf{P}_2$ certainly does not belong to $\mathscr{P}$ since we have $A(1) = 0$ instead of $A(1) = 1$.

given manifold need not itself belong to the manifold (except in the trivial case of linearity). In fact, it is sometimes impossible for it to do so (if $\mathscr{Q}$ is the boundary of a convex region).†

It is important to always bear in mind details of this kind and to reflect upon them. This is not only, and not mainly, because of their intrinsic importance—however notable this may be—but above all because one has to learn to free oneself from the ever present danger of confusing *prevision* and *prediction*. In a prediction any dependence should obviously be preserved (because it reduces to the choice of *a point* in $\mathscr{A}$, and not the barycentre of masses distributed over $\mathscr{Q}$). The type of argument which, in the examples given, turned out to be wrong if applied to prevision, would, on the other hand, be valid for a prediction. Despite knowing, and remembering, that the arguments do not hold for prevision, anyone (even You, even I) can inadvertently fall into error, applying them without sufficient thought in some particular problem, or in some small corner of the formulation of some particular problem.

There will be many and frequent occasions to warn against errors, misunderstandings, distortions, obscurities, contradictions and the other endless troubles which are so difficult to avoid when dealing with probability, and which are always essentially the result of ignoring the same warning: *prevision is not prediction*! It would not be a bad idea to imagine constantly in front of you an admonitory card—as is used by a certain well-known organization—bearing the message, 'Think!', but with an explanatory rider suited to the needs of probability theory and its applications:

> '*Think: prevision is not prediction!*'

There is an anecdote, concerning another such maxim, which may perhaps serve to make this recommendation more forceful. It reveals the fallacy of resorting to the self-deception of 'accepting for certain' the alternative on the basis of which one 'decides to act'; a vain attempt to replace a meaningful probability argument by an impossible translation of it into the inadequate logic of certainty. The anecdote is related by Grayson (on p. 52 of a book concerning which we shall have more to say: cf. Chapter 5, 5.5.3) in the following way:

> *Holes that are going to be dry shouldn't be drilled*

---

† If we wished to be precise, we should exclude the points on the boundary where one does not have strict convexity: in other words, those which are barycentres of other points; or, alternatively, those through which there is no hyperplane which leaves *all* the other points of $\mathscr{Q}$ on the same side.

'is printed on a sign hanging in one operator's office. This would truly be a "golden rule" if any oil or gas firm could live by it. Unfortunately, no one can—not even this particular operator who drilled 30 consecutive dry holes a few years ago.'

3.8 PROBABILITIES OF EVENTS

3.8.1. The properties of probabilities of events are simply special cases of the properties of previsions of random quantities. It will be sufficient to establish them quickly, and to illustrate their meaning within the form of representation that we have introduced.

The theorem of 'total probability'. This is the name given to the theorem that translates, into the field of probability, the additive property of prevision.

The case of incompatibility. If two events A and B are incompatible then, as we have already noted, their logical and arithmetic sums coincide: $E = A \lor B = A + B$, so that, if $\mathbf{P}$ is additive, $\mathbf{P}(E) = \mathbf{P}(A) + \mathbf{P}(B)$. The same result holds for the (logical and arithmetic) sum of any finite number of incompatible events: $E = E_1 \lor E_2 \lor \ldots \lor E_n = E_1 + E_2 + \ldots + E_n$, and hence

$$\mathbf{P}(E) = \mathbf{P}(E_1) + \mathbf{P}(E_2) + \ldots + \mathbf{P}(E_n).$$

We can state this formally:

Theorem. *In the case of incompatible events, the probability of the event-sum must be equal to the sum of the probabilities.*

The case of (finite) partitions. In particular, for a partition in which, in addition, the sum $E = 1$, and hence $\mathbf{P}(E) = 1$, one has the following:

Theorem. *In a (finite) partition the probabilities must sum to 1.*

In particular, for two complementary events E and $\tilde{E}$ (a partition with $n = 2$), it turns out that $\mathbf{P}(E) + \mathbf{P}(\tilde{E}) = 1$; that is to say, $\mathbf{P}(\tilde{E}) = 1 - \mathbf{P}(E) = \sim \mathbf{P}(E)$; or, in yet another form, if $\mathbf{P}(E) = p$, then $\mathbf{P}(\tilde{E}) = \tilde{p}$.
In words:

Theorem. *The probabilities of two complementary events must themselves be complementary.*

Recalling the properties of the constituents, one can state immediately the following:

Corollary. *In order that the probabilities of all the events E which are linearly dependent on $E_1, \ldots, E_n$ should be determined, it is necessary and sufficient to attribute probabilities to all the constituents $C_1 \ldots C_s$. These probabilities must sum to 1; the $\mathbf{P}(E)$ depend linearly upon them.*

3.8.2. *Sufficiency of the conditions.* The preceding statements tell us how 'we must'—or 'You must'—evaluate probabilities: in other words, they impose necessary conditions for coherence. In fact—with the obvious restriction that the probabilities be non-negative—they are also sufficient, in the sense that an evaluation satisfying them is coherent, no matter how You choose it. We have already seen this in general in Section 3.4; it may be useful to repeat the argument in this particular case where it is very simple and clear.

Suppose that to the events $E_1 \ldots E_n$ of a finite partition You have attributed non-negative probabilities $p_1 \ldots p_n$, summing to 1, and that I (thinking in terms of 'The first criterion' of Section 3.3) try to force You into a bet which assures me of certain gain. I have to fix the amounts c_i for the bets on the individual E_i in such a way that the resulting bet

$$X = c_1(E_1 - p_1) + c_2(E_2 - p_2) + \ldots + c_n(E_n - p_n)$$

is certainly positive; in other words,

$$c_1 E_1 + c_2 E_2 + \ldots + c_n E_n > c_1 p_1 + c_2 p_2 + \ldots + c_n p_n$$

no matter which of the E_i occurs. If E_i occurs, however, the left-hand side has the value c_i, and it is impossible for this to be always greater than the right-hand side, since the latter is itself a weighted average of the c_i.

3.8.3. *The case of compatibility; inequality.* For any arbitrary set of events, i.e. without making the assumption of incompatibility, we have

$$E = E_1 \vee E_2 \vee \ldots \vee E_n = 1 \wedge (E_1 + E_2 + \ldots + E_n) \leqslant E_1 + E_2 + \ldots + E_n$$

and hence

(6) $$\mathbf{P}(E) \leqslant \mathbf{P}(E_1) + \mathbf{P}(E_2) + \ldots + \mathbf{P}(E_n).$$

Stated formally:

Theorem. *The probability of the event-sum must be less than or equal to the sum of the probabilities.*

This is even more evident if one puts it in the form

$$\mathbf{P}(E) \leqslant \mathbf{P}(E_1 + E_2 + \ldots + E_n);$$

i.e. that the probability of the event-sum must be less than or equal to the prevision of the number of successes (one only has to consider that the latter takes into account multiplicities, whereas the former does not).

Expressions in terms of products. In the case of compatible events nothing can be said about $\mathbf{P}(E)$ other than the preceding inequality based just on the $\mathbf{P}(E_i)$. If we introduce other elements, and evaluate them, then, of course, things change. In terms of constituents, the only one we require is

$$C = \tilde{E}_1 \tilde{E}_2 \ldots \tilde{E}_n \quad \text{because} \quad E = \tilde{C}, \qquad \mathbf{P}(E) = 1 - \mathbf{P}(C).$$

Making use of the products of the E_i (two at a time, three at a time, etc.), and the expansion

(7) $$E = \sum_i E_i - \sum_{ij} E_i E_j + \sum_{ijh} E_i E_j E_h - \ldots \pm E_1 E_2 \ldots E_n,$$

we have at once the following:

Theorem. *For the probability of the event-sum we must always have*

(8) $$\mathbf{P}(E) = \sum_i \mathbf{P}(E_i) - \sum_{ij} \mathbf{P}(E_i E_j) + \sum_{ijh} \mathbf{P}(E_i E_j E_h) - \ldots \pm \mathbf{P}(E_1 E_2 \ldots E_n).$$

Observe that the expression is linear in the probabilities of the products. Note also the special cases of two and three events:

$$\mathbf{P}(A \vee B) = \mathbf{P}(A) + \mathbf{P}(B) - \mathbf{P}(AB),$$

$$\mathbf{P}(A \vee B \vee C) = \mathbf{P}(A) + \mathbf{P}(B) + \mathbf{P}(C) - \mathbf{P}(AB) - \mathbf{P}(BC) - \mathbf{P}(AC)$$
$$+ \mathbf{P}(ABC).$$

3.8.4. *Extensions.* The same formula serves to express the probability that out of n events a given h occur, and no others; and hence the probability that exactly h occur (no matter which ones). The occurrence of $E_1 E_2 \ldots E_h$, and no others, can be written as:

(9) $$E_1 E_2 \ldots E_h (1 - E_{h+1})(1 - E_{h+2}) \ldots (1 - E_n)$$
$$= E_1 E_2 \ldots E_h - \sum_i E_1 E_2 \ldots E_h E_{h+i} + \sum_{ij} E_1 E_2 \ldots E_h E_{h+i} E_{h+j} - \ldots$$
$$\pm E_1 E_2 \ldots E_n$$

(where, as can be seen, the sum with k indices is the sum of the products of $h + k$ events; the given h together with k of the others). The event $Y = h$, the number of successes $= h$, is the sum of $\binom{n}{h}$ events of the above kind; in other words, the sum of all the corresponding expressions. In this sum, the products h at a time appear only once, those $h + 1$ at a time appear $h + 1$ times (once for each combination h at a time of their $h + 1$ factors), and so on; in general, the products $h + k$ at a time each appear $\binom{h+k}{h}$ times. For this reason, denoting the sum of the products r at a time by $\sum^{(r)}$ for convenience, we have

(10) $$(Y = h) = \sum^{(h)} - \binom{h+1}{h} \sum^{(h+1)} + \binom{h+2}{h} \sum^{(h+2)} - \ldots \pm \binom{n}{h} \sum^{(n)}$$
$$= \sum_{r=h}^{n} (-1)^{r-h} \binom{r}{h} \sum^{(r)}.$$

If in place of the $\sum^{(r)}$ we substitute the sum *of the probabilities* of the products, $P(E_{i_1}E_{i_2}\ldots E_{i_r}) = p_{i_1i_2\ldots i_r}$, which we denote by S_r for short, the same formula gives the probability

(11) $$P(Y = h) = \sum_{r=h}^{n} (-1)^{r-h}\binom{r}{h} \sum p_{i_1i_2\ldots i_r} = \sum_{r=h}^{n} (-1)^{r-h}\binom{r}{h}S_r.$$

Note in particular:

$$P(Y = 0) = 1 - S_1 + S_2 - S_3 + \ldots \mp S_{n-1} \pm S_n$$

$$P(Y = 1) = S_1 - 2S_2 + 3S_3 - 4S_4 + \ldots \mp (n-1)S_{n-1} \pm nS_n$$

$$P(Y = 2) = S_2 - 3S_3 + 6S_4 - 10S_5 + \ldots \mp \binom{n-1}{2}S_{n-1} \pm \binom{n}{2}S_n$$

$$P(Y = n-1) = S_{n-1} - nS_n$$

$$P(Y = n) = S_n$$

(where $\pm$ stands for $(-1)^n$, and $\mp$ for $-(-1)^n$).

Example. A classical and instructive problem is that of *matching*, which lends itself to amusing formulations. If one has n letters and their respective envelopes, what is the probability that if *the letters are inserted into the envelopes at random* one has none, or one, or two,..., or n 'matchings'; i.e. letters in their own envelopes? The same problem arises if one pairs up at random right and left shoes from n pairs, or the husbands and wives of n couples, or the jackets and trousers of n suits, etc. Alternatively, if one gives back at random to n people their passports, the keys of their hotel rooms, hats left in the cloakroom, and so on. More standard versions are given by the matchings in position among playing cards from two identical decks (for instance by placing them at random in two rows), or between the number of the drawing from an urn of numbered balls and the number of the ball drawn.

The probability of a matching at any given position is obviously $1/n$, of two matchings at two given positions is $1/[n(n-1)]$, and, in general, of r matchings at r given positions is

$$\frac{1}{[n(n-1)\ldots(n-r+1)]} = \frac{(n-r)!}{n!}$$

(in fact: only one out of the n objects, or only one out of the $n(n-1)$ pairs,..., or only one out of the $n!/(n-r)!$ arrangements r at a time, is favourable).

The S_r are therefore the sum of $\binom{n}{r}$ terms all equal to $(n-r)!/n!$, so that $S_r = 1/r!$ (independent of n), from which, denoting the number of matchings by Y, we obtain

$$P(Y = 0) = \left\{1 - 1 + \frac{1}{2!} - \frac{1}{3!} + \ldots \pm \frac{1}{n!}\right\} = e^{-1} - R_n \cong e^{-1}\dagger$$

$$P(Y = h) = \left\{1 - 1 + \frac{1}{2!} - \frac{1}{3!} + \ldots \mp \frac{1}{(n-h)!}\right\}\bigg/ h! = (e^{-1} - R_{n-h})/h! \cong e^{-1}/h!$$

$\dagger$ R_n is the remainder of the series $\sum \pm 1/k!$ from the term $\pm 1/(n+1)!$ onwards: it is approximately equal to this first omitted term (which exceeds it in absolute value). With respect to e^{-1} it is practically negligible, except when n (respectively $n-h$) is very small (even for $n=10$ or $n-h=10$, the correction does not affect the decimal expression given for e^{-1}).

(in particular: $\mathbf{P}(Y = n - 1) = 0, \mathbf{P}(Y = n) = 1/n!$). Expressed numerically, $e^{-1} = 0.367879$.

In the limit, as n increases, the distribution tends to that in which

$$\mathbf{P}(Y = h) = e^{-1}/h!$$

(as we shall see later, Chapter 6, 6.11.2, this is the Poisson distribution with prevision $\mathbf{P}(Y) = 1$).

Observe that for the matching problem one could establish immediately by a direct argument that $\mathbf{P}(Y) = 1$ (i.e. that, in prevision, there is only one matching, whatever n is). We have only to note that it is given by $\mathbf{P}(Y) = n \cdot (1/n)$, since the prevision (probability) of a matching at any one of the n places is $1/n$.

Observe also that the relation $\mathbf{P}(Y = n - 1) = 0$ is obvious: in fact, $n - 1$ is not a possible value for Y because if we have matchings in $n - 1$ positions the last one cannot fail to give a matching (it is as well to point out this fact since it is easily overlooked!).

3.8.5. *Entropy.* Given a partition into events with probabilities $p_1, p_2, \ldots,$ p_n, we define the entropy to be the number

$$\sum_h p_h|\log_2 p_h| \qquad (\log_2 p_h = (\log p_h)/(\log 2)),$$

which represents the prevision of the number of YES–NO questions required to identify the true event.

This is immediate in the case of $n = 2^m$ equally probable events: m YES–NO questions are necessary and sufficient to know with certainty to which half, quarter, eighth, ..., of the partition the true event belongs, and finally to know precisely which one it is. If we have nine events with probabilities $\frac{1}{2}, \frac{1}{8}, \frac{1}{8}, \frac{1}{8}, \frac{1}{32}, \frac{1}{32}, \frac{1}{32}, \frac{1}{64}, \frac{1}{64}$, one question suffices if the first one is true; if not (with probability $\frac{1}{2}$) another two questions are sufficient to decide which one of the next three is true, or whether the true event is one of the remaining five; finally (with probability $\frac{1}{8}$), another two questions are necessary, plus (with probability $\frac{1}{32}$) a further one to decide between the last two events. The entropy in this example is therefore given by

$$1 + 2(\tfrac{1}{2}) + 2(\tfrac{1}{8}) + \tfrac{1}{32} = 2\tfrac{9}{32} = 2.28.$$

If it is not possible to proceed by successive halvings, some fraction is wasted (unless some device is available: for the time being, however, this brief introduction will suffice).

The unit of entropy is called a *bit* (contraction of '*binary digit*'): in the example above, the entropy was 2·28 bits; in the case of $1024 = 2^{10}$ equally probable cases it is 10 bits. For a given n, the entropy is maximized by an equipartition ($p_h = 1/n$): the reader might like to verify this as an exercise.

An item of information which leads to the exclusion of certain of the possible outcomes causes a decrease in entropy: this decrease is called the *amount of information*, and, like the entropy, is measured in bits (it is, in fact, the same thing with the opposite sign : some even call it *negative entropy*). We note that, for the time being, we are not in a position to provide a complete explanation of our assertion that an increase in information causes a decrease in entropy.

3.8.6. *Probability as measure or as mass.* In the set-theoretic interpretation of the events, it appears natural to think of probability—a non-negative, additive function taking the value 1 on the whole space—either as a *measure*, or as a *mass*.

The most widely used approach at the present time is the systematic identification of events as sets, and probability as measure (with all the advantages—as well as the risks!—which derive from a mechanical transposition of all the concepts, procedures and results of measure theory into the calculus of probability). To those reservations which we have already repeatedly expressed in connection with the systematic adoption of the set-theoretic interpretation of events, others must be added (in our opinion) concerning the further inflexibility introduced by the identification of probability with measure. This can, in fact, lead one to think that the representation in a space furnished with a measure binds events and random entities inseparably to a well-determined evaluation of probability. In the most elementary case, where we use Venn diagrams, the figures should be drawn in such a way that the area of each section be equal to its probability (taking the basic rectangle to have unit area). This, on the other hand, is in accordance with those points of view in which to each event (set) there corresponds an objectively (or, in any case, uniquely) determined probability.

If, instead, one wishes to distinguish between on the one hand the representation of the logical situation, and on the other the introduction of whatever coherent evaluation of probability one wants to make, it turns out to be preferable to think of probability as *mass*. The mass can in fact be distributed at will, without altering the geometric support and the 'measure', which might in that context appear more natural.† In the Venn diagram, without changing the figure in any way, there is no difficulty in imagining the possible ways of distributing a unit mass among the various parts (it does not matter if we put large masses on small pieces), and in imagining those ways that various individuals, real or hypothetical, would have chosen as their own opinion, or those we think they might choose.

Another advantage is the following: if one gives to the space of the representation the structure of the linear ambit, $\mathscr{A}$, then the well-known implications of the mechanical meaning of *mass* make clear all those

---

† It is not that different distributions of 'mass' could not equally well be called different 'measures'. It is, however, a fact that when talking in terms of measure one tends to make of it something fixed, with a special status, whereas when talking in terms of mass there is the physical feeling of being able to move it in whatever way one likes.

probabilistic properties which can be translated in terms of knowledge of the barycentre of a distribution (as we have already had occasion to see), or of moments of inertia, etc.†

We shall see shortly some particularly significant applications of this concept in the linear ambit determined by n events (in the sense explained in Chapter 2, Section 2.8, where the 'possible points' are finite in number, since they correspond to the constituents). Meanwhile, before concluding these comments on the set-theoretic interpretation, it is perhaps instructive to point out the simple, but not obvious, meaning that the expression of the probability of the event-sum acquires under this interpretation. We will consider the case of three events, where

$$E = A \lor B \lor C = A + B + C - AB - AC - BC + ABC.$$

In the Venn diagram, Chapter 2, Figure 2.1(b), the area of the union of the pieces A, B, C (or, alternatively, the mass contained in them) is calculated in the following way: first, we sum the areas of A, B and C; in this way, however, those of $A B \tilde{C}$, $A \tilde{B} C$, $\tilde{A} B C$ (doubly shaded) are counted twice, and that of ABC (triply shaded) is counted three times; subtracting those of AB, AC and BC, we re-establish the correct contribution for those originally counted twice; however, ABC is now counted three times less (since it belongs to AB and AC and BC) and therefore turns out to be ignored altogether; if we add it in, everything turns out as it should be.

3.9 LINEAR DEPENDENCE IN GENERAL

3.9.1. The straightforward theorems concerning 'total probability', which we established at the beginning of the previous section, certainly require no further explanations. It is, however, convenient to introduce the use of the representation with the spaces $\mathscr{A}$ and $\mathscr{L}$ by means of the simple cases, before proceeding to others of a less trivial nature.

We shall restrict ourselves, in general, to the three-dimensional case, which is the most obviously intuitive: the extension to n dimensions (which we shall occasionally mention) presents no difficulty for the reader who is familiar with such things, whereas for those who lack this familiarity it is better to be clear about the simpler case than to acquire confused and formal notions in a less accessible field.

---

† The suggestion has even been put forward that one could always think just in terms of the mass (or measures, or area) rather than in terms of the original meaning of probability: in this way we avoid the questions and doubts of a conceptual nature to which such a notion of probability can give rise. In general, however, in addition to removing the doubts this would also remove the raison d'être of the problems themselves (unless these only involve formal aspects, capable of being isolated from the context which provides them with meaning and content).

Let E_1, E_2, E_3 be three events (which, for the moment, we take to be logically independent; we shall introduce various assumptions as we go on), and let (x, y, z) be the cartesian reference system on which we superpose the linear ambit $\mathscr{A}$ and the linear space $\mathscr{L}$. The eight vertices of the unit cube

$$(0, 0, 0)\ (1, 0, 0)\ (0, 1, 0)\ (0, 0, 1)\ (0, 1, 1)\ (1, 0, 1)\ (1, 1, 0)\ (1, 1, 1),$$

thought of as points of $\mathscr{A}$, represent the constituents Q_i forming $\mathscr{Q}$;

$$Q_0 = \quad Q_1 = \quad Q_2 = \quad Q_3 = \quad Q_1' = \quad Q_2' = \quad Q_3' = \quad Q_0' =$$
$$\tilde{E}_1\tilde{E}_2\tilde{E}_3 \quad E_1\tilde{E}_2\tilde{E}_3 \quad \tilde{E}_1E_2\tilde{E}_3 \quad \tilde{E}_1\tilde{E}_2E_3 \quad \tilde{E}_1E_2E_3 \quad E_1\tilde{E}_2E_3 \quad E_1E_2\tilde{E}_3 \quad E_1E_2E_3$$

(where negations correspond to the *zeros*, affirmations to the *ones*); thought of as points (or vectors) of $\mathscr{L}$, they represent the random quantities

$$0 \quad E_1 \quad E_2 \quad E_3 \quad E_2 + E_3 \quad E_1 + E_3 \quad E_1 + E_2 \quad E_1 + E_2 + E_3$$

(where the presence of a summand corresponds to the *ones*).

The generic point (x, y, z), thought of as a point of $\mathscr{A}$, would mean that E_1 takes the value x, and similarly $E_2 = y$ and $E_3 = z$ (which is invalid, since the random quantities E_i cannot take on values other than 0, 1). This can be valid, however, as *prevision*, in the sense that $\mathbf{P}(E_1) = x$, $\mathbf{P}(E_2) = y$, $\mathbf{P}(E_3) = z$; in other words, (x, y, z) represents the prevision $\mathbf{P}$ which attributes to E_1, E_2, E_3 the probabilities $(p_1, p_2, p_3) = (x, y, z)$, and which is also expressible as the barycentre of the points Q_i with suitable weights (masses) q_i. Thought of as a point (or vector) of $\mathscr{L}$, (x, y, z) represents the random quantity $X = uE_1 + vE_2 + wE_3$ with coefficients $(u, v, w) = (x, y, z)$. Since $\mathbf{P}(X) = up_1 + vp_2 + wp_3 = ux + vy + wz$, $\mathbf{P}(X)$ can be interpreted as the inner product of the (dual) vectors $\mathbf{P}$ (or $P - 0$) of $\mathscr{A}$ and X (or $X - 0$) of $\mathscr{L}$; or, alternatively, as $\mathbf{P}(X) = (P - 0) \times (X - 0)$ in the metric space on which $\mathscr{A}$ and $\mathscr{L}$ have been superposed.

Until we state precisely the assumptions made concerning the E_i, i.e. establish which among the eight products are actually possible constituents, all this remains rather general and introductory in character; simply a repetition of things we know already, with a few additional details.

3.9.2. *The case of partitions.* If the E_i constitute a partition, there are three constituents: $Q_1 = (1, 0, 0)$, $Q_2 = (0, 1, 0)$, $Q_3 = (0, 0, 1)$. We know that the p_i can be any three non-negative numbers summing to 1. In other words, the admissible $\mathbf{P} = (x, y, z)$ belong to the plane $x + y + z = 1$. More precisely, they belong to the triangle having as its vertices the three possible points Q_1, Q_2, Q_3, and are, in fact, uniquely expressible as barycentres of these points, $P = q_1Q_1 + q_2Q_2 + q_3Q_3$, with weights $q_1 = x$, $q_2 = y$, $q_3 = z$. This triangle constitutes the space $\mathscr{P}$ of admissible previsions, and is precisely the convex hull of the set $\mathscr{Q}$ of possible outcomes (which reduces in this case to the three given vertices). Representing the triangle by a figure in the plane, one sees that the probabilities x, y, z, turn out to be the barycentric coordinates of the point P with respect to the Q_i. Since the triangle is equilateral, one has the standard 'ternary diagram' (as is used, for example, to indicate the composition of ternary alloys) in which x, y, z also have a more immediate interpretation as the *distances of the point from the sides*, taking as unity the height of the triangle (to which the sum of the three distances is always equal). It is also clear that a point outside of the triangle (not in the plane, or in the plane but outside the triangle) can be brought nearer to all the three vertices—i.e. to all the points of $\mathscr{Q}$—by transporting it into the triangle. This can be accomplished by projecting it onto the plane, and then, if the projection falls outside the triangle, by transporting it to the nearest point on the boundary. This is related to the 'second criterion', if we think of the penalty as being the square of the ordinary distance in this representation.

If we think in terms of $\mathscr{L}$, we could say, instead, that the point $(1, 1, 1)$ represents the random quantity which is certainly equal to 1, given that $E_1 + E_2 + E_3 = 1$. The fact that for the coordinates of P we must have $x + y + z = 1$ is then interpreted on the basis of the scalar product: $\mathbf{P}(1) = x \cdot 1 + y \cdot 1 + z \cdot 1 = 1$.

3.9.3. *The case of incompatibility.* If the E_i are incompatible (but not exhaustive) there are four constituents: the previous three and $Q_0 = (0, 0, 0)$; i.e. Q_0, Q_1, Q_2, Q_3. The above considerations still hold, except that we now have the relation $x + y + z \leqslant 1$ (instead of $= 1$). We still have P expressible uniquely as a barycentre, $P = q_0 Q_0 + q_1 Q_1 + q_2 Q_2 + q_3 Q_3$, of the Q_i, with weights $q_0 = 1 - x - y - z$, $q_1 = x$, $q_2 = y$, $q_3 = z$, and the space $\mathscr{P}$ (which was the triangle with vertices Q_1, Q_2, Q_3) is now the tetrahedron having in addition the vertex Q_0.

3.9.4. *The case of a product.* Let E_1 and E_2 be logically independent, and E_3 be their product: $E_3 = E_1 E_2$. The constituents are then the following four: $Q_0 = (0, 0, 0)$, $Q_1 = (1, 0, 0)$, $Q_2 = (0, 1, 0)$ and $Q'_0 = (1, 1, 1)$. The first three are in the plane $z = 0$, the last three are on $z = x + y - 1$; the other two groups of three are on $z = y$ and $z = x$, respectively. The space $\mathscr{P}$ is therefore the tetrahedron $z \geqslant 0$, $z \geqslant x + y - 1$, $z \leqslant x$, $z \leqslant y$, or, in other words, expressed compactly using $\wedge$ and $\vee$,

$$[\max(0, x + y - 1) =] \qquad 0 \vee (x + y - 1) \leqslant z \leqslant x \wedge y \quad [= \min(x, y)].$$

These are the restrictions under which one can arbitrarily choose the probabilities of two logically independent events and that of their product.

Here also, P is uniquely expressible as a barycentre

$$P = q_0 Q_0 + q_1 Q_1 + q_2 Q_2 + q_0' Q_0'$$

of the Q with weights $q_0 = 1 - x - y + z, q_1 = x - z, q_2 = y - z, q_0' = z$.

3.9.5. *The case of the event-sum.* This proceeds as above, except that $E_3 = E_1 \vee E_2$ (instead of $E_1 E_2$). Since the event-sum is $E_1 + E_2 - E_1 E_2$, this case reduces straightaway to the preceding ones. The constituents are Q_0, Q_1', Q_2', Q_0'; the inequalities for the tetrahedron $\mathscr{P}$ having these vertices are

$$[\max(x, y) =] \qquad x \vee y \leqslant z \leqslant 1 \wedge (x + y) \qquad [= \min(1, x + y)];$$

the weights which give $P = (x, y, z)$ as a barycentre in terms of the Q are

$$q_0 = 1 - z, \qquad q_1' = z - y, \qquad q_2' = z - x, \qquad q_0' = x + y - z.$$

Remark. In the preceding cases each P was derived as a barycentre of the Q with *uniquely determined weights* q; it is important to note (and we shall return to this later) that this circumstance is exceptional. To be more precise, this happens when and only when the Q are *linearly independent*—in the examples above we had, in fact, either 3 non-collinear, or 4 non-coplanar— or when they are (as events) *expressible as a linear combination of the given events*. In fact, they were, in the first case, E_1, E_2, E_3; in the second, $1 - E_1 - E_2 - E_3, E_1, E_2, E_3$; in the third, $1 - E_1 - E_2 + E_3, E_1 - E_3$, $E_2 - E_3, E_3$; and in the fourth, $1 - E_3, E_3 - E_2, E_3 - E_1, E_1 + E_2 - E_3$. In other words, the Q (as events) belonged in these cases to $\mathscr{L}$. Observe that the expressions for the Q in terms of the E are the same as those for the weights q in terms of x, y, z. *In the following examples this will no longer happen.*

3.9.6. *The case of exhaustivity.* If we specify *only* that E_1, E_2, E_3 are exhaustive, then there are seven constituents; the eight minus $Q_0 = (0, 0, 0)$, which is excluded. This latter vertex of the cube being missing, the convex hull $\mathscr{P}$ is the cube itself minus the tetrahedron defined by this vertex and the three adjacent ones; i.e. the part of the cube $0 \leqslant x, y, z \leqslant 1$ which satisfies the inequality $x + y + z \geqslant 1$. Each of its points P can be expressed—*in an infinite number of ways*—as a barycentre of points Q (unless the point coincides with a vertex, or belongs to an edge, or a triangular face, in which case the number of representations is finite). In fact, all we have to do is to choose non-negative weights q, summing to 1, such that

$$q_1 + q_2' + q_3' + q_0' = x, \qquad q_2 + q_1' + q_3' + q_0' = y,$$

$$q_3 + q_1' + q_2' + q_0' = z$$

(4 equations and 7 unknowns).

3.9.7. *The case where the negations are also exhaustive.* If we exclude both the extreme constituents, i.e. in addition to $Q_0 = (0, 0, 0)$ we also exclude $Q'_0 = (1, 1, 1)$, then six constituents remain. The cube has now had removed from it the two opposite tetrahedrons, and the remaining part $\mathscr{P}$ is that defined by the double inequality $1 \leqslant x + y + z \leqslant 2$.† Other considerations are as above.

A useful example is given by the comparisons between 3 random quantities, X, Y, Z; in other words, by considering the 3 events $E_1 = (X > Y)$, $E_2 = (Y > Z)$, $E_3 = (Z > X)$ (we assume excluded, or at least as practically negligible, the case of equality). By transitivity, the three events cannot turn out to be either all true or all false; there remain the other 6 constituents, corresponding to the $6 = 3!$ possible permutations. As an application, one might think, for example, of comparing the weights (or temperatures, etc.) of 3 objects.

Other cases. The following cases are similar (and are useful as exercises): $E_3 \subset E_1 E_2$ (5 constituents); E_1 and E_2 incompatible, $E_3 \subset (E_1 = E_2)$ (6 constituents), etc. Another example with 4 (independent!) constituents is given by $E_3 \equiv (E_1 = E_2)$.

3.9.8. *The case of logical independence.* All 8 constituents exist; $\mathscr{P}$ is the whole cube. This is the most complete and 'normal' case; there is little to say apart from thinking about it in the light of remarks concerning more elaborate cases.

The same case in any number of dimensions. If $E_1, E_2, \ldots, E_n$ are logically independent events, we will have 2^n constituents Q_i, the vertices of the unit hypercube; i.e. the points $(x_1, x_2, \ldots, x_n)$ in the linear ambit $\mathscr{A}$ with $x_i = 1$ or 0. The admissible previsions $\mathbf{P}$ are those of the cube $\mathscr{P}$, $0 \leqslant x_i \leqslant 1$, which is the convex hull of the set of the vertices $\mathscr{Q}$. The linear space $\mathscr{L}$ is formed by the random quantities $X = u_1 E_1 + u_2 E_2 + \ldots + u_n E_n$ which are linearly dependent (homogeneously, but it is easy to take into account separately an additive constant) on the events E_i. Conceptually, everything that has been stated for $n = 2$ and $n = 3$ also holds for arbitrary n (this saves us repeating everything in a more cumbersome notation and so making the exposition rather heavy going).

3.9.9. *General comments.* Each particular case differs from the final one by virtue of the exclusion of some of the constituents: instead of 2^n there are only $s < 2^n$. These determine a linear space of dimension $d (d \leqslant n, \log_2 s \leqslant d \leqslant s - 1)$; if $d < n$, the n events E_i are linearly dependent. In fact, if all

---

† The case $x + y + z = 2$ (with constituents Q'_1, Q'_2, Q'_3) is similar to that of the partition $(x + y + z = 1)$, and is obtained if $\tilde{E}_1, \tilde{E}_2, \tilde{E}_3$ form a partition.

the Q satisfy a linear relation $\sum_i x_i = \text{const.}$ the same holds for the E_i. For instance, in the above examples $x + y + z = 1$, $x + y + z = 2$ gives $E_1 + E_2 + E_3 = 1$ (or 2), so that we need only consider two events, for example E_1 and E_2, setting $E_3 = 1 - E_1 - E_2$ or $E_3 = 2 - E_1 - E_2$, respectively (this also holds in the general case). If we consider the unnecessary E_i as eliminated (since they are linearly dependent on the others), we can always arrange that $d = n$; in any case, $\mathscr{P}$ is the convex hull (d-dimensional polyhedron) having as vertices the points Q which form $\mathscr{Q}$.

Given some **P** (in $\mathscr{A}$), in other words, *having evaluated the probabilities* **P**(E_i) *of the given events*, **P** turns out to be determined for all those random quantities X which are linearly dependent on the E_i, and for no others; i.e. for those belonging to $\mathscr{L}$. In particular, the probability of an event E is determined if and only if E is one of these X.† This statement takes into account all the obvious cases: for example, the probability of $A \vee B$ is not determined by **P**(A) and **P**(B) (unless we assume incompatibility), but is determined if we include **P**(AB), since we have the relation $A + B = AB + A \vee B$. It is useful to see an example of how non-trivial events can be found among the X of $\mathscr{L}$ (i.e. the X that only have two possible values; which we can always represent as 0 and 1). We shall see then that, if E is not linearly dependent on the E_i, one can only say that $p' \leqslant \mathbf{P}(E) \leqslant p''$, where $p' = \sup \mathbf{P}(X)$ for the X of $\mathscr{L}$ which are certainly $\leqslant E$, and $p'' = \inf \mathbf{P}(X)$ for the X of $\mathscr{L}$ which are certainly $\geqslant E$.

3.9.10. *A non-obvious example of linear dependence.* Suppose that A, B, C, D, F, G are the participants in a competition, and that six other individuals each choose from among the participants their three 'favourites' (a prize being offered to all those who have included the winner among their 'favourites'). Suppose also that we know the choices to be: C, D, G for the first individual; B, C, G for the second; A, D, F for the third; B, F, G for the fourth; A, C, D for the fifth; D, F, G for the sixth. Finally, a seventh individual —suppose it is You—has chosen A, B, C. Is your guess, the event E say, linearly dependent on the events $E_1, \ldots, E_6$, which denote the guesses of the others, or not? This question might be important, for example, in the following situation: there is an expert in whom You have great confidence, so far as judging the competition and the participants is concerned, and whose opinion concerning your probability of winning is of interest to You. However, You do not know this directly (since You do not know what

---

† This does not exclude the possibility that for certain evaluations (limit cases in which an inequality reduces to an equality) **P**(E) can turn out to be determined for E which are not linearly dependent on the E_i we started with (and perhaps not even logically dependent). For instance, if neither of A and B is logically dependent on $A \vee B$, then knowing **P**($A \vee B$) is not sufficient to determine **P**(A) and **P**(B); if, however, **P**($A \vee B$) = 0, it follows necessarily that **P**(A) and **P**(B) are also zero.

probability of winning he attributes to each participant) but only indirectly (because You happen to know what probabilities he attributes to the guesses of the others turning out to be correct); is this enough?

We have the system of equations:

$$1 = A + B + C + D + F + G$$
$$E_1 = \qquad\qquad C + D \qquad + G$$
$$E_2 = \qquad B + C \qquad\qquad + G$$
$$E_3 = A \qquad\qquad + D + F$$
$$E_4 = \qquad B \qquad\qquad + F + G$$
$$E_5 = A \qquad + C + D$$
$$E_6 = \qquad\qquad\qquad D + F + G$$
$$E = A + B + C$$

(the first equation states, as we have seen, that the six cases are the only ones possible, and are incompatible). One could work out the determinant (and, by virtue of its being zero, could verify linear dependence), but instead we note (and leave the reader to verify it by working out the sum) that we have the relation

$$2E_1 - E_2 + E_3 - 3E_4 - 5E_5 + 5E_6 + 7E$$
$$= 3(A + B + C + D + F + G) = 3,$$

from which

$$E = \tfrac{1}{7}(3 - 2E_1 + E_2 - E_3 + 3E_4 + 5E_5 - 5E_6).$$

Hence, if I know the $p_i = \mathbf{P}(E_i)$ of the guesses, I can conclude that in the expert's opinion (assumed coherent) $p = \mathbf{P}(E)$ must be

$$p = \tfrac{1}{7}(3 - 2p_1 + p_2 + p_3 + 3p_4 + 5p_5 - 5p_6).$$

In a similar way one could, of course, see whether the $p_1 \ldots p_6$ are admissible (compatible with probabilities $\geqslant 0$ for the partition A, B, C, D, F, G, and if, in any case, they determine them, etc.). It may be a useful exercise to develop these questions in the context of this example (as it stands, or modifying it in some way).

3.10 THE FUNDAMENTAL THEOREM OF PROBABILITY

3.10.1. We turn now to proving and illustrating the general conclusion that we stated before, and which, in a more complete and precise form, constitutes the following:

Theorem. *Given the probabilities* $\mathbf{P}(E_i)$ $(i = 1, 2, \ldots, n)$ *of a finite number of events, the probability,* $\mathbf{P}(E)$, *of a further event* E, *either*

(a) *turns out to be determined (whatever* $\mathbf{P}$ *is) if* E *is linearly dependent on the* E_i *(as we already know); or*

(b) *can be assigned, coherently, any value in a closed interval* $p' \leqslant \mathbf{P}(E) \leqslant p''$ *(which can often give an illusory restriction, if* $p' = 0$ *and* $p'' = 1$, *or even, in limit cases for particular* $\mathbf{P}$, *give a well-determined result* $p = p' = p''$).

More precisely, p' *is the upper bound,* $\sup \mathbf{P}(X)$, *of the evaluations from below of the* $\mathbf{P}(X)$ *given by the random quantities* X *of* $\mathscr{L}$ *(i.e. linearly dependent on the* E_i) *for which we certainly have* $X \leqslant E$. *If* E *is not logically dependent on the* E_i, *observe that* $X \leqslant E$ *can be more usefully replaced by* $X \leqslant E'$, *where* E' *is the largest event logically dependent on the* E_i *contained in* E *(cf. Chapter 2, 2.7.3). The same can be said for* p'' *(replacing* $\sup$ *by* $\inf$, *maximum by minimum,* E' *by* E'', *and changing the direction of the inequalities, etc.).*

Proof. If $Q_1 \ldots Q_s$ denote the constituents, relative to $E_1 \ldots E_n$, and E is logically (but not linearly) dependent on the E_i, then the linear ambit $\mathscr{A}'$ obtained by the adjunction of E (i.e. by adding a new coordinate x to the preceding $x_1 \ldots x_n$) has the same constituents Q_h, but now placed at the vertices of a cube in $n + 1$ dimensions instead of n. Each $Q = (x_1, x_2, \ldots, x_n)$ is either left as it was (with $x = 0$), or moved onto the parallel S_n ($x = 1$), becoming either $(x_1, x_2, \ldots, x_n, 0)$ or $(x_1, x_2, \ldots, x_n, 1)$, according as Q is contained in $\tilde{E}$ or in E. The convex hull $\mathscr{P}'$ in S_{n+1} (in $\mathscr{A}'$) has as its projection onto the preceding S_n ($\mathscr{A}$) the preceding $\mathscr{P}$. For each admissible $\mathbf{P}$ in the latter (with coordinates $p_i = \mathbf{P}(E_i)$), the admissible extensions in $\mathscr{A}'$ are the points $\mathbf{P}'$ which project onto $\mathbf{P}$ and belong to $\mathscr{P}'$; i.e. belong to the segment $p' \leqslant x \leqslant p''$ which is the intersection of the ray $(p_1, p_2, \ldots, p_n, x)$ with $\mathscr{P}'$. The extreme points $(x = p', x = p'')$ are on the boundary of $\mathscr{P}'$, i.e. on one of the hyperplanes (in n dimensions) which constitute its faces (they could be on more than one—vertices, edges, etc.—but this does not affect the issue). Suppose the hyperplane is given by $\sum u_i x_i + ux = c$; in other words, suppose that the relation $\sum u_i E_i + uE = c$ holds on it, i.e. that $E = (c - \sum u_i E_i)/u$: then the X in $\mathscr{L}$ defined by the right-hand side has the given property, and yields $p' = \mathbf{P}(X)$. Similarly for p''.

3.10.2. *Applications.* Let us generalize some of the examples considered previously in S_3. Those concerning the number of successes,

$$Y = E_1 + E_2 + E_3,$$

now become the consideration of $Y = E_1 + E_2 + \ldots + E_n$, and we can look at various subcases. Suppose that either Y is known, $Y = y$ $(0 \leqslant y \leqslant n)$ (as in the previous cases where $Y = 1$ and $Y = 2$), or certainly lies between two given extreme values y' and y'' $(0 \leqslant y' < y'' \leqslant n)$ (as in the previous cases, where $1 \leqslant Y \leqslant 2$). The interpretation of this last example, as given in 3.9.7, will now be extended (in different ways) to comparisons between n objects: finally, the case of the event-sum will require all the products.

3.10.3. *Knowledge about frequency*. This first example is noteworthy in that it constitutes the first and most elementary link in the long chain of conclusions which, as we proceed, will clarify and enrich our insight into the relationship which holds between *probability* and *frequency*. This is important both for what the conclusions do say and, perhaps even more so (in some situations at least), in order to get used to not interpreting them as saying something which they do not say.

The simplest case is that in which the number of successes, $Y = E_1 + E_2 + \ldots + E_n$, is known (for certain); i.e. the *frequency* Y/n is known (for certain). Let $Y = y$, so that $Y/n = y/n$. The following are possible examples: in an election, out of n candidates we know that y are to be elected; in an examination, y candidates out of n passed (but we are still ignorant of which ones); in a drawing of the lottery, out of $n = 90$ numbers $y = 5$ will be drawn; at $n = 90$ successive drawings of all the balls in Bingo, all the $y = 15$ numbers on your card will come out.

As an extension, we have the case in which we know the limits between which Y must lie; $y' \leqslant Y \leqslant y''$ (and hence that the frequency must be between y'/n and y''/n). In the preceding examples: it may be that the electoral system allows the number elected to vary between y' and y''; that on the basis of partial information about the examinations one knows that at least y' have passed and at least $n - y''$ have not; if we consider 10 drawings of the lottery instead of one (for instance, all the 10 'wheels' on the same day), then of the $n = 90$ numbers the total of different numbers drawn can vary between $y' = 5$ (all the sets of five identical) and $y'' = 50$ (no number repeated).

It is obvious that, as in the case $n = 3$, the sum of the $\mathbf{P}(E_i)$, i.e. $\mathbf{P}(Y)$, must give in the first case y, and in the second a value $y' \leqslant \mathbf{P}(Y) \leqslant y''$. Put more forcefully; dividing by n, *the probabilities $\mathbf{P}(E_i)$ must be such that their arithmetic mean coincides with the known frequency y/n, or falls between the extreme values, y'/n and y''/n, that the frequency can assume* (end-points included). This is all that can be said on the basis of the given information. In general, one might say more: for example, that each number in the lottery has probability $\frac{5}{90}$ of coming up in a given drawing, and not different probabilities with mean $\frac{5}{90}$. This could only be done, however, on the basis of additional knowledge or considerations which must be kept separate.

3.10.4. *The linear ambit of events logically dependent on n given events*. For the purpose in hand, it is obviously sufficient to consider the linear ambit, let us call it $\mathscr{A}^*$, generated by the s constituents Q_h (these form a partition, and so the dimension is actually $s - 1$, given the identity

$Q_1 + Q_2 + \ldots + Q_s = 1$). We could also generate it by means of the E_i and their products (two at a time, three at a time, etc.). We saw, in 3.8.3, that in this way one can express the event-sum linearly, and we shall now see that it is possible to express all the constituents linearly, and hence all the events which are logically dependent on the E_i. We will suppose that the E_i are logically independent, so that $s = 2^n$; in the other case, the treatment is equally valid, except that the constituents and the products which turn out to be impossible have to be omitted.

Let us illustrate the situation by referring to the case of three logically independent events and their products; for convenience we denote the three events by A, B, C (instead of E_1, E_2, E_3) and their products by $F = AB$, $G = AC$, $H = BC$ and $E = ABC$. We have 7 events which are linearly independent because there exists only one linear relation between the $2^3 = 8$ constituents (the sum $= 1$). Some inequalities (implications) hold among them, however; for instance, $A \geqslant AB \geqslant ABC$ so that $A \geqslant F \geqslant E$ (as is obvious if one considers that of the $2^7 = 128$ vertices of the cube in 7 dimensions only the 8 corresponding to the constituents relative to A, B, C, are possible).

We list the constituents, giving their coordinates in the ambit $\mathscr{A}^*$, and the linear expressions in the dual space $\mathscr{L}^*$:

$$ABCFGHE = (1, 1, 1, 1, 1, 1, 1) = E,$$

$$AB\tilde{C}F\tilde{G}\tilde{H}\tilde{E} = (1, 1, 0, 1, 0, 0, 0) = F - E,$$

$$A\tilde{B}C\tilde{F}G\tilde{H}\tilde{E} = (1, 0, 1, 0, 1, 0, 0) = G - E,$$

$$\tilde{A}BC\tilde{F}\tilde{G}H\tilde{E} = (0, 1, 1, 0, 0, 1, 0) = H - E,$$

$$A\tilde{B}\tilde{C}\tilde{F}\tilde{G}\tilde{H}\tilde{E} = (1, 0, 0, 0, 0, 0, 0) = A - F - G + E,$$

$$\tilde{A}B\tilde{C}\tilde{F}\tilde{G}\tilde{H}\tilde{E} = (0, 1, 0, 0, 0, 0, 0) = B - F - H + E,$$

$$\tilde{A}\tilde{B}C\tilde{F}\tilde{G}\tilde{H}\tilde{E} = (0, 0, 1, 0, 0, 0, 0) = C - G - H + E,$$

$$\tilde{A}\tilde{B}\tilde{C}\tilde{F}\tilde{G}\tilde{H}\tilde{E} = (0, 0, 0, 0, 0, 0, 0) = 1 - A - B - C + F + G + H - E.$$

These expressions, and the analogous ones for each of the events logically dependent on A, B, C, are obtained as shown in the following example:

$$\tilde{A}B\tilde{C} = (1 - A)B(1 - C) = B - AB - BC + ABC = B - F - H + E.$$

The necessary and sufficient condition for coherence is that the probabilities of the constituents are non-negative (they automatically turn out to sum to 1), and therefore the following inequalities (where, for simplicity, we denote the probability of an event by the corresponding lower case letter) are necessary and sufficient:

$$e \geqslant 0, \qquad f, g, h \geqslant e, \qquad a \geqslant f + g - e, \qquad b \geqslant f + h - e,$$

$$c \geqslant g + h - e, \qquad (a + b + c) - (f + g + h) + e \leqslant 1.$$

3.10.5. *A canonical expression for random quantities.* By analogy, we indicate here how, in the same manner, each random quantity

$$X = c_0 + c_1 E_1 + c_2 E_2 + \ldots + c_n E_n,$$

linearly expressible in terms of the events E_i, can be put in a meaningful canonical form by reducing it to a linear combination

$$X = x_1 C_1 + x_2 C_2 + \ldots + x_s C_s$$

of the constituents C_h (the x_h are the possible values of X, assumed in correspondence to the occurrence of the C_h). As an example: if we denote two logically independent events by A and B, and the constituents by $Q_1 = AB$, $Q_2 = A\tilde{B}$, $Q_3 = \tilde{A}B$, $Q_4 = \tilde{A}\tilde{B}$, where $1 = Q_1 + Q_2 + Q_3 + Q_4$, $A = Q_1 + Q_2$, $B = Q_1 + Q_3$, we have, for instance, for $X = 3 - 4A + B$:

$$X = 3(Q_1 + Q_2 + Q_3 + Q_4) - 4(Q_1 + Q_2) + (Q_1 + Q_3)$$

$$= 0 \cdot Q_1 + (-1) \cdot Q_2 + 4 \cdot Q_3 + 3 \cdot Q_4.$$

X assumes the possible values $-1, 0, 3, 4$, corresponding to Q_2, Q_1, Q_4, Q_3.

3.10.6. *Comment.* The above considerations are intended to familiarize the reader (in the case of events) with the crucially important idea of the relations of linearity and inequality, and to stress *a fact* and *a criterion* which will be of use in what follows, and more generally.

The *fact* is the *possibility* of expressing all that can legitimately be said by arguing solely in terms of the events (and random quantities) whose prevision is known. That is to say, without leaving the linear ambit determined by the latter, without imagining already present a probability distribution over larger ambits, those in which the extension is possible, albeit in an infinite number of ways.

The *criterion* lies in the *commitment* to systematically exploiting this fact; the commitment considered as the expression of a fundamental methodological need in the theory of probability (at least in the conception which we here maintain). All this is not usually emphasized.

These considerations should go some way to excusing the length of the exposition, which is certainly excessive in comparison with what would be desirable if this topic were well enough known in general to permit us to restrict ourselves to a few brief remarks.

3.10.7. *The case of an infinite number of events* (*or random quantities*). The fundamental theorem of probability (and prevision), given in 3.10.1, permits us—even in countably infinite or non-denumerable cases where, of course, the number of choices is infinite—to proceed to attribute to all the events and random quantities that we wish, one after the other, probabilities and previsions coherent with the preceding ones. The arguments presented do not become invalid when we pass to the infinite case, because the conditions of coherence always refer just to finite subsets: cf. Appendix, Section 15.

This demonstrates the theorem of the *unconditional existence and extendibility of coherent previsions* of events and random quantities in any (open)† field. In other words:

If, within the field in which they are made, the previsions do not already give rise to incoherence, no incoherence arises to prevent the existence of coherent previsions in any field whatever, coinciding with the preceding ones whenever these apply.

3.11 ZERO PROBABILITIES: CRITICAL QUESTIONS

3.11.1. In both the criteria put forward in order to define probability there was a point whose clarification we held over to the sequel. It was the same point in both cases; the wherefore of the precaution taken in excluding the possibilities of gains being *all uniformly negative*, but not that of gains being all negative (without the 'uniformity' condition). Another matter, connected with this, is the removal of the reservations regarding the prevision of unbounded random quantities.

We are dealing with critical questions, and, if we only wished to consider those aspects relating to applications, they could be omitted, or confined to the Appendix. This, however, is not possible. In Chapter 6, we have to study distributions, and to throw light on the conceptual differences and their wherefores, introduced in accordance with the present viewpoint, it is better to focus right from the beginning on those aspects which will play a fundamental rôle.

The fact is that a logical construction is such in so far as it is a whole in which 'tout se tient'; otherwise, it is nothing of the sort. Questions which are seemingly completely otiose and insignificant can have, and do have, interconnections with all the rest, and are essential for an understanding of them. To ignore them, or merely to mention them in passing, is dangerous,

---

† 'Open' is meant in the sense of not being preconstituted, not constrained, not a 'Procrustean bed', not a Borel field, not consisting of events, etc., which have a given meaning or structure, but a field in which we can, at any moment, insert whatever might come to mind.

especially when they impinge on delicate and controversial matters: too many ideas then remain rather vague, and give rise to an accumulation of doubts.

For this reason, having reached the end of Chapter 3, we shall now consider the questions of a critical nature which have arisen; we shall do the same at the end of Chapter 4, coming back to these same questions under a new guise; finally, at the end of Chapter 5, we shall arrive at the same kind of considerations, although with respect to topics which are less technical and more general. We shall attempt to confine ourselves to the minimum necessary discussion, expressed as simply as possible. The few additional clarifications or examples will be recognizable 'at a glance' by virtue of the small print.

3.11.2. It would not be accurate to say that all the problems reduce to the presence of zero probabilities, but, in order to have a guideline to follow, it is convenient to think in these terms (just as it is not only suggestive but also appropriate to mention them in the section heading).

It seems impossible that there is anything at all to be said about zero probabilities. Instead, we have the following basic questions:

(I) can a possible event have zero probability? If so:

(II) is it possible to compare the zero probabilities of possible events (to say if they are equal, or what their relation is, etc.)?

(III) can a union of events with zero probabilities have a positive probability (in particular, can it be the certain event)?

(IV) are there any connections with problems concerning random quantities, and in particular with the problem of prevision for unbounded random quantities?

Question (II) crops up again within the topics of Chapter 4 and will be discussed there; we had to mention it, not only to put it in its natural position as a 'question', but also to give prior warning that any incidental comments that we make here for convenience, will be clarified at the appropriate place: we will draw attention to this by writing '(II!)'.

Questions (I) and (III) can be bracketed and discussed together straightaway; afterwards we shall pass on to (IV). However, there was a reason for putting the two questions (I) and (III) separately. Question (III), which evidently requires to be put in the context of infinite partitions, might lead one to think and state that one can only have possible events with zero probability *if they belong to infinite partitions* (!). This is monstrous. If E has probability p (in particular $= 0$) it *is* an event with probability p (in particular with zero probability) both when considered in itself, or in the dichotomy E and $\tilde{E}$, or in any other partition into few, many or an infinite numbers of events, obtained by partitioning in any way whatsoever. Unfortunately, this propensity to see each event embedded in some scheme, together with others usually studied with it, gives rise to serious confusions both in theoretical matters (as is the case here) and practically (as in the examples in Chapter 5, 5.8.7).

This having been said as an appropriate warning, we can pose question (III) once again by asking *whether in an infinite partition one can attribute zero probability to all the events.* In this form, the question becomes essentially equivalent to that concerning the different types of additivity: *finite,* only for a finite sum; *countable,* for the denumerable case; *perfect,* if the additivity always holds.

There are precisely three answers, corresponding to these three types (with a variation, which is related to (I)):

A = Affirmative, N = Negative (N' and N''), C = Conditional

(and in what follows, we shall denote them and the corresponding points of view with the initials A, N and C, or, if necessary, A, N', N'' and C).

A : Yes. Probability is finitely additive. The union of an infinite number of incompatible events of zero probability can always have positive probability, and can even be the certain event.

N : No. Probability is perfectly additive. In any partition there is a finite, or countable, number of events with positive probabilities, summing to one: the others have zero probability both individually and together.

C : It depends. The answer is NO if we are dealing with a countable partition, because probability is countably additive; the sum of a countable number of zeroes is zero. The answer is YES if we are dealing with an uncountable infinity,† because probability is not perfectly additive: the sum of an uncountable infinity of zeroes can be positive.

In the case of the answer N, there are however two subcases to be distinguished with reference to question (I) (for which, in cases A and C, the answer can only be YES).

N' : Probability zero implies impossibility. What has been said above is a consequence of this identification.

N'' : Probability zero does not imply impossibility. However, the behaviour is the same: even if we take the union of them all, the events of probability zero form an event with zero probability.

---

† I do not know whether this corresponds exactly to the conception of the supporters of this thesis (often one only talks about the case of the continuum).

3.11.3. Let me say at once that the thesis we support here is that of *A, finite additivity;* explicitly, *the probability of a union of incompatible events is greater than or equal to the supremum of the sums of a finite number of them.* Apart from the present author, it would seem that only B. O. Koopman (1940) has systematically adopted and developed this thesis. Others, like Good (1965), admit only finite additivity as an axiom, but do nothing to follow up this observation. Others again, like Dubins and Savage (1965), make use of finite additivity for special purposes and topics.

The thesis *N* is supported, as far as I know, only by certain logicians, such as Carnap, Shimony and Kemeny (as a consequence of a definition of 'strict coherence').†

The thesis *C* is the one most commonly accepted at present; it had, if not its origin, its systematization in Kolmogorov's axioms (1933). Its success owes much to the mathematical convenience of making the calculus of probability merely a translation of modern measure theory (we shall say a lot more about this in Chapter 6). No-one has given a real justification of countable additivity (other than just taking it as a 'natural extension' of finite additivity); indeed, many authors do also take into account cases in which it does not hold, but they consider them separately, not as absurd, but nonetheless 'pathological', outside the 'normal' theory.

3.11.4. Let us review, briefly, the main objections to the various theses (we number them: $A1, A2, \ldots; N1, N2, \ldots; C1, C2, \ldots$). Our point of view is, of course, represented by the objections to *N* and *C*, and by the answers $(A1a, A1b, \ldots; A2a, A2b, \ldots)$ to the objections raised against *A*. We will also interpolate some examples $(E1, E2, \ldots)$.

A1 This is an objection from the standpoint of *N* (or rather *N′*): *it is not sufficient to exclude as inadmissible those bets with gain X certainly negative* $(\vdash X < 0$: *weak coherence); it is necessary to exclude them if the gain is certainly non-positive* $(\vdash X \leqslant 0$: *strict coherence). This means that 'zero probability' is equivalent to 'impossibility'.*

The most decisive reply will be objection *N2*, but it is better not to evade a reply which clarifies the points (perhaps persuasive) put forward in *A1*; this reply will constitute a preliminary refutation of *N* (*N1*).

A1a It should be unnecessary to point out that the inadmissibility of a bet is always relative to the set of choices offered by a given scheme. It is obvious that if among the possible choices there was the choice 'do not make a bet at all', nobody would choose an alternative which could only lead to losses (this, however, means nothing).

---

† In addition to these serious authors, there is no point in mentioning the large number who refer to zero probability as impossibility, either to simplify matters in elementary treatments, or because of confusion, or because of metaphysical prejudices.

A1b In the simplest scheme, let $X = -E$ (loss $= 1$ if E occurs; e.g. the risk we are facing), and consider the appropriateness of insuring oneself by paying a premium p. Let us suppose that one is willing to pay $(\frac{1}{2})^n$ (and no more) if $p = (\frac{1}{2})^n$; e.g. $E =$ all heads in n tosses. If $E =$ all heads in an infinite number of tosses, I will not be willing to pay more than zero (every $\varepsilon > 0$ is $> (\frac{1}{2})^n$ for sufficiently large n, and would be too much even if the risk were infinitely greater). The lesser evil, therefore, is not to insure oneself; in other words, to act in this respect (*but not in others*) as if E were impossible.

A1c There is more, however. The condition of coherence is and must be (as we established in 3.3.5 and 3.3.6) *even weaker*† than the one criticized in *A*1, allowing in addition bets in which one can *only lose*! Let us suppose that an individual is subjected to a certain loss of a sum $1/N$ (where N is an 'integer chosen at random', with equal—and therefore zero—probabilities for each value, and hence for each finite segment $N \leqslant n$ (II!)). There is no advantage in paying a sum ε (however small) to avoid this certain loss, because it would always be practically certain that the loss avoided would be very much smaller.

N1 = A1d Summarizing and concluding, we have the following. The variants (from the weakest to the strongest) consist in excluding X if

$$\sup X < 0, \qquad \sup X \leqslant 0 \quad \text{with } X = 0 \text{ impossible}, \qquad \sup X \leqslant 0.$$

Objection *A*1 criticizes the middle statement, and supports the last one. In *A1c* we explained why, on the contrary, we think it necessary to support the first one.

N2 The variant N' is logically absurd unless one excludes the possibility of considering a partition with an uncountable infinity of possible cases (e.g. the continuum). In the denumerable case objections arise which also apply to C ($C3 = N4$, and so on).

N3 The variant N'' does away with *N*2: nevertheless, *the meaning of zero probability is still exceptionally restrictive* (much more so than in C, and even there it is too restrictive; cf. *C*4).

In fact, one should be able to define $E^* =$ the union of all events with zero probabilities = the maximal event with zero probability (let us call it 'the catastrophe'). Under the non-catastrophic hypothesis (with probability = 1) one goes back to N'; only in the opposite cases are events with zero probability no longer impossible (and, consequently, (II!) can have any probability whatsoever).

---

† If we wished to give this condition a name, we might call it *sufficient coherence* (in contrast to *weak* and *strict* coherence).

3.11.5. *C*1 *C appears to be less logically plausible than A and N*—we suspect 'Adhockery for mathematical convenience'—because the distinction between finite and infinite has without doubt a logical and philosophical relevance, whereas it might seem strange to draw the crucial distinction between finite and non-denumerable on the one hand, and countable on the other.

*C*2 A difficulty which derives from this is the following: *given a partition (e.g. whose cardinality is that of the continuum) into events of zero probability,* what happens if as a consequence of additional information one believes that *only a countable infinity remain possible*? In particular, if one assumes them (II!) equally probable? Or under the most general hypothesis?

*E*1 Initially, X has a uniform distribution over the real numbers between 0 and 1 (all points equally probable (II!)). Additional information reveals that X is rational.

*E*2 It seems obvious (but recall (II!)) that in this case—i.e. *E*1 after the given 'additional information'—*the values which remain possible, i.e. the rational values of* $[0, 1]$, *are (still) equally probable* (they define a 'random choice' from the original set).

If one thought of actually interpreting the problem geometrically, one might perhaps doubt the judgment of all the rationals as equally probable, considering as 'rather special' the end-points, mid-point, fractions with small denominator, decimal fractions with only a few figures, etc.

This effect is lessened if one thinks of taking the 'distance between two points chosen at random' (the first minus the second; if negative add 1, take the result mod 1).

It disappears altogether if one thinks in terms of a circle obtained by rolling up the segment without indicating which is the 'zero' point.

*C*3 = *N*4 Objection *C*2 can also be raised in the countable case (and then it also concerns *N*). *Suppose that we have a countable infinity of possible cases, one with* $p = 1$ *(and the others therefore with* $p = 0$*); assume we know that the first one has not occurred.*

*E*3 Let N be the number of passages through the origin in a random walk for which $\mathbf{P}(N > n) = 1$ for all n (an example is Heads and Tails); information: $N \neq \infty$.†

---

† This information could only be given by somebody who had explored the world as it appears after the end of time . . . . Objections to 'lack of realism' would, however, be out of place here as it is merely a question of logical compatibility. Where they are appropriate (and usually insufficiently dealt with), the exigencies of realism will be examined here (especially in the Appendix), perhaps at greater length than hitherto, and perhaps more than is reasonable. One cannot refute the exact nature of a conclusion based on the examination of a 'pathological' curve (e.g. that of Helge van Koch) by the pretext that there exist neither pencils, nor sheets of paper, nor hands, by means of which it could be drawn.

*E*4 In general, in such cases it is plausible to say that the

$$p_h = \mathbf{P}(N = h|N \neq \infty)$$

are all zero, and (II!) each is infinitely greater than the preceding one. We limit ourselves to a mere statement of this in order to be able to refer to this example without examining it deeply.

*C*4 *The meaning of p* = 0 *is too restrictive even in C* (although much less so than in *N* ; cf. *N*3). Expressed in a vague form, but one which corresponds exactly to the state of things, this is the 'essence' of those considerations and examples already given (*C*2, *C*3, *E*1, *E*2), and of those to come. The fact that, whereas, for any finite *n*, uniform partitions are allowed (all *p* = 1/*n*), in the countable case only extremely unbalanced partitions are allowed (under *C* and *N*), may serve as a 'symptom', which makes this 'restrictiveness' appear pathological.

We shall see, on the one hand, just how unbalanced they are, and, on the other, the objections to which this gives rise from a realistic point of view. The latter, of course, will vary according to the conception one holds.

*C*5 = *N*5 By taking the sum of probabilities to be = 1 (suppose we denote the probabilities by $p_1, p_2, \ldots, p_i, \ldots$, in decreasing order), one necessarily has an inequality such that for any $\varepsilon > 0$, however small, a finite number of events—the first n_ε—together have probability $> 1 - \varepsilon$, and the infinity of the others together have probability $< \varepsilon$. (In such circumstances, I am tempted to say that the events 'are not countably infinite' but 'a finite number—up to trifles').

*E*5 The point made in *C*5 = *N*5 appears even more strange if we take as an example the following observation.

> If, instead of the whole infinity of events, one only had the first $N = n/\varepsilon$ (where ε and $n = n_\varepsilon$ are as in the preceding case), there would be nothing to prevent one judging them equally probable (or almost so) in accordance with some assumed reasons or opinions. The total probability of the first *n* would then have been ε instead of $1 - \varepsilon$. Of course, even the infinity of probabilities could have all been taken $< 1/N$, but the enormity of the inequality would reappear if we took some $n' = n'_\varepsilon$ and $N' = n'/\varepsilon$ to start with.

From a mathematical standpoint this is obvious What is strange is simply that a formal axiom, instead of being *neutral* with respect to the evaluations (or, for those who believe in them, with respect to the objective reasons), and only imposing formal conditions of coherence, on the contrary, imposes constraints of the above kind without even bothering about examining the possibility of there being a case against doing so.

3.11.6. Let us try to better imagine the reactions of individuals with different points of view.

$C6 = N6$ Suppose we are given a countable partition into events E_i, and let us put ourselves into the subjectivistic position. An individual wishes to evaluate the $p_i = \mathbf{P}(E_i)$; he is free to choose them as he pleases, except that, if he wants to be coherent, he must be careful not to inadvertently violate the conditions of coherence.

Someone tells him that in order to be coherent he can choose the p_i in any way he likes, so long as the sum $= 1$ (it is the same thing as in the finite case, anyway!). The same thing?!!! You must be joking, the other will answer. In the finite case, this condition allowed me to choose the probabilities to be all equal, or slightly different, or very different; in short, I could express any opinion whatsoever. Here, on the other hand, the *content* of my judgments enter into the picture: I am allowed to express them only if they are unbalanced to the extent illustrated in $C5$–$N5$–$E5$. Otherwise, even if I think they are equally probable—as I would do in the case of $E2$—I am obliged to pick 'at random' a convergent series which, however I choose it, is in absolute contrast to what I think. If not, you call me *incoherent*! In leaving the finite domain, is it I who has ceased to understand anything, or is it you who has gone mad?

$C7 = N7$ In the same situation, an objectivist of the classical school finds himself facing case $E2$ (for him 'in conditions of symmetry all possible cases are equally probable').

This much is obvious: the infinite number of cases are equally probable and therefore they all have probability $1/\infty = 0$ (perhaps—he may think—I am not expressing myself in an orthodox fashion; the conclusion, however, is this one). To the objection of the teacher who wants a series with sum $= 1$, and who is not worried if one asks him whether he really wants an opinion so unbalanced as to give rise to the points raised in $E5$, he too will cry out: 'Is it I who has ceased to understand anything, or is it you who has gone mad?' And he will explain: 'I swear that I find myself in the ideal conditions of complete ignorance, with the absence of any reason to doubt whether any point has objective probability greater than that of any other one. In no other case can I be so sure of being able to state with precision that the objective probabilities are equal, because it is only in this case, where I cannot even see or distinguish the rational points, that I have reached the final sublime peak of total and unsurpassable ignorance. And now, what is the use of it? What are the objective probabilities I must give the various points, and how do I know which of them must be assigned a large probability, a small one, or a very small one?'.

$C8 = N8$ For the frequentist, this is even easier. If he thinks of a sequence of experiments (an ideal version of roulette, reduced to a point-ball which can stop at any rational point of the circle of $E2$) he will be in doubt as to whether a point will appear just a few times, or many times, or even infinitely many times. It is unlikely, however, that he will think for a moment that some point—and especially one which can be individuated right from the beginning —will appear so often as to have a limit-frequency different from zero.

$C9 = N9$ Here is a new and genuine mathematical objection to countable additivity: for those who conceive of probabilities as limit-frequencies (over a sequence, or, in von Mises' terminology, a 'Collective'), the fact that *limit-frequencies must satisfy finite additivity, but not countable additivity*, should be decisive.

(So far as I know, however, none of them has ever taken this observation into account, let alone disputed it; clearly it has been overlooked, although it seems to me I have repeated it on many occasions).

3.11.7. *C*10 A probability which is countably (but not perfectly) additive cannot be defined on the power set of the infinite set of events under consideration.

Therefore, it is necessary:

(a) either to introduce restrictions which only allow one to refer to events given by certain 'subsets', excluding the others (in this case the logical justifications are not obvious, and the mathematical ones, which require the creation of special events by endowing the 'space' with topological properties, seem merely to have the status of 'Adhockeries for mathematical convenience');

(b) or to accept perfect additivity, i.e. *N*, which appears *more logical* than *C*, for this reason in addition to that already given in *C*1 (but one encounters *N*2, and abandons any treatment in the continuum, even by means of the measure-theoretic model which is the actual aim of *C*);

(c) or to accept finite additivity; i.e. *A*.

3.11.8. Do there exist objections to *A* (besides *A*1, which we have examined already)? In all honesty—and I shall willingly change my mind if any contrary evidence is brought to my attention—it seems to me that one should in general refer to prejudices and habits, rather than to objections. Independently of the discussion of specific aspects of the real problem (which are always neglected), it is these habits and prejudices which lead one to consider as 'natural', or 'absurd', those things in other branches of mathematics which are more or less customary, more or less up-to-date, and, above all, more or less 'convenient'. We refer to those fields where, in the absence of an intrinsic meaning, already existing and imposed from the outside onto the possible translations into mathematical definitions and axioms, it is admissible to choose those concepts and hypotheses which are most convenient, to choose them 'for mathematical convenience'.

We shall see something of these aspects and attitudes in Chapter 6 and in the Appendix. (It is often difficult to analyse them because they are more psychological than mathematical in character, and because one usually has to deduce things from odd comments rather than from explicit and systematic explanations.) If one wants to pick out an example of a sufficiently concrete position, having some validity,† I merely point to the following.

---

† I hope that the reader can himself demolish the frequent attempts to 'prove' countable additivity under the tacit assumption of the validity of some property equivalent to it.

*A*2 It seems to many people that a countable partition which is not unbalanced (i.e. not reducing to cases 'finite up to trifles', as we jokingly called them in *C*5) is 'not feasible'. A positive integer *N*, unknown (random) and capable of taking on any value (between 0 and ∞, which is excluded), is always, in any practically or conceptually imaginable example, almost certainly not too large (and an upper bound is not given solely in order to avoid a more or less arbitrary choice). A partition of a set whose cardinality is that of the continuum, e.g. an interval, into a countably infinite number of (*L*-) measurable sets, is necessarily such that all the measure (except an arbitrarily small residual) is given by a finite number of them. They can be overlapping (as in the Vitali case) but then they are not measurable, and therefore not even 'mentionable', and not even susceptible of a constructive description independently of the axiom of choice.

It is necessary to reply to this from various viewpoints.

*A*2*a* From the subjectivistic point of view—since, subject to the conditions of coherence, one has complete freedom of choice in evaluating the probabilities—one can perfectly well assign greater probability to a set with only one point than to a set which has very large measure, or is non-measurable. Conversely, can this line of argument justify attributing large probability to sets consisting of a single point and with small measure, and negligible probability to the large sets, leaving out the intermediate cases?

*A*2*b* Do not these examples themselves (although in a slightly more sophisticated manner) reveal the prejudice of assuming the measure-theoretic model as the universal one?

*A*3 Another plausible objection : all these examples and counter-examples are artificial, with no practical interest ; there is no reason to prefer a less convenient theory simply because it allows us to take account of them.

*A*3*a* The examples have a critical function ; to test the logical consistency of the various points of view. To accept the point of view which (I hope) they reveal to be the logically correct one does not imply that one has to occupy oneself with matters of this nature,† but only to avoid expressing oneself in a way that appears to be incorrect (albeit with reference to 'pathological' examples).

*A*3*b* Indeed, in practice, it will probably turn out to be advisable to limit oneself to *even simpler* ideas, sticking to the more elementary ambit (Jordan–Peano measure, Riemann integral) where the conclusions are unexceptionable, rather than passing to the more 'modern' set-up (Borel or Lebesgue

---

† Let us recall that the critical examples which Peano inserted into Genocchi's lecture notes, in order to show that certain 'theorems' did not always hold in 'pathological' cases, met with an exactly similar attitude of disapproval and incomprehension.

measure, Lebesgue integral), given that the usual extension is based on a convention which is inadmissible as a general axiom, and difficult to justify in a realistic way as a particular hypothesis for individual practical cases. It seems to me that it is difficult to justify not only its validity, but even that possible interpretations and applications to actual and practical problems are not illusory.

A3c If we are going to talk about which theory is 'less convenient', we must distinguish the sense in which 'convenient' is to be understood. The theory given by *C* is, in general, more convenient to handle, and is convenient because it provides a well-determined answer in many cases where *A* just gives bounds. From the standpoint of *A*, it is wrong to substitute an exact answer in place of these bounds (and, anyway, inconvenient, since it forces us to exclude all those examples that might appear artificial, but which are not absurd). From some points of view, *A* is even more tractable; e.g. every limit of a probability distribution is, in *A*, a probability distribution (possibly not proper): this is not true in *C*. It is, in any case, a question of things which are logically relevant, not one of mathematical convenience.

A4 One more objection (a little premature as far as the applications it refers to are concerned, but not in terms of its formal meaning, nor for the understanding of example *E6* below).

Proofs made in the spirit of *A* in order to invalidate the interpretations of *asymptotic results* (not yet discussed) as *limit-results* (deduced in accordance with conception *C*) often make use of the device of introducing a number *N*, which is 'chosen at random' (zero probability for each single *n* and finite segment $N \leqslant n$), assuming that from *N* onwards a certain process proceeds in a different way from that foreseen in the scheme of description.

This said, the objection is: *That's a different story: if the scheme changes, if there is a violent change, then the conclusions established under the assumption that the scheme remains unaltered, without foreseeing any possibility of a violent change, will certainly break down.*

A4a Statements of this kind do not take account of the situation. The 'scheme', as usually described, does not explicitly foresee the possibility of a violent change, but it does not exclude it either: it is entirely neutral. It is therefore improper to refer to a 'violent change': the question of a violent change arises only when one adds to the mathematical scheme something more in the way of interpretation, which would be difficult to express. Indeed, if it were expressed, it would render trivial the result, which is beautiful and true only if one assumes that countable additivity is less restrictive than would appear from the following kind of example.

E6 As in *E2*, we can imagine 'choosing at random' a rational number in [0, 1] with a finite number of decimal places (all with the same probability

(II!)),† the number of places being itself random, and not preassigned. If we think of a selection of the successive decimals (or of their successive deciphering or calculation, if they have been 'drawn all at once' and can be worked out successively, as for π), the process is clearly identical to that of drawing any real number whatsoever. At each drawing, all 10 figures have the same probability $\frac{1}{10}$, whatever the previous results may have been.‡

If by 'catastrophe' we mean the exceeding of the last non-zero figure, it is certain that sooner or later this will happen. *But it will not be a catastrophe :* we will not be able to realize it ; nothing will change in the described scheme. Even after 100 or 1,000,000 or 10^{1000} consecutive zeroes, provided we have no gift of divination, the probability that the next figure will be zero is $\frac{1}{10}$, as for any other figure; the probability that the next 100 figures will all be zero is 10^{-100}, as for any other 100-figure number; the probability that the figures will continue to be zero for evermore is zero, exactly as it is at any other instant, and after any arbitrary sequence of figures.

In this example, all the probabilistic assumptions explicitly stated for the process hold exactly ; these lead to the conclusion that, with probability $= 1$, the 10 figures will each show up with limit-frequency $\frac{1}{10}$ (whereas, the limit-frequency is here $=1$ for the figure 0, and $=0$ for the others). The only assumption which does not hold is that of countable additivity, but if anyone considers it as an axiom, instead of a particular restriction (not valid in our example), he has the right (?) to omit its explicit statement and to check whether it holds.

3.11.9. *Conclusion (for the time being).* I do not know whether, and to what extent, the arguments put forward here have been persuasive. On the other hand, it is premature to accept or reject them before encountering other aspects of them and having seen their implications (in Section 3.12 following, at the end of Chapter 4, and in Chapter 6 and elsewhere, more or less incidentally). In view of this, however, I would like to have succeeded in convincing the reader of one thing ; that we are dealing with a complex of problems, connected and meaningful, concerning which there are many things to be discussed under various headings : the conceptual, the mathematical, the practical. It is not just, as might seem logical at first sight,

---

† If one wishes, instead of choosing from this set one can imagine the choice of any rational whatsoever, as in *E*2. The rationals can be put together in 'equivalence classes' (where two numbers differ by a bounded decimal fraction; i.e. they coincide from some point on) and in each class *an identifiable representative can be chosen*; the one which is *periodic right from the beginning*. Every rational uniquely determines the components $r = p + d$ (p periodic, d decimal), and the sets I_d (of the r with the same d) give rise to a partition of the rationals into a countable number of sets superposable by translations (mod 1). To choose r is therefore a way of choosing d.

The partition is similar to that of Vitali for the reals, but here, fortunately, an infinite number of choices is not required.

‡ We should refer to stochastic independence, but we shall come to it in the next chapter, Chapter 4, and here content ourselves with just mentioning the idea.

a question of arbitrary conventions for the subtleties involved, having no connection with real problems.

3.12 RANDOM QUANTITIES WITH AN INFINITE NUMBER OF POSSIBLE VALUES

3.12.1. The above considerations obviously also apply to the case in which there are an infinite number of possible values for a random quantity X. Some new features also arise, however. We shall not concern ourselves with the general case until Chapter 6, but in the meantime it is necessary to mention certain refinements, although only for the more elementary case (elementary in a certain sense, at least) of a countable infinity of possible values $x_h (h = 1, 2, \ldots)$. To these will correspond—or rather can be attributed by the person who evaluates them—probabilities p_h, either positive or zero (they might even all be zero), with

$$\sum_h p_h = 1 - p^* \leqslant 1, \qquad (0 \leqslant p^* \leqslant 1).$$

For any interval or set I, one could say, knowing only the x_h and p_h, that $\mathbf{P}(X \in I) = \sum_h p_h(x_h \in I)$ if the set contains a finite number of points, but only that

$$\sum_h p_h(x_h \in I) \leqslant \mathbf{P}(X \in I) \leqslant \sum_h p_h(x_h \in I) + p^*$$

if it contains an infinite number (given that the probability p^* can always be imagined as deriving solely from these).

3.12.2. In particular, if x is an accumulation point of the x_h (it does not matter whether it is one of them or not), we can have non-zero *adherent* probabilities, the latter defined to be the limit of $\mathbf{P}(x - \varepsilon < X < x)$ or $\mathbf{P}(x < X < x + \varepsilon)$ as $\varepsilon \to 0$ ($\varepsilon > 0$), and their sum (if we wish to distinguish, we refer to adherent from the left, adherent from the right). The adherent probabilities (or masses) cannot exceed p^*; not even if we take them all together, or even include those possibly adherent (from the left) to $+\infty$ and (from the right) to $-\infty$.† The adherent probabilities could not only have total

---

† One can either allow $+\infty$ and $-\infty$ to also appear among the possible values, or one can exclude them. Including them would entail thinking of X as a random point on the completed real line (compactified) with the adjunction of the 'extremes' $+\infty$ and $-\infty$. There is nothing absurd about this, although it is not usual to do and there is no point in insisting upon it. Every now and again we will make brief mention of such eventualities, but without entering into any obligation to observe case by case whether what is said is valid there also.

  On the other hand, we must note a certain conflict of interest. As far as prevision is concerned (and here the inequalities are essential), the values $+\infty$ and $-\infty$ are distinct and very far apart (in fact, opposite). From an analytic point of view, however, it would be more natural to consider them as a single value (except for looking at it in terms of approaching from the left and right), thinking, for instance, of the complex sphere (and, in that context, of the circle of real numbers) and of functions which are 'continuous' there, like $y = 1/x$ at $x = 0$ (cf. *Matematica logico-intuitiva*, 3rd Ed., pp. 124–133).

probability $< p^*$, but also zero (in other words, non-existent), although p^* was positive, or even $p^* = 1$. As an example: $X = $ rational between 0 and 1, with the probability of each subinterval equal to its length (the uniform distribution).

3.12.3. The argument concerning the prevision $\mathbf{P}(X)$ is new and specific to this case. It is unnecessary to note that whatever one says concerning $\mathbf{P}(X)$ holds for any $\mathbf{P}(\gamma(X))$, where $Y = \gamma(X)$ is any function of X, whose possible values are $y_h = \gamma(x_h)$ with probabilities p_h (except that, if one of these values corresponds to an infinite number of the x_h, its probability may be, if $p^* > 0$, greater than the sum of the p_h instead of being equal to it).

What does the knowledge of the possible values x_h and their probabilities p_h allow us to say concerning $\mathbf{P}(X)$? Or rather, expressing ourselves in terms of what the question means in a (subjective) probabilistic sense, what restrictions does the knowledge of the x_h and an existing evaluation of the p_h (which we wish to remain coherent) impose on us when it comes to evaluating the prevision of X?

It is convenient to begin with the case of a *bounded* random quantity X, and to consider directly the minimum and the maximum of the accumulation points, which we denote by x' and x''; we therefore have

$$-\infty < \inf X \leqslant x' \leqslant x'' \leqslant \sup X < +\infty.$$

Let us prove that if $p^* = 0$ (i.e. if $\sum_h p_h = 1$, as it is if countable additivity holds) we must have the unique result $\mathbf{P}(X) = \sum_h p_h x_h$, as in the finite case. Apart from this special case we can only say that

$$\sum_h p_h x_h + p^* x' \leqslant \mathbf{P}(X) \leqslant \sum_h p_h x_h + p^* x''.$$

Thus, if we are not in the above case, $p^* = 0$, $\mathbf{P}(X)$ turns out to be uniquely determined if and only if $x' = x''$; in other words, if the x_h have a unique accumulation point, hence a limit to which they converge.

Proof. For a given $\varepsilon > 0$, take N sufficiently large so that we have

$$\sum_h p_h(h \geqslant N) < \varepsilon,$$

and put $X = X_1 + X_2 + X_3$ with

$X_1 = X = x_h$ if $h < N$, and otherwise $= 0$,

$X_2 = X = x_h$ if $h \geqslant N$ and $x_h < x' - \varepsilon$ or $x_h > x'' + \varepsilon$, and otherwise $= 0$,

$X_3 = X = x_h$ if $h \geqslant N$ and $x' - \varepsilon \leqslant x_h \leqslant x'' + \varepsilon$, and otherwise $= 0$.

We have

$$\mathbf{P}(X_1) = \sum_h p_h x_h (h < N) \to \sum_h p_h x_h \ (x_h \text{ bounded!});$$

$$\varepsilon \inf X \leqslant \mathbf{P}(X_2) \leqslant \varepsilon \sup X,$$

because there are at most a finite number of possible values between $\inf X$ and $x' - \varepsilon$, and the same for those between $x'' + \varepsilon$ and $\sup X$, and the total probability of those between them with $h \geqslant N$ is the sum of a finite number of the p_h for which the sum of the series is $< \varepsilon$. Finally, we have

$$p^*(x' - \varepsilon) \leqslant \mathbf{P}(X_3) \leqslant p^*(x'' + \varepsilon).$$

All this holds for every ε and hence, as $\varepsilon \to 0$, one obtains the given bounds.

Remark. It is most instructive and important to observe that these bounds *cannot be improved on;* in other words, it is actually admissible to evaluate $\mathbf{P}(X)$ by giving it any value whatsoever between the two end-points (inclusive). The p^* resulting from infinite zero probabilities (distributed on the possible x_h; it does not matter if these already have positive probabilities $p_h > 0$ or instead have $p_h = 0$) could well be considered as deriving from an infinite number of the x_h converging towards x', or towards x'', and in any intermediate way.

(In addition, one notes that the proof neither presupposes nor establishes countable additivity: it holds here—as it may hold elsewhere—by virtue of additional assumptions implicit in the definition of the particular case.)

3.12.4. We pass from the case of bounded X to that of X *unbounded.* The case of one-sided unboundedness must be considered separately, and we therefore begin with the case of X unbounded from above (obviously, the analysis holds also for the other case); the general case follows as a corollary.

We also suppose that with certainty $X \geqslant 0$ (i.e. $\inf X \geqslant 0$); in the general case it is sufficient to put $X = X_1 - X_2$, $X_1 = 0 \vee X$, $X_2 = |0 \wedge X|$, in order to reduce everything to random quantities which are certainly non-negative.

Moreover—in order not to complicate the exposition by encountering anew the circumstances already seen in the finite case—we suppose that there do not exist finite accumulation points. We can therefore suppose the x_h to be increasing, and tending to $+\infty$ as h tends to infinity.†

---

† It is clear that the conclusions of this special case are essentially valid in general if one considers that $X' \leqslant X \leqslant X''$, where we set $X' = $ (the smallest integer $\leqslant X$), $X'' = X + 1$ (and the unit of measurement can be taken as small as we please); X' and X'' are automatically of the type considered (but to pursue this would introduce things which we reserve for the treatment of the continuous case).

If $X_\infty = +\infty$ exists among the possible values, it is not necessary that the finite possible values be unbounded (and not even that they be infinite in number) in order for us to be in the unbounded case.

Under these conditions, putting

$$P_n = \sum_{h=1}^{n} p_h, \qquad P = \lim P_n, \qquad p^* = 1 - P, \qquad S_n = \sum_{h=1}^{n} p_h x_h,$$

$$S = \lim S_n,$$

we have

$$P_n = \mathbf{P}(X \leqslant x_n), \qquad 1 - P_n = \mathbf{P}(X > x_n),$$

$p^* = $ the mass adherent from the left at $+\infty$, or placed at $x = +\infty$, or some here, some there;

$$S_n = \mathbf{P}\{X(X \leqslant x_n)\}, \qquad S_n + x_n(1 - P_n) = \mathbf{P}(X \wedge x_n)$$

(the previsions of X either 'amputated' or 'truncated' at x_n; i.e. replaced, if X exceeds x_n, either by 0 or by x_n, respectively).

Since each 'truncated X' is always $\leqslant X$, we necessarily have

$$\mathbf{P}(X) \geqslant S_n + x_n(1 - P_n) \quad \text{for some } n, \text{ and hence}$$

$$\mathbf{P}(X) \geqslant S + x_n(1 - P) = S + x_n p^* \quad \text{for some } n$$

(because, if we let n increase in S_n and P_n, while keeping x_n fixed, the expression increases, but less than it would if x_n also were allowed to vary, and tends to the given limit).

It necessarily follows straightaway from this that $\mathbf{P}(X) = \infty$ if $S = \infty$ (the series $\sum_h p_h x_h$ diverges), or if $p^* \neq 0$ (there exists a probability placed at, or adherent to, $+\infty$), or both.

In the opposite case, $p^* = 0$ and S finite (the series of the p_h having sum $= 1$, and the series of the $p_h x_h$ being convergent), admissible evaluations of $\mathbf{P}(X)$ are given by

$$\mathbf{P}(X) = S = \sum_{h=0}^{\infty} p_h x_h, \quad \text{or any greater value, including } +\infty.$$

This is proved *by continuity* (and in the next section—Section 3.13—we briefly discuss that property of continuity which we shall make use of here).

First of all, we set $X'_n = X \ (X \leqslant x_n)$ (X amputated) with $p'_h = p_h$ for $h \leqslant n$, $p'_h = 0$ for $h > n$, and $p'_0 = \sum p_h(h > n) = \mathbf{P}(X'_n = 0)$; as n increases, all the $p'_h = \mathbf{P}(X'_n = h)$ tend to p_h, but $\mathbf{P}(X'_n) = S_n \to S$.

We then set $X''_n = X'_n + a_n(X > n)$, in other words, X''_n (like X'_n) coincides with X if the latter does not exceed x_n, but when it does we replace it with a_n instead of with 0; a_n denotes the first of the x_h for which $x_h p_0 \geqslant n$.† The value a_n already gives a contribution $\geqslant n$, hence we certainly have

$$\mathbf{P}(X''_n) \geqslant n \to \infty.$$

_____

† The argument, with a simple modification, also holds in the case in which $p_h = 0$ for all possible x_h from a certain $h = N$ on, so that $p_0 = 0$. One could, for instance, let $p_0 = (\frac{1}{2})^n$, taking this probability away from one or more of the p_h (for instance, from p_1 if $p_1 = 0$, starting from that n for which $(\frac{1}{2})^n < p_1$).

We repeat the conclusions in a schematic form:

$$\text{in the case} \begin{cases} p^* > 0 & \mathbf{P}(X) = +\infty\,; \\ & \\ p^* = 0 & \begin{cases} S = +\infty & \mathbf{P}(X) = +\infty\,; \\ S < +\infty & S \leqslant \mathbf{P}(X) \leqslant +\infty. \end{cases} \end{cases}$$

3.12.5. If X is unbounded from above and below, $\mathbf{P}(X)$ is completely undetermined. This is obvious straightaway from the fact that we could always have '$\infty - \infty$'; one can obtain this more rigorously by a passage to the limit in the previous cases (suitably balancing the positive and negative terms).

However, one might consider as *special* the evaluation which consists in taking, both for the positive part $0 \vee X$ and for the negative part $0 \wedge X$, the minimum (in absolute value) admissible prevision—denoting it by $\hat{\mathbf{P}}$—and setting in general

$$\hat{\mathbf{P}}(X) = \hat{\mathbf{P}}(0 \vee X) - \hat{\mathbf{P}}(|0 \wedge X|) \quad \text{(or, briefly, } \hat{S} = S^+ + S^-\text{)}.$$

'Special' is *not used in general sense* but if, and so long as, one can consider that, in a given case, the unbounded X is a theoretical schematization substituted for simplicity in place of an actual X, which is in reality bounded, but whose bounds are very large and imprecisely known.

This asymptotic prevision (as we shall call it for this reason) turns out to be:

$$\hat{S} = S^+ + S^- \begin{cases} \textit{finite,} \text{ if } S^+ \text{ and } S^- \text{ are;} \\ \\ \textit{infinite,} \text{ if one of the components is: } +\infty \text{ if } S^+ = +\infty\,; \\ \quad -\infty \text{ if } S^- = -\infty\,; \\ \\ \textit{undefined,} \text{ if both are infinite.} \end{cases}$$

3.13 THE CONTINUITY PROPERTY

The property says (and we shall make this precise and prove it) that *coherence is preserved in a passage to the limit.* The property does not hold (without further conditions) when we impose countable additivity. This turns out to be very useful as a tool in proofs of admissibility like the ones just given above (3.12.4).

Theorem. Let $\mathbf{P}_n(E)$ be the evaluations of (coherent) probabilities defined over the same field of events $\mathscr{E}$ (or over different fields of events having $\mathscr{E}$ in common), and put $\mathbf{P}(E) = \lim \mathbf{P}_n(E)$ when it exists (letting $\mathscr{E}' \subseteq \mathscr{E}$ be the set of the E for which the limit exists). In this field the $\mathbf{P}(E)$ itself constitutes a (coherent) evaluation of probability.

Remark. In place of the (more 'familiar') formulation above, it would be (mathematically) preferable to substitute that in which one speaks of the prevision of random quantities rather than the probability of events, and hence of linear spaces (with appropriate definitions and convergence) rather than 'fields'.

Proof. The conditions of coherence are expressed by linear equations (or inequalities) involving a finite number of elements (events, or random quantities); in the passage to the limit these are preserved.

Remark. In a more expressive formulation (and more precise, so long as one recalls that the meaning of 'convergence' is that given above): an evaluation of probability **P** adhering to a set $\mathscr{P}$ of coherent evaluations is coherent.

CHAPTER 4

Conditional Prevision and Probability

4.1 PREVISION AND THE STATE OF INFORMATION

We have all at times insisted on making clear the fact that every prevision, and, in particular, every evaluation of probability, is conditional; not only on the mentality or psychology of the individual involved, at the time in question, but also, and especially, on the state of information in which he finds himself at that moment.

> Those who would like to 'explain' differences in mentality by means of the diversity of previous individual experiences, in other words—broadly speaking—by means of the diversity of 'states of information', might even like to suppress the reference to the first factor and include it in the second. A theory of this kind is such that it cannot be refuted, but it seems (in our opinion) rather meaningless, being untestable, vacuous and metaphysical: in fact, since two different individuals (even if they are identical twins) cannot have had, instant by instant, the same identical sensations, any attempt at verification or refutation assumes an absurd hypothesis. It is like asking whether or not it is true that had I lived in the Napoleonic era and had participated in the Battle of Austerlitz I would have been wounded in the arm.

As long as we are just referring to evaluations relative to the same individual and state of information, there is no need to make any explicit mention of it; e.g. instead of $\mathbf{P}(E)$, writing something like $\mathbf{P}(E|H_0)$, where H_0 stands for 'everything that is part of that individual's knowledge at that instant'. Indeed, something which in itself is so obvious, and yet so complicated and vague to put into words, is clearer if left to be understood implicitly rather than if one thinks of it condensed into a symbol, like H_0.

Naturally, things change if we want to combine previsions which are relative to different states of information, and we shall see later that one cannot do without this. In precise terms, we shall write $\mathbf{P}(E|H)$ for the *probability 'of the event E conditional on the event H'* (or even the *probability 'of the conditional event E|H'*), which is the probability that You attribute to E if You think that in addition to your present information, i.e. the H_0 which we understand implicitly, *it will become known to You that H is true (and nothing else)*. This H, on the other hand, may be a combination of

134

'simpler' events (this is obvious, but it is better to point it out explicitly); in other words, it can denote, in a condensed manner, a whole complex of new information, no matter how extensive (so long as it is well delimited).

The above explanations may be useful as a preliminary guide to the meaning of the concept of *conditional probability*, $P(E|H)$—and, more generally, of *conditional prevision*, $P(X|H)$—which we are about to introduce. We ought to warn the reader, however, against an overhasty acceptance of these initial explanations, which, of necessity, skipped over certain important details, a discussion of which would have been premature (cf. the *Remarks* given in Chapter 11, 11.2.2). Think, instead, in terms of the definition that we are now going to give.

The definition is based on the same concepts and criteria that we met previously (cf. Chapter 3), except for the additional assumption that *any agreement made*—i.e. any *bet* or *penalty clause*—will remain *without effect if H does not turn out to be true:* in other words, everything is *conditional on the 'hypothesis' H.* (Concerning the terminology 'hypothesis', cf. Section 4.4.2.)

The 'first criterion' provides an intuitive explanation, which we exploit only to anticipate the meaning of the 'theorem of compound probabilities'. By paying the price $P(HE)$, I can be sure of receiving one lira if HE occurs; but I can obtain the same result by paying $P(E|H)$ only if I know H is true, and I can arrange for this amount, $S = P(E|H)$, in the case of the occurrence of H by paying $S . P(H)$ now; hence

(1) $$P(HE) = P(H) . P(E|H).$$

The same is true if, instead of an event E, I consider an arbitrary random quantity X; it is sufficient to observe that HX coincides with X, or is zero, according as H is true or false, and the extension of the preceding argument to this case becomes obvious.

4.2 DEFINITION OF CONDITIONAL PREVISION (AND PROBABILITY)

In order to give definitions of *conditional probability* and *conditional prevision*, and as a foundation for rigorous proofs, we choose to base ourselves on the 'second criterion'.

Definition. Given a random quantity X and a *possible* event H, suppose it has been decided that You are subject to a penalty

$$L = H\left(\frac{X - \bar{x}}{k}\right)^2$$

(k fixed arbitrarily in advance), where $\bar{x}$ is the value which You are at liberty to choose as You like. (Note: we have $L = 0$ if $H = 0 = false$; $L = [(X - \bar{x})/k]^2$ if $H = 1 = true$.)

$\mathbf{P}(X|H)$, *the prevision of X conditional on H* (in your opinion), is the value $\bar{x}$ which You choose for this purpose.

In particular, if X is an event, E, then $\mathbf{P}(E|H)$, so defined, is called *the probability of E conditional on H* (in your opinion).

Coherence. It is assumed that (in normal circumstances) You do not prefer a given penalty if You can choose a different one which is *certainly* smaller.

A *necessary and sufficient condition* for coherence in the evaluation of $\mathbf{P}(X|H)$, $\mathbf{P}(H)$ and $\mathbf{P}(HX)$, is compliance with the relation

(2) $$\mathbf{P}(HX) = \mathbf{P}(H) \cdot \mathbf{P}(X|H),$$

in addition to the inequalities $\inf(X|H) \leqslant \mathbf{P}(X|H) \leqslant \sup(X|H)$, and $0 \leqslant \mathbf{P}(H) \leqslant 1$; in the case of an event, $X = E$, relation (1),

$$\mathbf{P}(HE) = \mathbf{P}(H) \cdot \mathbf{P}(E|H),$$

is called the *theorem of compound probabilities*, and the inequality for $\mathbf{P}(X|H)$ reduces to $0 \leqslant \mathbf{P}(E|H) \leqslant 1$ (being $= 0$, or $= 1$, in the case where EH, or $\tilde{E}H$, respectively, is impossible).

By $\inf(X|H)$ and $\sup(X|H)$, we denote the lower and upper bounds of the possible values for X which are *consistent* with H; such values are simply the possible values of HX, with the proviso that the value 0 is to be included only if $X = 0$ is compatible with H (i.e. if HX can come from $H = 1, X = 0$, and not only, as is necessarily the case, from $H = 0$, with X arbitrary).

4.3 PROOF OF THE THEOREM OF COMPOUND PROBABILITIES

Let us consider first the case of events, and denote by x, y, z the values we suppose to be chosen, according to the given criterion, as evaluations of $\mathbf{P}(E|H), \mathbf{P}(H), \mathbf{P}(HE)$. In this case, the theorem is expressed by (1), and, with the above notation, it states that $z = xy$.

The penalty (taking the coefficient $k = 1$) turns out to be

$$L = H \cdot (E - x)^2 + (H - y) + (HE - z)^2,$$

i.e., in the three cases to be distinguished,

$$HE \ (H = E = HE = 1), \qquad H\tilde{E} \ (H = 1, E = HE = 0)$$

$$\text{and} \quad \tilde{H} \ (H = HE = 0),$$

we have

$$HE: \quad L = u = (1 - x)^2 + (1 - y)^2 + (1 - z)^2$$
$$H\tilde{E}: \quad L = v = \quad x^2 \quad + (1 - y)^2 + \quad z^2$$
$$\tilde{H} \ : \quad L = w = \qquad\qquad y^2 \quad + \quad z^2$$

Geometrically (interpreting x, y, z as cartesian coordinates: see Figure 4.1), the penalties u, v, w, in the three cases, are the squares of the distances of the point (x, y, z) from, respectively, the point $(1, 1, 1)$, the point $(0, 1, 0)$, and the x-axis (i.e. from the point $(x, 0, 0)$, the projection of (x, y, z) onto the axis). The four points lie in the same plane if a fifth one, $(x, 1, z/y)$, does also (this is the intersection of the line joining the last two with the plane $y = 1$), and this therefore must coincide with $(x, 1, x)$—which is on the line joining the first two points. In order for this to happen, we must have $z = xy$, i.e. the point (x, y, z) must lie on this paraboloid (and, of course, inside the unit cube): in this case, it is not possible to simultaneously shorten the three distances; in other cases this is possible.†

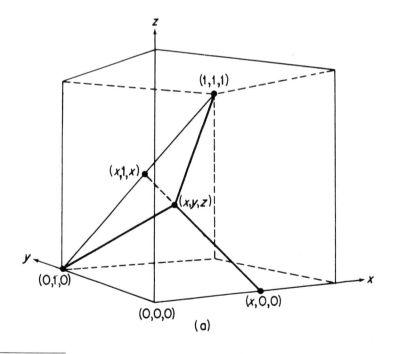

(a)

---

† A more detailed discussion can be found in B. de Finetti, 'Probabilità composte e teoria delle decisioni', *Rendic. di Matematica* (1964), 128–134. An English translation of this appears in B. de Finetti, *Probability. Induction and Statistics*, Wiley (1972).
N.B. The coordinates y and z are there interchanged with respect to their usage here

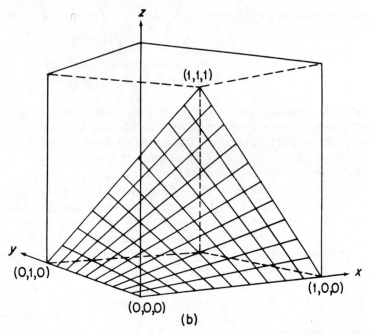

Figure 4.1   The two diagrams illustrate, in two stages, the argument given in Section 4.3. (a) shows why the prevision-point (x, y, z) must lie on a generator of the paraboloid $z = xy$ (presenting visually the argument of the text). (b) shows the set of all possible prevision-points (the part of the paraboloid inside the unit cube)

Turning to the general case of an arbitrary random quantity X, let us again use the notation $x = \mathbf{P}(X|H)$, $y = \mathbf{P}(H)$ and $z = \mathbf{P}(HX)$, and observe that the previous representation is still valid, except that, instead of the two points $(1, 1, 1)$ and $(0, 1, 0)$ on the line $y = 1$, $z = x$, we must consider all the points whose abscissae x are possible for X and compatible with H. In fact, expanding (in canonical form) we have,

$$L = H \cdot (X - x)^2 + (H - y)^2 + (HX - z)^2$$

$$= H[(X - x)^2 + (1 - y)^2 + (X - z)^2] + (1 - H)(y^2 + z^2).$$

If (x, y, z) were not on the paraboloid $z = xy$ (i.e. not in the plane through the line $y = 1$, $z = x$ and the point $(x, 0, 0)$), one could, as before, make it approach, simultaneously, both the x-axis and each point of the given line. In order that this should not be possible, it is necessary, in addition, to restrict oneself to the area (a quadrilateral bounded by the straight lines generating the paraboloid) given by

$$0 \leqslant y \leqslant 1 \quad \text{and} \quad \inf(X|H) \leqslant x \leqslant \sup(X|H).$$

The convenience of substituting $y = 0$ and $y = 1$ for any $y < 0$, or $y > 1$, respectively, is obvious; that x must not be outside the bounds for $(X|H)$ becomes clear (without spending time on the calculations) if one observes, in mechanical terms, that in order to cancel out a force acting at the point (x, y, xy) directed towards $(x, 0, 0)$—i.e. tending to make it approach the x-axis—it is necessary to have a force directed towards $(x, 1, x)$, which is opposite (or, alternatively, more than one, directed towards points which are on both sides of this point on the line $y = 1, z = x$). If the possible points were all on one side (and only in this case) all distances could be shortened by moving towards the nearest bound.†

4.4 REMARKS

4.4.1. Let us note first of all that, as we have already seen in passing, in questions concerning the conditioned event, $E|H$, the event E itself does not actually enter the picture: the cases to be distinguished are, in fact, $HE, H\tilde{E}, \tilde{H}$. Since H is called the 'hypothesis' of the conditioned event, HE could be called the 'thesis', $H\tilde{E}$ the 'antithesis', and $\tilde{H}$ the 'anti-hypothesis'. Every conditioned event $E|H$ could then be written in the *reduced* form 'thesis'| 'hypothesis', $HE|H$ (in fact, it does not matter whether one bets that if H occurs E does, or that if H occurs both H and E do). One might consider $E|H$ as a tri-event with values $1|1 = 1, 0|1 = 0, \quad 0|0 = 1|0 = \varnothing$, where $1 = true$, $0 = false$, $\varnothing = void$, according as it leads to a *win* or a *loss* or a *calling off* of a possible conditional bet. More generally, for a conditioned (random) quantity, $X|H$, one could put $X|1 = X, X|0 = \varnothing$ (if $\varnothing$ is thought of as outside the real field, inf$(X|H)$ and sup$(X|H)$ automatically acquire the desired meaning, introduced previously as a convention). The systematic use of algorithms based on this set of ideas does not seem sufficiently worthwhile to compensate for the bother of introducing them; however, this brief mention may suggest a few arguments for which it might turn out to be suitable.

4.4.2. As far as the use of the term 'hypothesis' for H is concerned, it should be unnecessary to point out that it refers only to the position of H in $E|H$ (or in $X|H$), and that, apart from this, H is any event whatsoever. We say this merely to avoid any possible doubts deriving from memories of obsolete terminologies (like 'probability of the hypotheses' or, even worse, 'of the causes', a notion charged with metaphysical undertones.

---

† This conclusion might fail to hold if the possible points were all on the same side of $(x, 1, x)$, but having this point as a bound (lower or upper). We will dwell upon detailed considerations of this kind in the sequel.

4.4.3. This being so, together with $E|H$ one can always consider $H|E$ as well (where E becomes the 'hypothesis'); indeed, since $EH = HE$, we obtain immediately the relationship between the probabilities of these two conditional events:

$$\mathbf{P}(EH) = \mathbf{P}(E)\mathbf{P}(H|E) = \mathbf{P}(H)\mathbf{P}(E|H),$$

which implies that

(3) $\mathbf{P}(E|H) = \mathbf{P}(E)\dfrac{\mathbf{P}(H|E)}{\mathbf{P}(H)}$ (provided $\mathbf{P}(H) \neq 0$);

this last formula is *Bayes's theorem*, whose fundamental rôle will be seen over and over again. Observe, however, that it is merely a different version, or corollary, of the theorem of compound probabilities.

The fact that relationships of this kind are of interest, also shows why it is not convenient (contrary to appearances) to consider systematically the reduced form, $HE|H$ (i.e. $E|H$ with $E\tilde{H} = 0$), which would simply give

$$\mathbf{P}(E|H) = \mathbf{P}(E)/\mathbf{P}(H).$$

4.4.4. Anyway, on the basis of the theorem of compound probabilities, one can deduce (provided $\mathbf{P}(H) \neq 0$) that

(4) $\mathbf{P}(E|H) = \mathbf{P}(HE)/\mathbf{P}(H)$;

this shows that, from a formal standpoint, and assuming coherence, conditional probability is not a new concept, since it can be expressed by means of the concept of probability which we already possess. This observation is, in fact, made use of in the axiomatic treatments; however, using this approach, one obtains the formula, not the meaning. For this reason (and also so as not to leave out the case, albeit a limit-case, where $\mathbf{P}(H) = 0$) we have considered it necessary to start from the *essential* definitions and *prove* the *theorem* of compound probabilities (instead of reducing it to a definition, which could appear arbitrary).

4.5   PROBABILITY AND PREVISION CONDITIONAL ON A GIVEN EVENT H

4.5.1. Let us examine how, for all the events E, and random quantities X, of interest, one passes from probabilities $\mathbf{P}(E)$, and previsions $\mathbf{P}(X)$ (we will call them *actual*, in order to distinguish them), to those *conditional* on a given event H. We already know that $\mathbf{P}(E|H) = \mathbf{P}(HE)/\mathbf{P}(H)$—let us suppose that $\mathbf{P}(H) \neq 0$—and, in general, that $\mathbf{P}(X|H) = \mathbf{P}(HX)/\mathbf{P}(H)$, but it is useful to think about this and give some illustrations, and in the meantime to observe also that $\mathbf{P}(\cdot\,|H)$ is additive, etc.; i.e. it is an admissible $\mathbf{P}$ (an element of $\mathscr{P}$ in the linear ambit we started with). In fact,

$$\mathbf{P}(X + Y|H) = \mathbf{P}(HX + HY)/\mathbf{P}(H) = \mathbf{P}(HX)/\mathbf{P}(H) + \mathbf{P}(HY)/\mathbf{P}(H);$$

in particular, for events A and B, $\mathbf{P}(A + B|H) = \mathbf{P}(A|H) + \mathbf{P}(B|H)$, and in the case of incompatibility the same holds for $\mathbf{P}(A \vee B|H)$; we therefore have

$$\mathbf{P}(\tilde{E}|H) = 1 - \mathbf{P}(E|H), \quad \text{etc.}$$

4.5.2. Decomposing E into $EH + E\tilde{H}$ (incompatible parts, constrained to be in H and in $\tilde{H}$, respectively) one sees immediately that it is the first part which gives rise to the value $\mathbf{P}(E|H) = \mathbf{P}(EH)/\mathbf{P}(H)$ (i.e. it increases in the same ratio as $\mathbf{P}(H)$ to 1, and the same is true for H, which goes from $\mathbf{P}(H)$ to $\mathbf{P}(H|H) = 1$), whereas the contribution of the second part is zero

$$(\mathbf{P}(E\tilde{H}|H) = \mathbf{P}(E\tilde{H}H|H) = \mathbf{P}(0|H) = 0).$$

Interpreting the events as sets, and the probability as mass, one obtains for this case a more effective and instructive image; considering the probability conditional on H implies

making all masses outside the set H ('hypothesis') vanish,

normalizing the remaining masses (i.e. altering them, proportionately, so that the total mass is again 'one').

The same rule holds for $\mathbf{P}(X|H)$, and could also be interpreted within this same framework (but in a less obvious form and, for the time being anyway, unintuitively).

4.5.3. Mentioning this is not only convenient from the point of view of having the rule of calculation easily at hand, but, as we have said, it is conceptually instructive. If these obvious considerations are well understood, confusions which are often irremediable will be avoided. The acquisition of a further piece of information, H—in other words, *experience*, since experience is nothing more than the acquisition of further information— acts always and only in the way we have just described: *suppressing the alternatives that turn out to be no longer possible* (i.e. leading to a more strict limitation of expectations). As a result of this, the probabilities are the $\mathbf{P}(E|H)$ instead of the $\mathbf{P}(E)$, but *not because experience has forced us to modify or correct them, or has taught us to evaluate them in a better way* (even if statements of this kind might perhaps appear tolerable at the level of a crude popularization): the probabilities are the same as before—even if in complicated cases this is less evident and perhaps, at first sight, not even believable —*except for the disappearance of those which dropped out and the consequent normalization of those which remained.*

4.6 LIKELIHOOD

4.6.1. *Bayes's theorem*—in the case of events E, but not random quantities X—permits us to write $\mathbf{P}(\cdot \,|H)$ in the form we met above, a form which is often more expressive and practical:

(5) $$\mathbf{P}(E|H) = \mathbf{P}(E)\mathbf{P}(H|E)/\mathbf{P}(H) = K \cdot \mathbf{P}(E)\mathbf{P}(H|E),$$

where the normalizing factor, $1/\mathbf{P}(H)$, can be simply denoted by K, and, more often than not, can be obtained more or less automatically without calculating $\mathbf{P}(H)$. For this reason, it is often convenient to talk simply in terms of *proportionality* (i.e. by considering $\mathbf{P}(\cdot \,|H)$ only up to an arbitrary, non-zero, multiplicative constant, which can be determined, if necessary, by normalizing).

One could say that $\mathbf{P}(\cdot \,|H)$ is proportional to $\mathbf{P}(\cdot)$ and to $\mathbf{P}(H|\cdot)$, where the dot stands for E, thought of as varying over the set of all the events of interest. More concisely, this is usually expressed by saying that

'final probability' $= K$ 'initial probability' $\times$ 'likelihood', where $= K$ denotes proportionality, and we agree to call:

the *initial* and *final* probabilities those not conditional or conditional on H, respectively (i.e. evaluated before and after having acquired the additional knowledge in question, H), and

the *likelihood* of H given E, the $\mathbf{P}(H|E)$ thought of as a function of E (and possibly multiplied by any factor independent of E, e.g. $1/\mathbf{P}(H)$, the use of which would allow the substitution of '$=$' for '$= K$', or anything resulting from the omission of common factors, more or less cumbersome, or constant, or dependent on H). The term 'likelihood' is to be understood in the sense that a larger or smaller value of $\mathbf{P}(H|E)$ corresponds to the fact that the knowledge of the occurrence of E would make H either more or less probable (our meaning would be better conveyed if we spoke of the 'likelihoodization' of H by E).

4.6.2. This discussion leads to an understanding of how it should be possible to pass from the initial probabilities to the final ones through intermediate stages, under the assumption that we obtain, successively, additional pieces of information $H_1, H_2, \ldots, H_n$ (giving, altogether, $H = H_1 H_2 \ldots H_n$). In fact, one can also verify analytically that

$$\mathbf{P}(E|H_1 H_2) = \mathbf{P}(EH_1 H_2)/\mathbf{P}(H_1 H_2)$$

$$= [\mathbf{P}(E)\mathbf{P}(H_1|E)\mathbf{P}(H_2|EH_1)]/[\mathbf{P}(H_1)\mathbf{P}(H_2|H_1)]$$

$$= K \cdot \mathbf{P}(E) \cdot \mathbf{P}(H_1|E) \cdot \mathbf{P}(H_2|EH_1)$$

$= $ (the probability of E) $\times$ (the likelihood of H_1 given E)
$\times$ (the likelihood of H_2 given EH_1).

In general,

$$\mathbf{P}(E|H) = \mathbf{P}(E|H_1H_2\ldots H_n)$$
$$= K \cdot \mathbf{P}(E) \cdot \mathbf{P}(H_1|E) \cdot \mathbf{P}(H_2|EH_1) \cdot \mathbf{P}(H_3|EH_1H_2)\ldots\mathbf{P}(H_n|EH_1H_2\ldots H_{n-1}).$$

Although the introduction of the term 'likelihood' merely gives a name to a factor in Bayes's formula, which refers to its rôle in the formula (in addition to the existing term, conditional probability, and apart from the indeterminacy we agreed to by defining it up to multiplicative factors), it has the advantage of emphasizing this factor, which will be present in various forms in more and more complicated problems.

4.7 PROBABILITY CONDITIONAL ON A PARTITION $\mathscr{H}$

Let us consider a (finite†) partition $\mathscr{H} = (H_1, H_2, \ldots, H_s)$, and the probabilities, $\mathbf{P}(E|H_j)$, of an arbitrary event E conditional on each of the H_j. Since $EH_1 + EH_2 + \ldots + EH_s = E(H_1 + H_2 + \ldots + H_s) = E \cdot 1 = E$, and $\mathbf{P}(EH_j) = \mathbf{P}(H_j)\mathbf{P}(E|H_j)$, one has

$$(6) \qquad \mathbf{P}(E) = \sum_j \mathbf{P}(H_j)\mathbf{P}(E|H_j):$$

in words, it is the weighted average, with weights $\mathbf{P}(H_j)$, of the probabilities of E conditional on the different H_j. In particular, it lies between them:

$$(7) \qquad \min \mathbf{P}(E|H_j) \leqslant \mathbf{P}(E) \leqslant \max \mathbf{P}(E|H_j)$$

(and it coincides with them if they are all equal). We shall call this property (which is not always valid for infinite partitions) the *conglomerative property* of conditional probability (and prevision).

If we consider as a random quantity, and denote by $\mathbf{P}(E|\mathscr{H})$, the quantity whose value is $\mathbf{P}(E|H_1)$ if H_1 occurs, etc., in other words, in formulae,

$$\mathbf{P}(E|\mathscr{H}) = H_1\mathbf{P}(E|H_1) + H_2\mathbf{P}(E|H_2) + \ldots + H_s\mathbf{P}(E|H_s) = \sum_{H\in\mathscr{H}} H \cdot \mathbf{P}(E|H),$$
(8)

we can write the expression above as

$$(9) \qquad \mathbf{P}(E) = \mathbf{P}[\mathbf{P}(E|\mathscr{H})].$$

More generally, we have, of course,

$$(10) \qquad \mathbf{P}(X) = \mathbf{P}[\mathbf{P}(X|\mathscr{H})].$$

The procedure displayed above, obtaining a prevision by decomposing it into previsions conditional on the alternatives in a partition (which may

---

† This restriction cannot be removed without further conditions (see later: Section 4.19).

often be chosen in such a way as to make the task easier, either through mathematical convenience, or through psychological judgement), is very helpful in many cases. We shall see this in *ad hoc* examples, and even more so in the frequent references we make to it in what follows.

4.8 COMMENTS

The idea of considering $\mathbf{P}(E|\mathcal{H})$ as a random quantity requires some further comment.

4.8.1. As we have said, a random quantity X is a quantity which is well-defined, in an objective sense, although unknown. Does this mean then that, taking $X = \mathbf{P}(E|\mathcal{H})$ with the meaning that $X = \mathbf{P}(E|H_j) = x_j$ if H_j occurs, under such a hypothesis it is objectively true that the value of the above-mentioned probability is x_j? Certainly not; but the possibility of this doubt must be removed. The problem is meaningful only after a particular evaluation of the probabilities $\mathbf{P}(E|H_j)$ has been taken into consideration; whether this is a subjective evaluation of a given individual, or a hypothetical evaluation. Given this, independently of the fact that the x_j have been determined as a result of these actual or hypothetical evaluations, instead of by measuring magnitudes or by choosing them at random, they are objectively determined numbers. That the value of X turns out to be x_j when H_j occurs is true in the sense that x_j is the value that by definition has been associated with H_j. The fact that the association is as an evaluation of $\mathbf{P}(E|H_j)$, made at a certain moment, by a certain individual, may or may not be of interest, but is irrelevant to the definition.

4.8.2. For equation (9) (or (10)) to be true, it is of course necessary that $\mathbf{P}$ always refers to the same individual: the average of the $\mathbf{P}_1(E|H_j)$ of one individual weighted by the $\mathbf{P}_2(H_j)$ of another does not give the $\mathbf{P}(E)$ of either of them; neither $\mathbf{P}_1(E)$ nor $\mathbf{P}_2(E)$.

4.8.3. The idea of considering $\mathbf{P}(E|\mathcal{H})$ as a random quantity often leads to a temptation which one should be warned against: this is the temptation of saying that we are faced with an '*unknown* probability', which is either x_1 or $x_2 \ldots$ or x_s, but we do not know which is the *true* value, x_j, until we know which of the hypotheses H_j is the *true* one. At any moment, the probability is that relative to the information one has; it can refer, for convenience, to different hypothetical pieces of information that can be arbitrarily chosen in an infinite number of ways, thus obtaining an infinite number of different conditional probabilities. None of them, and likewise none of the possible hypotheses, has any special status entitling them to be regarded as more or less 'true'. Any one of them could be 'true' if one had the information corresponding to it; in the same way as the one corresponding to one's present information is true at the moment.

4.8.4. In those cases in which it turns out to be *convenient* to refer to a partition—and these are the only cases in which the temptation meets needs which are essentially meaningful—it is a question, as we have just made clear above, of 'probabilities conditional on unknown objective hypotheses'. As usual, by 'convenient' we are referring to making an evaluation easier by taking one step at a time, and by choosing the easiest steps.

Probability is the result of an evaluation; it has no meaning until the evaluation has been made and, from then on, it is known to the one who has made it.† For this obvious reason alone, the phrase 'unknown probabilities' is already intrinsically improper, but what is worse is that the improper terminology leads to a basic confusion of the issues involved (or reveals it as already existing). This is the confusion that consists in thinking that the evaluation of a probability can only take place in a certain 'ideal state' of information, in some *privileged* state; in thinking that, when our information is different (as it will be, in general), more or less complete, in part more so, in part less so, or different in kind, we should abandon any probabilistic argument (and, perhaps, rely on adhockeries).

4.8.5. On the contrary, there are innumerable possible partitions, which might appear more or less special in character. In order to restrict ourselves to a single example, let us assume that we have to make a drawing from an urn containing 100 balls. We do not know the respective numbers of white and black balls, but, for the sake of simplicity, let us suppose that we attribute equal probabilities to symmetric compositions, and equal probability to each of the 100 balls: the probability of drawing a white ball is therefore $= \frac{1}{2}$. Someone might say, however, that the *true* probability is not $\frac{1}{2}$ but $b/100$, where b denotes the (unknown) number of white balls: the true probability is thus unknown, unless one knows how many white balls there are. Another person might observe, on the other hand, that 1000 drawings have been made from that urn, and, happening to know that a white ball has been drawn B times, one could say that the *true* probability is $B/1000$. A third party might add that both pieces of information are necessary since the second one could lead him to deviate slightly from attributing equal probabilities to all the balls (accepting it, in the absence of any facts, as a frequency, somewhat divergent from the actual composition). A fourth person might say that he would consider the knowledge of the position of each ball in the urn at the time of the drawing as constituting complete information (in order to take into account the habits of the individual doing the drawing; his preference for picking high or low in the urn): alternatively, if there is an automatic device for mixing them up and extracting one, the knowledge of the exact initial positions which would allow him to obtain the result by calculation (emulating Laplace's demon).‡

Only in this case (given the ability) would one arrive, at last, at the true, special partition, which is the one in which the theory of probability is no

---

† *For me, someone else's* evaluation may be unknown, etc.; however, it is for me an objective fact (an evaluation), independently of the subjective reasons which, within him, have led to its determination.

‡ In practice, the various partitions which may present themselves as 'reasonable' are, in fact, much more numerous than in this example, which is already quite 'traditional' in itself.

longer of any use because we have reached a state of certainty. The probability, 'true but unknown', of drawing a white ball is 100% under the hypothesis that the ball to be drawn is white, and 0% under the hypothesis that it is black.

But uncertainty is what it is; information is the information that one actually has (until we can obtain more, and so reduce uncertainty). If one wants to make use of the theory of probability one can only apply it to the actual situation; if one wants to make a plaything of it, little problems can be invented on which it is imagined that one can pin the label 'objective' in a facile fashion; one must not mix up the two things, however: even Don Quixote did not consider venturing forth upon the world astride a rocking-horse.

4.9 STOCHASTIC DEPENDENCE AND INDEPENDENCE; CORRELATION

4.9.1. The probability of E conditional on H, $\mathbf{P}(E|H)$, can be either equal to $\mathbf{P}(E)$, or greater, or less. This means that the knowledge (or the assumption) that H is true either does not change our evaluation of probability for E, or leads us to increase it, or to diminish it, respectively. In the first case, one says that E is *stochastically independent* of H (or *uncorrelated* with H); in the other cases, E is said to be *stochastically dependent* on H; more precisely, either *positively* or *negatively correlated* with H.

We observe straightaway that the property is symmetrical: the theorem of compound probabilities enables us to write down immediately (for $\mathbf{P}(E)$ and $\mathbf{P}(H)$ non-zero)

$$(11) \qquad \frac{\mathbf{P}(E|H)}{\mathbf{P}(E)} = \frac{\mathbf{P}(H|E)}{\mathbf{P}(H)} = \frac{\mathbf{P}(EH)}{\mathbf{P}(E).\mathbf{P}(H)},$$

and hence it turns out that *the ratio by which the probability of E increases or decreases when conditioned on H is the same as that for H conditioned on E, and it is also equal to the ratio between the probability of EH and the product of the probabilities of E and H.* Obviously, in the case of stochastic independence, this product is $\mathbf{P}(EH)$; in fact,

(12) $\mathbf{P}(EH) = \mathbf{P}(H).\mathbf{P}(E|H) = \mathbf{P}(H)\mathbf{P}(E)$ assuming $\mathbf{P}(E|H) = \mathbf{P}(E)$.

Therefore, we may also say, in a symmetric form, that two events *are* stochastically independent (uncorrelated) or are negatively or positively correlated (with each other). It is clear that if E and H are positively correlated the same is true for $\tilde{E}$ and $\tilde{H}$, whereas the reverse is true for E and $\tilde{H}$, and for $\tilde{E}$ and H: if one of the pairs is stochastically independent (uncorrelated) the same is true in all four cases (verify this as an exercise).

Remarks. This symmetry in behaviour between *positive correlation* and *negative correlation* no longer holds, however, when more than two events are considered. Although positive correlations, however strong, are always possible, negative correlations are not possible unless they are very weak (at least on average), the more so the greater the number of events.

The proof will be given (for the general case of random quantities) in 4.17.5: at the present time we do not even have the concepts required to express the statement, except in the informal way given above. At this juncture, it was necessary to point out the conceptually significant aspects of the matter rather than leaving it until the technical exposition to which we referred. In that exposition, Figures 4.3(a) and (b) reveal the reason, in an intuitive fashion, by means of the following analogy: *it is possible to imagine as many vectors as we wish forming arbitrarily small angles, but not forming angles which are all 'rather' obtuse.†*

4.9.2. For more than two events, $E_1, E_2, \ldots, E_n$, say, we could, of course, consider pairwise stochastic independence, $\mathbf{P}(E_i E_j) = \mathbf{P}(E_i)\mathbf{P}(E_j)$, $i \neq j$, but, in fact, they are termed *stochastically independent* only if

(13) $$\mathbf{P}(E_{i_1}E_{i_2} \ldots E_{i_k}) = \mathbf{P}(E_{i_1})\mathbf{P}(E_{i_2}) \ldots \mathbf{P}(E_{i_k})$$

holds for any arbitrary product of the events E_i: this condition is, as we shall see later, more restrictive. This property, if it holds for the E_i, also holds if some of them are replaced by their negations $\tilde{E}_i$, as we have already observed in the case of two events. We therefore have, *for stochastically independent events* E_i, whose probabilities are denoted by p_i, that the probability of a product, such as $\tilde{E}_1 \tilde{E}_2 E_3 E_4 \tilde{E}_5$, is obtained by simply writing p in place of E; thus $\tilde{p}_1 \tilde{p}_2 p_3 p_4 \tilde{p}_5$, i.e. $(1 - p_1)(1 - p_2)p_3 p_4 (1 - p_5)$. More generally, *for any event E, which is logically dependent on the E_i, and expressed arithmetically in terms of them in canonical form* (with $+, .$ and $\sim$), *the probability is expressible in terms of the p_i by the same formula.‡* For example, if

$$E = (E_1 \vee E_2 E_3)(\tilde{E}_4 \vee E_5 \tilde{E}_6),$$

expanding, we obtain

$$E = (E_1 + E_2 E_3 - E_1 E_2 E_3)(\tilde{E}_4 + E_5 \tilde{E}_6 - \tilde{E}_4 E_5 \tilde{E}_6)$$

---

† This sentence is rather vague, but rather than make it complicated it is preferable to ask the reader to accept it for now, simply as a reference to what we shall see in more detail shortly.

‡ The reduction to canonical form is not necessary: it is only required to draw attention to the fact that, when we expand, powers E_i^k, with $k > 1$, do not appear formally; to these would correspond probabilities p_i^k instead of p_i, as must be the case by virtue of the idempotence of the E_i, $E_i^k = E_i$. For example, if $E = (E_1 \vee E_2)(E_1 \vee E_3) = (E_1 + E_2 - E_1 E_2)(E_1 + E_3 - E_1 E_3)$, and we substituted straightaway, we would wrongly obtain $\mathbf{P}(E) = (p_1 + p_2 - p_1 p_2)(p_1 + p_3 - p_1 p_3) = p_1^2 \tilde{p}_2 \tilde{p}_3 + p_1(\tilde{p}_2 p_3 + p_2 p_3) + p_2 p_3$, whereas, in place of the first factor, p_1^2, we should have p_1. As a general rule, one might consider substituting the p_i for the E_i, suppressing the exponents at the end: this procedure could be dangerous, however, since if the p_i were equal, for example, and were replaced straightaway by p, one would make a mistake in the opposite direction.

etc., and finally one could substitute p for E. In fact, since no E appears repeated in both parentheses, we can substitute straightaway (without arriving at a single sum of products) and write

$$\mathbf{P}(E) = (p_1 + p_2 p_3 - p_1 p_2 p_3)(\tilde{p}_4 + p_5 \tilde{p}_6 - \tilde{p}_4 p_5 \tilde{p}_6).$$

4.9.3. A particular, celebrated case, and one which has been extensively studied, is that of *stochastically independent and equally probable events*, $p_i = p$; this is the *Bernoulli scheme*, also referred to as that of 'repeated trials'. For every E, logically dependent on n such events, the probability $\mathbf{P}(E)$ turns out to be expressed by a polynomial in p (of degree at most n); for example, the E considered above (depending on the 6 events $E_1 \ldots E_6$) would have the probability

$$\mathbf{P}(E) = (p + p^2 - p^3)(\tilde{p} + p\tilde{p} - p\tilde{p}^2) = p(1 + p - p^2)(1 - p + p^2 - p^3)$$

$$= p - p^3 + p^4 - 2p^5 + p^6.$$

Less obvious algebraically, but more meaningful, would be the analogous expression as a homogeneous polynomial of degree n in the two variables p and $\tilde{p} = (1 - p)$; it is obtained, in an obvious fashion, by multiplying each term by a suitable power of $(p + \tilde{p}) = 1$. In the previous example, operating in the two factors right from the beginning, one has, for example,†

$$\mathbf{P}(E) = p[(p + \tilde{p})^2 + p(p + \tilde{p}) - p^2]\tilde{p}[(p + \tilde{p})^2 + p(p + \tilde{p}) - p\tilde{p}]$$

$$= p\tilde{p}^5 + 5p^2\tilde{p}^4 + 9p^3\tilde{p}^3 + 8p^4\tilde{p}^2 + 2p^5\tilde{p}.$$

The significance of this lies in the following: the coefficients denote the number of constituents of E corresponding to the different frequencies of the E_i. In precise terms, the coefficient of $p^h \tilde{p}^{n-h}$ is the number of constituents in which h of the E_i occur, and $n - h$ do not: in other words, with h factors of the form E_i and $n - h$ of the form $\tilde{E}_i$. In the example given, one sees that there is one constituent with a single occurrence (i.e. $(1 \vee 0 . 0)(\tilde{0} \vee 0 . \tilde{0})$), 5 with two, 9 with three, 8 with four and 2 with five (this is easily verified because the two factors each have 5 favourable constituents, of which those containing 0, 1, 2, 3 occurrences number, respectively, 0, 1, 3, 1 and 1, 2, 2, 0).

4.9.4. An even more special case is that in which $p = \frac{1}{2}$. This is usually referred to as the case of *Heads and Tails* (although we could also think in terms of any other interpretation and application, and although the case of Heads and Tails is an exceptional one, where some 'objective circumstance' forces us to adopt this evaluation of probability). In this case, each constituent

---

† By introducing the *ratio*, $r = p/\tilde{p}$ (cf. Chapter 5), we have $p^h \tilde{p}^{n-h} = \tilde{p}^n r^h$, and therefore the polynomial in p and $\tilde{p}$ can be written as $\tilde{p}^n \times$ a polynomial in r: in the example given, we would have $\mathbf{P}(E) = \tilde{p}^6(r + 5r^2 + 9r^3 + 8r^4 + 2r^5)$.

has probability $p^h \tilde{p}^{n-h} = (\frac{1}{2})^n$, and $\mathbf{P}(E) = (\frac{1}{2})^n \times$ the sum of the coefficients of the polynomial in p and $\tilde{p}$ (or in r), which is, in other words, the ratio between the number of constituents (or cases) which are favourable to E, and the total number (2^n) of constituents.

4.10 STOCHASTIC INDEPENDENCE AMONG (FINITE) PARTITIONS

4.10.1. There is an obvious and immediate extension of the notion of stochastic independence from the case of events to that of (finite) partitions; in other words, if one wants to use such terminology, to multi-events, like $E' = (E_1', E_2', \ldots, E_{m'}')$ and $E'' = (E_1'', E_2'', \ldots, E_{m''}'')$, and, in particular, to random quantities with a finite number of possible values. It will simply imply that every event of a partition is stochastically independent of every event of the other one: $\mathbf{P}(E_h' E_k'') = \mathbf{P}(E_h')\mathbf{P}(E_k'')(h = 1, 2, \ldots, m'; k = 1, 2, \ldots, m'')$, and, in particular, for random quantities X and Y it will mean that

$$(14) \quad \mathbf{P}[(X, Y) = (x_h, y_k)] = \mathbf{P}[(X = x_h) . (Y = y_k)] = \mathbf{P}(X = x_h) . (Y = y_k).$$

And so on for three or more partitions or random quantities (referring always to the finite case).

4.10.2. Let us now prove that pairwise stochastic independence is, as we said, a necessary but not sufficient condition for the stochastic independence of n events (and, *a fortiori*, of n partitions): two examples will suffice.

Let A, B, C, D be the events of a partition, to each of which we attribute probability $\frac{1}{4}$. The events $E_1 = D + A, E_2 = D + B, E_3 = D + C$ are pairwise independent ($E_i E_j = D, \mathbf{P}(E_i E_j) = \frac{1}{4} = \mathbf{P}(E_i)\mathbf{P}(E_j) = \frac{1}{2}.\frac{1}{2}$), but are not so when taken three at a time, since $E_1 E_2 E_3 = D$, and the probability of the product of all three of them is still $\frac{1}{4}$ instead of $\frac{1}{8}$.

Similarly, considering $A + B, B + C, C + A$, the products two at a time would have probability $\frac{1}{4}$, but the product of all three is impossible and therefore has probability zero and not $\frac{1}{8}$.

More generally, one can have stochastic independence up to a given order, 'm by m' say, but not beyond this, as the following example (a generalization of the previous ones) shows. Let $E_1, E_2, \ldots, E_m$ be stochastically independent events each of probability $\frac{1}{2}$ (i.e. every 'constituent' has probability $(\frac{1}{2})^m$), and let E be the event which consists in the fact that among the E_i there are an odd number of false ones: $E = (\tilde{E}_1 + \tilde{E}_2 + \ldots + \tilde{E}_m = \text{odd})$. It is clear that E is logically dependent on the E_i (by definition, and, on the other hand, $EE_1 \ldots E_m = 0$ with certainty, since either some of the E_i are 0, or all of the $\tilde{E}_i$ and their sum is 0, hence not odd, so that $E = 0$), but is stochastically independent of $m - 1$ of them (conditionally on any results of these, E coincides either with the omitted event or with its negation).

4.10.3. Suppose we have two partitions, into m' events $E'_1 \ldots E'_{m'}$ and into m'' events $E''_1 \ldots E''_{m''}$, respectively. To say that in each of them the probabilities of the different events are equal (to $p' = 1/m'$ and $p'' = 1/m''$, respectively), and that they are stochastically independent, implies that the $m = m'm''$ events $E'_h E''_k$ of the product-partition all have the same probability, $p = p'p'' = 1/(m'm'') = 1/m$; conversely, this property implies the two previous ones. The same obviously holds for three or more partitions. We shall come back to this fact, which is the basis for many applications of the combinatorial type.

4.10.4. If we have different partitions, or multi-events, which are stochastically independent and have equally distributed probability (e.g. successive drawings with replacement from an urn, with fixed probabilities of drawings for balls of m different colours, $p_1 + p_2 + \ldots + p_m = 1$), we have an extension of the Bernoulli scheme given above; 'repeated trials' for multi-events. It is clear how the considerations made in the previous case could be generalized: for every event E which is logically dependent on n m-events, the probability $\mathbf{P}(E)$ can be expressed as a polynomial $\sum c_{h_1 h_2 \ldots h_m} p_1^{h_1} p_2^{h_2} \ldots p_m^{h_m}$ (the sum being over all m-tuples of non-negative integers with sum $= n$). The coefficients give the number of favourable constituents containing the ith result h_i times ($i = 1, 2, \ldots, m$). In the case of equal probabilities ($p_1 = p_2 = \ldots = p_m = 1/m$), a generalization of Heads and Tails ($m = 2$), the probabilities are

(15) $\mathbf{P}(E) = (1/m^n) \times$ the sum of the coefficients of the polynomial
       $=$ the ratio of the number of constituents (or cases) favourable to E and the total number (m^n) of all constituents (possible cases).

4.11 ON THE MEANING OF STOCHASTIC INDEPENDENCE

4.11.1. It is absolutely essential to continue to underline the fact that the notion of stochastic independence does not belong to the domain of the logic of certainty, but to that of prevision, and that therefore—like probability and prevision—it has a *subjective* meaning. After presenting the necessary details in an abstract setting, we shall need to dwell upon the various considerations required to illustrate them in practice. This is of paramount importance if one takes into account that people usually seem to think—or, at least, allow it to be thought, since objections are rarely put forward—that the meaning of stochastic independence is self-evident and objective, and that this property always holds, except for special cases of interdependence. So much so, that in applications to many practical problems† one often comes across notions and formulae which are valid if the hypothesis of stochastic independence is adopted, but where this hypothesis does not turn out to be justified and is not, in fact, introduced explicitly, but only tacitly, and perhaps inadvertently. The habit of simply saying 'independence', as if it were a unique notion, plays a part in obscuring the special nature of the notion of stochastic independence. For the sake of brevity, we shall also adopt this habit when there is no ambiguity, or when it is not required to underline the sense: we shall only do it, however, after having given warning of this, and of the existence of other notions which are, in a certain sense, similar. We have already met those of *linear* and *logical* independence (whose meaning resides within the logic of certainty), and the notion of things being *uncorrelated* (which, in the case of events, is synonymous with pairwise stochastic independence, but which, in the case of random quantities, will turn out to be different, as we shall shortly see).

4.11.2. The definition of stochastic independence depends on the evaluation of probability; i.e. on the choice of a particular $\mathbf{P}$. If A and B are two *logically independent* events, an individual can evaluate $\mathbf{P}(A)$, $\mathbf{P}(B)$ and $\mathbf{P}(AB)$ in any way whatsoever, provided that (cf. Chapter 3, 3.9.4) $\mathbf{P}(AB)$ turns out to be not less than $\mathbf{P}(A) + \mathbf{P}(B) - 1$, and not greater than either of $\mathbf{P}(A)$ and $\mathbf{P}(B)$ (which, in any case, are all numbers between 0 and 1). The ratio $\mathbf{P}(AB)/\mathbf{P}(A)\mathbf{P}(B)$ can therefore assume all non-negative values, depending on the appraisal of the person making the evaluation.‡

Even if, for the sake of brevity, we shall occasionally say that two events (or partitions, etc.) *are* stochastically independent, it must be remembered that this is 'with respect to a given $\mathbf{P}$'; in other words, 'according to the opinion of the person who has chosen the evaluation $\mathbf{P}$' is to be understood. In particular, in the case of *logically independent* events or partitions, however the probabilities are evaluated, the evaluation extended on the basis of the hypothesis of independence is coherent. If, on the other hand, *we do not*

---

† As H. Bühlmann observes (in a report at the ASTIN Congress in Trieste, 1963), the condition of independence is often understood and assumed to be valid when it is not valid at all. He refers to the field of insurance and actuarial mathematics (but what he says is unfortunately true in many other fields). Sometimes, rather than tacitly stating, or considering as obvious, the condition of independence, one considers that 'not knowing much about the interdependence' provides a justification for it. This is tantamount to saying that if we do not know much about the behaviour of a function we can argue as if we knew that it were a constant.

‡ After having evaluated $\mathbf{P}(A) = a$ and $\mathbf{P}(B) = b$, the ratio $\mathbf{P}(AB)/\mathbf{P}(A)\mathbf{P}(B)$ can still assume all non-negative values if $a + b \leqslant 1$, and all values not less than $1 - (\tilde{a}\tilde{b}/ab)$ otherwise. In any case, the three cases of positive, zero and even negative correlation (since this minimum is always less than 1) remain possible.

have logical independence, i.e. some product is impossible, e.g. $E = E_i' E_j'' E_h'''$ (three elements of three partitions), we necessarily have $P(E) = 0$: we can have the relation $P(E) = P(E_i')P(E_j'')P(E_h''')$ if at least one of the factors is zero, the relations $P(E|E_i'E_j'') = P(E_h'''|E_i'E_j'') = P(E_h''')$ (and similar ones) only if all the factors are zero. In other words, the given arithmetic conditions of stochastic independence *cannot hold*, except in the limit cases mentioned above, which do not fall within the definition given in the form of a product, and the more extreme cases, which do not even fall within the definition given in terms of conditional probability. Rather than accept this anomaly, it is preferable to eliminate it by including logical independence as a prerequisite for the definition of stochastic independence. The justification of this is that it is equivalent to taking into account the difference between possible events to which zero probability is attributed, and impossible events. This is the same distinction as that between empty sets and non-empty sets of measure zero; a much more fundamental distinction than that between non-empty sets with zero or non-zero measure.

Given these considerations about limit-cases, we can now say (in the case of finite partitions) that *stochastic independence* presupposes *logical independence* (but certainly not vice-versa). As far as *linear* dependence is concerned, we recall that it is a particular form of logical dependence, and therefore it excludes stochastic independence.

In order to complete this hierarchy of notions, let us say at this point that absence of correlation will be a subjective notion weaker than stochastic independence (but when applied under more and more restrictive conditions it may lead to it).

4.12 STOCHASTIC DEPENDENCE IN THE DIRECT SENSE

Let us now illustrate some of the kinds of factors which may often influence our judgments of whether events are stochastically independent or dependent. It is necessary to learn how to think carefully about the presence of these factors in order to avoid assuming too readily the hypothesis of stochastic independence, a practice we have already criticized. In putting forward these few cases, we are not attempting an exhaustive treatment, and the mention of these cases is not meant to correspond to a classification having any theoretical value (indeed, the distinctions which we shall make, with the sole aim of drawing together a few examples, might become empty, nebulous abstractions if taken too seriously).

Anyway, without any intention of becoming theoretical, let us call, informally, stochastic dependence *in the direct sense*, the case that arises in the most evident form, and in the most obvious and common examples in treatments from all conceptual viewpoints. This is the case in which the occurrence of an event changes the circumstances surrounding the occurrence of another one (in a way considered relevant to the evaluation of the probability). Standard examples are : drawings from an urn without replacement (where the drawing of a white ball decreases the percentage of white balls for the next drawing); contagious diseases (where a diseased individual increases the probability that people close to him catch the illness); the breakdown of machines, etc. (where the difficulties caused by a breakdown of one of them precipitates the breakdown of others); the outcomes of successive trials in a competition (where, due to the initial results, the objective conditions for the succeeding trials change; e.g. the height of the bar in a high-jump competition), and so on.

Examples of this kind draw attention to dependence 'in one direction'—chronologically (dependence of what happens afterwards on what has happened before). This corresponds to the interpretation—often, in fact, referred to when considering cases of this kind—based on the idea of '*cause*'. That this is irrelevant is seen by observing that the relationship of dependence or independence is symmetric. Anyway, we take this opportunity of remarking that, for 'conditional' bets too, it is of no importance whether the 'fact' refers to the future or the past, and, in particular, whether, chronologically, it follows or precedes the other 'fact' assumed as the hypothesis for the validity of the bet. One could very well bet on the occurrence of a certain event today, stipulating that the bet will be effective only if some other event takes place in a month's time.

Our desire to discuss this case of 'direct' dependence was not so much because it needed attention drawing to it, but, on the contrary, to make the reader subsequently aware of the incompleteness of discussions which mention only this form of dependence, and lead one to believe that, apart from such cases, there is no reason to depart from the formulation in terms of stochastic independence. We therefore proceed now to consider certain other examples.

4.13 STOCHASTIC DEPENDENCE IN THE INDIRECT SENSE

By this we mean, in an informal way, as above, those cases in which the occurrence of an event has no influence on the occurrence of another one, but in which there are some circumstances that can influence both events. In other words—if one wishes to speak in terms of 'causes'—there is a 'cause' common to these events, but there is no direct 'causal' relationship between them. For example, in considering (the possibility of) two ships both being wrecked in the same area, on the same day (even without assuming collisions or any direct interference of this kind), one might rightly imagine a positive correlation, since both probabilities are influenced in the same way by common circumstances (like the state of the sea; calm or stormy). The same holds true for the deaths of two individuals during next winter, since, if it is very cold, the probability of death will increase for both of them. In the same way, if we ask whether two participants in a competition will achieve better results than some other participant, the result obtained by the latter will influence the two events in the same way, even if one judges the three results to be stochastically independent. This latter example can also be given an interpretation in terms of a game of chance in which A and B 'win' if they obtain a greater score than the 'bank' does. Interpreting the score as that obtained by throwing a die, then, in terms of the 'score' obtained by the 'bank', the probabilities of wins for A or B, or both, are given by

the 'bank's' score (H):	1	2	3	4	5	6
$P(A\|H) = P(B\|H) =$	5/6	4/6	3/6	2/6	1/6	0
$P(AB\|H) =$	25/36	16/36	9/36	4/36	1/36	0

and averaging (assuming that each of the six cases has probability $= 1/6$)

$$P(A) = P(B) = 15/36 = 5/12 = 41 \cdot 67\%$$

$$P(A)P(B) = 25/144 = 75/432 = 17 \cdot 36\%$$

$$P(AB) = 55/216 = 110/432 = 25 \cdot 45\% > P(A)P(B).$$

This example shows that conditional on each of the possible hypotheses for the 'bank's' score, $H = ($'points' $= h)$ with $h = 1, 2, \ldots, 6$, the two events are stochastically independent, but that this independence conditional on each event of a partition *does not imply stochastic independence*. We will return shortly to an explicit consideration of this notion and this result, to which the case of indirect dependence essentially reduces.

There is one case, however, which derives even less from 'objective' circumstances.

4.14 STOCHASTIC DEPENDENCE THROUGH AN INCREASE IN INFORMATION

If it is true (as it is, in fact) and if one can justify (as we have, for the moment, simply assumed) that the probability of an event is often evaluated on the basis of observed frequencies of more or less similar events, then this fact implies a stochastic dependence. In fact, observed events provide a certain amount of experience capable of modifying, as time goes on, the evaluations of probabilities based on frequencies. Indeed, it is precisely the analysis based on these present considerations that will lead later (Chapter 11) to an explanation of why and under what conditions such a criterion of evaluation turns out to be justified.

The situation to which we refer is obviously relevant in the case of 'new' phenomena; i.e. those about which there is little past experience: think, for instance, of the success or failure of the first space launches; of the first trials employing a new drug, or something of that kind; of the probability of death in a species of animal never before observed; of the risks attached to nuclear experimentation, and so on. Putting on one side the hypothesis of 'new', the situation does not change in essence, but does change quantitatively since a few, or even many, trials cannot produce any substantial alteration of a frequency arrived at after a great many previous trials. This is so unless one is led to behave as if faced with a 'new' phenomenon: thinking, for instance, that because of a change in circumstances (or for whatever other reason) the future frequency of an 'old' phenomenon (like mortality, fire, hail, or anything else) will closely resemble the frequency suggested by a small number of recent experiences, rather than the frequency observed in a large number of less recent experiences.†

In a certain sense, the situation is the same as that of drawings with replacement from an urn of unknown composition: the probabilities of white balls at successive drawings turn out to be interdependent because the results, as they are obtained, make one's ideas about the composition of the urn more precise (and the smaller the past experience, the greater the influence it has on our ideas). This case could really have been included among the previous examples of indirect dependence (dependence on the unknown composition of the urn); the only difference—an irrelevant one—is the fact that here the composition is an unknown but pre-existent datum, whereas in the other examples we were dealing with the influence of future events, uncertain at the moment when the question was posed. Instead, in the given examples of 'new phenomena' our disposition to review the evaluation was not attributed to ignorance of circumstances, or of specific, objectively determined magnitudes, but, in a general way, to a lack of familiarity with the phenomenon. There may be those who would like to say that such an 'objective magnitude' is the 'constant, but unknown, probability'. We have explained many times, however, that it is not admissible to speak in this way, and we shall also see that it is unnecessary, because, by arguing in a sensible way about meaningful notions, one comes to the same conclusions as would be obtained by meaningless arguments, introducing meaningless notions. Anyway, this means that none of the cases present any essential differences, neither conceptually nor mathematically, notwithstanding the external differences which required us to look at them separately in order to avoid an over-restricted view.

---

† This is the problem studied by American actuaries under the heading of 'Credibility Theory'; cf. the two lectures by A. L. Mayerson and B. de Finetti containing information and discussion about this topic: *Giorn. Ist. Ital. Attuari* (1964).

The temptation to proceed further with these considerations, which could not be completed here, is best resisted: we recall that their purpose was simply to persuade the reader that, *in a certain sense, it is stochastic independence which constitutes a rather idealized limit-case*, and that dependence is the norm, rather than the contrary (whose acceptance is the bad habit referred to by Bühlmann; cf. 4.11.1, footnote).

4.15   CONDITIONAL STOCHASTIC INDEPENDENCE

4.15.1. In the previous examples, we have encountered the notion of conditional stochastic independence (conditional on an event, on a partition); it is necessary to add something more systematic in this connection.

We shall say that $E_1 \ldots E_n$ are stochastically independent with respect to H (or with respect to each H_j of a partition) if they are such with respect to the function (or in general the functions) $\mathbf{P}$ of the type $\mathbf{P}(\cdot) = \mathbf{P}(\cdot|H)$ (i.e. $\mathbf{P}(E_1 E_2|H) = \mathbf{P}(E_1|H) \cdot \mathbf{P}(E_2|H)$, etc.).

In the example (of beating the 'bank' when throwing dice), we found that A and B, stochastically independent with respect to a partition, turned out to be positively correlated; $\mathbf{P}(AB) > \mathbf{P}(A)\mathbf{P}(B)$. We now want to examine the question in general, beginning with a very simple example (less restrictive than the previous one, in the sense that the probabilities of the two events are not assumed to be equal). Let us consider just two hypotheses, H and $\tilde{H}$, with probabilities c and $\tilde{c}$; let the events A and B have probabilities a' and b' conditional on H, and a'' and b'' conditional on $\tilde{H}$. The probability of AB will be

$$(16) \qquad \mathbf{P}(AB) = c \cdot \mathbf{P}(AB|H) + \tilde{c} \cdot \mathbf{P}(AB|\tilde{H}) = ca'b' + \tilde{c}a''b'',$$

whereas, in order that A and B be independent, it should have been

$$\mathbf{P}(AB) = \mathbf{P}(A) \cdot \mathbf{P}(B) = (ca' + \tilde{c}a'')(cb' + \tilde{c}b'')$$

$$= c^2 a'b' + c\tilde{c}(a'b'' + a''b') + \tilde{c}^2 a''b'';$$

the difference is

$$(17) \qquad \mathbf{P}(AB) - \mathbf{P}(A)\mathbf{P}(B) = (c - c^2)a'b' - c\tilde{c}(a'b'' + a''b') + (\tilde{c} - \tilde{c}^2)a''b''$$

$$= c\tilde{c}(a'b' + a''b'' - a'b'' - a''b') = c\tilde{c}(a' - a'')(b' - b'').$$

One therefore has stochastic independence only in the trivial cases: $c = 0$ or 1, or $a' = a''$, or $b' = b''$; in other words, if the two hypotheses do not have zero probability, only if A (or B) is stochastically independent of them:

$$\mathbf{P}(A) = \mathbf{P}(A|H) = \mathbf{P}(A|\tilde{H}).$$

If this does not happen, one has positive or negative correlation according as the probabilities of A and B vary in the same or the opposite sense when conditional on H rather than $\tilde{H}$. This is what we would have expected.

4.15.2. The same problem, with a partition into s hypotheses $H_1 \ldots H_s$ instead of two, with probabilities $c_1 \ldots c_s$, and with

$$\mathbf{P}(A|H_j) = a_j, \qquad \mathbf{P}(B|H_j) = b_j,$$

gives:

(18)
$$\mathbf{P}(A) = a = \sum c_j a_j, \qquad \mathbf{P}(B) = b = \sum c_j b_j, \qquad \sum c_j = 1,$$
$$\mathbf{P}(AB) = \sum c_j a_j b_j = \sum c_j[a + (a_j - a)][b + (b_j - b)]$$
$$= ab + \sum c_j(a_j - a)(b_j - b),$$
$$\mathbf{P}(AB) - \mathbf{P}(A)\mathbf{P}(B) = \sum c_j(a_j - a)(b_j - b).$$

One can easily see directly from this expression that if when the a_j increase the b_j increase as well, the difference is positive; i.e. A and B turn out to be positively correlated (negatively if the change is in the opposite direction): this generalizes the previous conclusion. In particular, if (conditional on each H_j) A and B have equal probabilities, $a_j = b_j$, they are positively correlated (so that the conclusion of the example concerning the die and the bank was necessary, not just incidental). More generally, once we have defined correlation between random quantities, we shall see that the expression obtained above will correspond to the following statement: A and B are positively or negatively correlated, or uncorrelated, according to the sense in which the random quantities $X = \mathbf{P}(A|\mathscr{H})$ and $Y = \mathbf{P}(B|\mathscr{H})$ are correlated; in other words, according as $\mathbf{P}(XY) \gtreqless \mathbf{P}(X)\mathbf{P}(Y)$.

4.15.3. The case of conditional stochastic independence gives rise to a particularly interesting case of inductive argument; i.e. of determining the probabilities of the different possible hypotheses conditional on the information regarding the outcomes of any events which are judged to be *stochastically independent of each other, conditionally on each of the above mentioned 'hypotheses'*.

This is—to refer to the standard example of the classical variety—the case of drawings with replacement from an urn of unknown composition: the hypotheses are the different compositions of the urn (e.g. percentages of white and black balls), the events are the drawing of a white ball on given trials. On the other hand, in order to demonstrate the importance of this in less academic examples, this is often the form of argument used to evaluate the probability of the two hypotheses of the guilt or innocence of an accused man on the basis of the ascertainment of a certain number of facts having the status of 'circumstantial evidence', or 'proof'. If the latter facts differ as much as possible they can therefore be taken as stochastically independent of each other, conditional on both hypotheses, and with different probabilities conditionally on the two hypotheses.

It goes without saying that jurors and magistrates would reject with horror the idea of a verdict as an evaluation of probability: in order to have their feet on solid ground, they feel obliged to present as the 'truth', or as a 'certainty', some version which, through the procedures provided, has qualified as the official and compulsory version (and which, therefore, cannot be open to correction, even if an individual who was officially murdered many years ago shows up looking very much alive†). It is sad, to say the least, to see such an unconscientious preference for a 'certainty', which is almost always fictitious, rather than a responsible and accurate evaluation of probability. Perhaps the saddest thing, however, is the thought that the world will probably remain for quite some time at the mercy of a mentality so distorted and arrogant that it neither retracts nor hesitates even when faced with the most grotesque absurdities.‡

One more example: Heads and Tails using a coin that we think may be 'imperfect' (i.e. it may 'favour' one side more than the other). As different 'hypotheses' in this case, one often considers the 'hypothesis of an imperfection giving rise to a probability p of heads', a different 'hypothesis' for each value of p, or for a certain number of values p_h; e.g., in order to simplify matters, increments of 1 %. This formulation is not very satisfactory, because the definition of a hypothesis on the basis of an evaluation of probability is a nonsense; however, before seeing (in Chapter 11) the way in which an equivalent, and correct, formulation can be given, based on the notion of 'exchangeable events', without speaking of such 'hypotheses', one can accept this image, for the time being, as a 'temporary formulation'. This is acceptable on account of the above observation that it is equivalent in its actual conclusions to the correct formulation, even if it is, strictly speaking, meaningless.

4.15.4. Formally, the particular case we are referring to reduces to the obvious simplification introduced in the expression for $\mathbf{P}(E|H)$ (given in

---

† As happened recently in Sicily.

‡ Some even assert that in the absence of proofs sufficient for conviction the accused should always be discharged 'for not having committed the crime'. On the other hand, it can well happen that it is certain that one of two suspects is guilty, e.g. one or other, or both, of a married couple (like in the 'Bebawi case', Rome 1966). Judicial wisdom, which ignores common sense, and, therefore, probability, would then have to assert, in effect, that all the inhabitants of the world are under suspicion apart from two people, one of whom is the murderer, who are officially free and protected from any possibility of suspicion.

Translators' note. The Bebawis were a married couple appearing in a murder trial, who were each accusing the other of the murder. They were both acquitted on the grounds that the cases against them were insufficiently proved.

Section 4.6.2), if the items of information H_i, which make up H, are stochastically independent of each other conditional on the events E. Then, in fact, $P(H_2|EH_1)$ reduces to $P(H_2|E)$, $P(H_3|EH_1H_2)$ reduces to $P(H_3|E)$, etc., and, finally, the likelihood for the information $H_1H_2\ldots H_n$ (the product of the H_i) is nothing other than the product of the likelihoods for the single H_i, so that:

(19) $P(E|H) = P(E|H_1H_2\ldots H_n) = KP(E)P(H_1|E)P(H_2|E)\ldots P(H_n|E).$

In a form which is sometimes more expressive, given two events E (E_h and E_k, say) we can write

(19′)
$$\frac{P(E_h|H)}{P(E_k|H)} = \frac{P(E_h)}{P(E_k)} \cdot \frac{P(H_1|E_h)}{P(H_1|E_k)} \cdot \frac{P(H_2|E_h)}{P(H_2|E_k)} \cdots \frac{P(H_n|E_h)}{P(H_n|E_k)}.$$

In other words: the ratio of the final probabilities (of any two events E) is given by the ratio between their initial probabilities times the ratios of the likelihoods for each item of information H_j. One should note the particular case in which, in place of E_k, we substitute the negation $\tilde{E}_h$ of E_h: put more succinctly, $E_h = E$ and $E_k = \tilde{E} = 1 - E$, and then one obtains a relationship between the initial and final ratios $P(E)/P(\tilde{E})$, and the ratios $P(H_j|E)/P(H_j|\tilde{E})$, which we might call *ratios of probability* and *ratios of likelihood*, respectively: we shall talk about this explicitly in Chapter 5, 5.2.4–5.2.5.

This result expresses—at least in the Bayesian version†—the 'Likelihood Principle':

'*For the purpose of inferences concerning the events E, the information obtained from the occurrence of the H_j can be arrived at from the knowledge of the likelihoods $P(E_h|H_j)$ (or of their ratios).*'

It is, however, necessary (in order to avoid possible misunderstandings) to underline that this is true *only if* the conditions specified above hold; we will discuss this in greater detail in Chapter 11.

In the meantime, let us point out a qualitative and expressive formulation of one particular conclusion which corresponds to many practical situations:

'*Suppose a thesis (e.g. the guilt of an accused man) is supported by a great deal of circumstantial evidence of different forms, but in agreement with each other; then even if each piece of evidence is in itself insufficient to produce any strong belief, the thesis is decisively strengthened by their joint effect.*'

This statement is known as 'Cardinal Newman's principle', since it was he (taking it over from previous authors) who made it famous as the basis of his mode of argument in his work the 'Grammar of Assent'.

---

† The reservation expressed by this parenthetical clause is due to the fact that some people believe that the sense in which this 'principle' is understood by non-Bayesian authors, and in particular by Allan Birnbaum who has written about it and supported it, is different. Thus far, I have been unable to discover what these supposed essential differences are (apart from the interpretation; subjectivistic or non-subjectivistic).

4.15.5. *Remarks*. In the case of *independence* also we find ambiguity, as already illustrated in Section 4.8. There, it was a question of considering as the 'true' probability not that relative to the actual state of information, but a different one, unknown, conditional on some idealized form of unacquired information. Here, it is a question of calling 'independent' those events which are such conditional on a certain 'ideal' partition. Again, a typical example is that of drawings from an urn of unknown composition, which are independent conditional on the knowledge of the composition (or on any assumption about it), but are not independent for someone ignorant of the composition.† Precisely because of the interdependence induced by this ignorance, the successive information about the outcomes of the drawings serves to modify the evaluations of probability (in the sense of Section 14). In the case of independence, all such information would, by definition, have no effect.‡

4.15.6. The previous example takes on an even more 'paradoxical' air (for those who cannot distinguish dependence and conditional independence, or, at any rate, do not always remember that everything is relative to a given state of information) if the drawings are made without replacement.

> This is the case of a 'lucky-dip': N tickets are for sale (and before being sold their markings are unknown), n of them are winning tickets (and one checks this by examining each ticket one has bought) which give one the right to a prize: we suppose, to avoid complications, that the prizes are identical. Conditional on the knowledge of the number of prizes, n, for a given number of tickets sold one's probability of buying a winning ticket is *less*, the more prizes that have been won. If, initially, one were very uncertain about the percentage of winning tickets (i.e. distributed the probability to be attributed to the various hypotheses over a wide range, e.g., as a limit-case, gave equal probabilities to all the hypotheses $n = 0, 1, 2, \ldots, N$), the more frequent the occurrence of winning tickets, the more one's probability increases for the tickets yet to be sold. Under the intermediate assumption, which consists in knowing that the number n has been determined by casting a die N times and taking $n =$ the number of times a '6' occurs, the probability would remain constant ($= \frac{1}{6}$) independently of any information concerning tickets sold and prizes won. (This is obvious; it is the same thing as actually playing dice: in any case, it would be a useful exercise to check the conclusion without using this direct argument.)

Examples of this kind (dice, urns, roulette, . . .) are convenient because they are reduced to standard schemes. Precisely for this reason, however, they

---

† An even better way of putting it is to say that they are 'exchangeable': we will talk about this in Chapter 11.

‡ Lindley (in the 2nd volume of *Probability and Statistics*), in order not to diverge too much from existing terminology, chose to continue to talk of *independence* (without, in cases of this kind, adding '*conditional*'). He told me that a student once objected: '*How, then, can an experience be informative?*'. This means (I observed) that your teaching is so good that it leads people to a correct understanding despite the incorrect terminology. However, it is better to use the correct terminology in order that nobody becomes confused, or has to make a strenuous mental effort in order not to be confused.

have little use or significance, and, hence, it is desirable to give a more concrete and practical interpretation of the same example.

From a box containing 1000 specimens of a certain gadget, about 100 were drawn and used: 15 of them did not work properly (whereas, according to the standard specification, this should have been around 5). Should one use the others or throw them away (assuming, for example, that if more than 10% were defective their use would cause more damage than the cost of throwing them away)? We shall limit ourselves to the conceptual aspects: the exact calculations, with precisely specified hypotheses, could be made now, but we shall reserve this until Chapters 11 and 12.

The data given says nothing except in relation to what we know, or imagine, regarding systems of production and packing. If, for packing them into boxes, the gadgets are chosen at random, there is no reason to be less (or more) confident about the remaining articles: the fact of them being together with other articles which are defective in a greater or lesser percentage is purely fortuitous. If, on the other hand, one believes that the contents of a box come from the production of a given machine at a given time, the conclusion may be different, in either sense. If one thinks that the defects are due to a machine being temporarily out of adjustment, then the usual attitude of fearing that the high percentage of defectives might also be found in the rest of the box is reasonable. If, instead, one thinks that there is a periodic cause (in an extreme case, that the 7th article in every series of 20 turns out to be defective), it is almost certain that each box contains almost exactly 50 defective pieces (at any rate, with less imprecision than under the first hypothesis). The conclusion is then the opposite one: having already removed 15 defective articles, instead of 5, it is to be expected that 35 remain, rather than 45 (and the bad initial outcomes improve the prospects for the remainder, rather than making them worse).

4.16 NON-CORRELATION; CORRELATION (POSITIVE OR NEGATIVE)

4.16.1. The condition $P(AB) = P(A)P(B)$ for events was referred to as both the condition for stochastic independence and the condition for non-correlation; in the case of two random quantities, X and Y, the same condition $P(XY) = P(X)P(Y)$ will still be called the condition for non-correlation (or of positive or negative correlation if either $>$ or $<$ is substituted for $=$), whereas by stochastic independence one implies a more restrictive condition, which, for the time being has only been introduced for the case of random quantities with a finite number of possible values.

One can show straightaway that the above-mentioned condition is more restrictive; in other words, that stochastic independence implies non-correlation (but not conversely, *except in the case of two random quantities with only two possible values, and hence, in particular, for events*). Let x_i ($i = 1, 2, \ldots, m'$) denote the possible values for X, and $p_i' = P(X = x_i)$ their probabilities; similarly, let y_j and p_j'' denote the m'' possible values and probabilities for Y. We denote the probability of the pair (x_i, y_j) by p_{ij}; i.e. $p_{ij} = P[(X = x_i)(Y = y_j)]$, and we observe that the p_{ij}, given the p_i' and p_j'', can be any of the $m'm''$ values (lying in $[0, 1]$) satisfying the $m' + m'' - 1$

linear conditions $\sum_j p_{ij} = p'_i$, $\sum_i p_{ij} = p''_j$ (one of which is superfluous, since $\sum p'_i = \sum p''_j = 1$). They are therefore determined up to

$$m'm'' - (m' + m'' - 1) = (m' - 1)(m'' - 1)$$

degrees of freedom (except in boundary cases, where some of the p'_i or p''_j are $= 0$). The condition for non-correlation gives a further equation in the p_{ij}:

$$\mathbf{P}(XY) - \mathbf{P}(X)\mathbf{P}(Y) = \sum_{ij} x_i y_j (p_{ij} - p'_i p''_j) = 0,$$

which is clearly satisfied in the case of stochastic independence (we always have $p_{ij} = p'_i p''_j$), and still allows $(m' - 1)(m'' - 1) - 1$ degrees of freedom. In other words, it permits infinitely many other solutions—i.e. schemes of non-correlation without stochastic independence—unless $m' = m'' = 2$; q.e.d.

4.16.2. As for the statement that by 'strengthening' non-correlation one can obtain stochastic independence, we were referring to the possibility of considering, besides the non-correlation between X and Y, the same relation between arbitrary functions of X and Y, $X' = \alpha(X)$ and $Y' = \beta(Y)$, say: $\mathbf{P}(X'Y') = \mathbf{P}(X')\mathbf{P}(Y')$, i.e. $\mathbf{P}[\alpha(X)\beta(Y)] = \mathbf{P}[\alpha(X)]\mathbf{P}[\beta(Y)]$. In the case of X and Y with a finite number of possible values (the only case for which we have so far defined stochastic independence) it is obvious that such a relation holds, whatever the functions α and β are, if X and Y are stochastically independent (with the above notation, if $p_{ij} = p'_i p''_j$, we have $\sum p_{ij}\alpha(x_i)\beta(y_j) = \sum p'_i p''_j\alpha(x_i)\beta(y_j)$. Conversely, it follows that $(m' - 1)(m'' - 1) - 1$ suitable (i.e. linearly independent), additional conditions of this kind will suffice to imply stochastic independence. For the general case (an infinite number of possible values), similar conclusions will hold, except that we shall require the adjunction of *infinitely many* conditions of this kind, and, in addition, clarification of the meaning of the definition by means of suitable critical considerations (cf. Chapter 6).

4.16.3. If, for $X_1, X_2, \ldots, X_r$, we not only have

$$\mathbf{P}(X_i X_j) = \mathbf{P}(X_i)\mathbf{P}(X_j)$$

but also

$$\mathbf{P}(X_i X_j X_h) = \mathbf{P}(X_i)\mathbf{P}(X_j)\mathbf{P}(X_h), \quad \text{etc.},$$

we could, of course, define, and look at, *non-correlation of order three (or greater)*, for any arbitrary distinct X. Equivalently (and perhaps more simply), we can say that, when $\mathbf{P}(X_i) = 0$, non-correlation of order k means that $\mathbf{P}(Z) = 0$ for each Z which is the product of $h \leqslant k$ distinct factors X_i; the general case can be reduced to this one by saying that it implies non-correlation of order k of the $X_i - \mathbf{P}(X_i)$. However—with a convention opposite to that for stochastic independence—when we simply say 'non-correlation', 'pairwise' should always be understood. This is both because this is the case of most frequent interest, and in order to be able to use, in the case of events, the two convenient and easily distinguishable terms, '(stochastically) independent' and 'uncorrelated', without having to specify 'independent, that is to say, independent *of every order*' and 'uncorrelated, that is to say, *pairwise* uncorrelated', respectively.

4.16.4. Pairwise non-correlation (unlike independence) has, in fact, an autonomous and fundamental meaning, no matter how many random quantities are being considered together. More generally, a *measure* of correlation is of interest, and this will be provided by the *correlation coefficient*, $r(X, Y)$, between two random quantities (to be defined by equation (24) in Section 4.16.6). In the same way as knowledge of the previsions $\mathbf{P}(X_i)$ was sufficient in order to know the prevision of every linear function of the X_i, $X = \sum a_i X_i$, knowledge of the prevision of the squares, $\mathbf{P}(X_i^2)$ (in addition to that of the $\mathbf{P}(X_i)$), and of the correlation coefficients $r_{ij} = r(X_i, X_j)$, is sufficient to determine the prevision of every quadratic function of the X_i:

$$X = \{\text{a second-degree polynomial in the } X_i\} = \sum_{ij} a_{ij} X_i X_j + \sum_i a_i X_i + a_0,\dagger$$

$$(20) \qquad \mathbf{P}(X) = \sum_{ij} a_{ij}\mathbf{P}(X_i X_j) + \sum_i a_i\mathbf{P}(X_i) + a_0.$$

Knowledge of the *second-order previsions* is often sufficient for the solution of many problems (if not completely, by giving some bounds). If one thinks of the image (still not made precise, but intuitively clear) of *probability as distribution of mass*, the knowledge of the previsions is equivalent to the knowledge of the *barycentre*, and that of the second-order previsions (or second-degree characteristics of the distribution) is equivalent to knowledge of the *moments of inertia*.

The reasons for the importance of such knowledge, albeit limited, of the distribution in the calculus of probability (as in statistics), are, essentially, the same as those which determine their importance in mechanics (although, in general, not as precisely as is the latter case, due to the connection with energy, etc.).

4.16.5. *Separations and deviations.* It is often convenient to write

$$X = x + (X - x)$$

where $x = m = \mathbf{P}(X)$, or some other special value (like the *median* or the *mode*, which we shall discuss in Chapter 6, 6.6.6), or even with a generic x

---

† The first summation will suffice if we include the index 0 corresponding to the fictitious random quantity $X_0 \equiv 1$ (cf. Chapter 2, 2.8.3); in this case, a_i becomes $a_{i0} + a_{0i}$ and a_0 becomes a_{00}.

Moreover, it is, of course, irrelevant whether we take as zero the a_{ij} with $i > j$, or conversely with $i < j$, or instead take $a_{ij} = a_{ji}$, or whatever, according to the circumstances: the only relevant thing is $a_{ij} + a_{ji}$.

(representing an arbitrary given number). We shall call the difference $X - x$ the *separation* (of X from x); if we take the absolute value (as is often useful), $|X - x|$ is called the *deviation*.

As far as the second-order previsions are concerned, it is clear that, in general, it is convenient to take them relative to the barycentre, $x_i = m_i = \mathbf{P}(X_i)$, the point with respect to which the moments are smallest.

$$\mathbf{P}(X - x)^2 = \mathbf{P}[(X - m) - (x - m)]^2 = \mathbf{P}(X - m)^2 + (x - m)^2$$
$$- \{2(x - m)\mathbf{P}(X - m)\},$$

but the final term vanishes $(\mathbf{P}(X - m) = m - m = 0)$ and we have the following result, well-known in mechanics: the moment with respect to a point x is the moment about the barycentre (the first term) plus the square of the distance from the barycentre (the second term: here the mass $= 1$), and clearly the minimum is at $x = m$.

$\mathbf{P}(X - m)^2$ is called the *variance* of X, and its square root (in mechanics, the *radius of gyration*; the distance at which the mass should be concentrated in order to preserve the moment of inertia†) is called the *mean standard deviation* or, more briefly, the *standard deviation*. It is denoted by

$$(21) \qquad \sigma(X) = \sqrt{[\mathbf{P}(X - m)^2]} = \sqrt{[\mathbf{P}(X^2) - m^2]} \qquad (m = \mathbf{P}(X)),\ddagger$$

or sometimes σ_X (or simply σ if there is no ambiguity). The variance will be denoted by $\boldsymbol{\sigma}^2(X)$, σ_X^2 or σ^2.

The separation (and the deviation) from m, divided by the standard deviation, are called the *standardized separation*, $(X - m)/\sigma$, and the *standardized deviation* $|X - m|/\sigma$.

In this way, we can express the square terms of Section 4.16.4 by means of previsions and variances (i.e. by means of previsions and standard deviations):

$$(22) \qquad \mathbf{P}(X_i X_i) = \mathbf{P}(X_i^2) = \boldsymbol{\sigma}^2(X_i) + \mathbf{P}^2(X_i) = \sigma_i^2 + m_i^2$$

(where $\mathbf{P}^2(X) = [\mathbf{P}(X)]^2$),

and similarly the cross-product terms, $\mathbf{P}(X_i X_j)$ with $i \neq j$;

$$(23) \qquad \mathbf{P}(X_i X_j) = m_i m_j + \mathbf{P}[(X_i - m_i)(X_j - m_j)] = m_i m_j + \sigma_{ij},$$

where σ_{ij}, so defined, is called the *covariance* of X_i and X_j,§ and, writing $\sigma_{ij} = \sigma_i \sigma_j r_{ij}$, we arrive at the introduction of the correlation coefficient, as mentioned above.

---

† This is an example of a mean according to Chisini's definition! Cf. Chapter 2, 2.9.2.
‡ σ is *boldface* when it is an operator (and the same holds for r).
§ In particular, for consistency, $\sigma_{ii} = \sigma_i^2$.

4.16.6. In order to define the *correlation coefficient* we denote by X and Y the two random quantities, and suppose that $\mathbf{P}(X) = \mathbf{P}(Y) = 0$; then setting

$$\mathbf{P}(XY) = \sigma(X)\sigma(Y)\mathbf{r}(X, Y),$$

we have, by definition,

$$(24) \qquad\qquad \mathbf{r}(X, Y) = \frac{\mathbf{P}(XY)}{\sigma(X)\sigma(Y)}.$$

It was clear from the very beginning that the correlation coefficient would be zero, positive or negative, according as X and Y are uncorrelated, positively correlated, or negatively correlated. It is equally obvious that if $Y = X$, then $r = 1$, and that if $Y = -X$, then $r = -1$, and it is also clear that multiplying X and/or Y by constants does not change r, except possibly in sign:

$$\mathbf{r}(aX, bY) = \pm\mathbf{r}(X, Y),$$

$+$ or $-$, according to the sign of ab. If $a = 0$, or $b = 0$, then $aX = 0$ or $bY = 0$ and r has no meaning; the previous observation can therefore be completed by saying that if $Y = aX$, then $\mathbf{r}(X, Y) = \pm 1$ (sign of a).

It is already intuitively obvious from the above that r can assume all values between ± 1, but no others, and we shall now prove this: it will suffice to restate the standard argument about quadratics. We always have $(Y - tX)^2 \geqslant 0$ (or zero, in the limit case where for some $t = t_0$ we have the identity $Y = t_0 X$), and hence $t^2 X^2 - 2tXY + Y^2 \geqslant 0$; taking its prevision, $t^2\mathbf{P}(X^2) - 2t\mathbf{P}(XY) + \mathbf{P}(Y^2) \geqslant 0$, and so, since the discriminant must be negative, $|\mathbf{P}(XY)|^2 < \mathbf{P}(X^2)\mathbf{P}(Y^2)$; q.e.d.

In order to extend the definition to the case in which we do not have $\mathbf{P}(X) = \mathbf{P}(Y) = 0$, it suffices to observe that the separations from the prevision, $X - m_X$ and $Y - m_Y$, must be substituted for X and Y, and $\mathbf{P}(XY)$ therefore replaced by

$$\mathbf{P}[(X - m_X)(Y - m_Y)] = \mathbf{P}(XY) - m_X m_Y.$$

It is useful to remark that a different extension of the definition could have been obtained by leaving $\mathbf{P}(XY)$ as the numerator, and changing the denominator to $\mathbf{P}_Q(X)\mathbf{P}_Q(Y)$, where $\mathbf{P}_Q(X) = \sqrt{\mathbf{P}(X^2)} =$ quadratic prevision of X. The same properties and proofs would hold, but the meaning would be different: if we denote this alternative coefficient (temporarily) by $\hat{r}$, $\hat{r} = 0$ would imply $\mathbf{P}(XY) = 0$, instead of $= m_X m_Y$, and $\hat{r} = \pm 1$ would follow from $Y = aX$ instead of from $Y - m_Y = a(X - m_X)$.

The meaning of all this will be clear under the geometric interpretation which we are now about to introduce.

Remarks. We cannot (as a rule) say that in order to have $\mathbf{P}_Q(X) = 0$ we must have $X = 0$, but only that all the probability must be at least *adherent* to 0. To have $\mathbf{P}_Q(X) = 0$, we must obviously have $\mathbf{P}(|X| \geqslant \varepsilon) = 0$ for all $\varepsilon > 0$ (if this were equal to $p > 0$, we would in fact have $\mathbf{P}_Q^2(X) > p\varepsilon^2$), but this does not exclude the possibility of $\mathbf{P}(X \neq 0)$ being > 0 or even $= 1$ (e.g. if the only possible values are the sequence $x_n = 1/n$, each with zero probability). Anyway, we shall say, if $\mathbf{P}_Q(X) = 0$, that X *coincides* with 0, and write $X \doteq 0$; similarly, we say that X and Y coincide, $X \doteq Y$, if $X - Y \doteq 0$.

4.17   A GEOMETRIC INTERPRETATION

4.17.1. We have already considered (Chapter 2, 2.8.1) the linear space $\mathscr{L}$ of random quantities X: it is an affine vector space (whose origin is the 'random' quantity which is identically $= 0$) in which each X is represented by a vector (and linear combinations by linear combinations). We also agreed to denote by X_0 the 'random' quantity whose value is identically 1, and by x_0 the axis on which the 'certain' (constant) quantities lie.

Once we have introduced a prevision $\mathbf{P}$, we know that $\mathbf{P}(X)$ is a linear function of the vector X, with $\mathbf{P}(cX_0) = c$ (on the axis representing certainty, coinciding with the abscissa c). To give $\mathbf{P}$ is to give the plane of the *fair* random quantities (with $\mathbf{P}(X) = 0$): to find $\mathbf{P}(X) = m$ means, in fact, to find that m for which $\mathbf{P}(X - m) = 0$; in other words, to decompose X into $m + (X - m)$, the sum of a vector mX_0, known with certainty $(m = mX_0)$, and a fair vector. One might prefer to think of $x_0 = m$ as the point of intersection of the axis of certainty with the plane parallel to the fair plane, passing through the point X (where 'the point X' is short for $O + X$, the end point of the vector X which starts from O).

Functions of the second degree in random quantities belonging to $\mathscr{L}$—i.e. arbitrary numbers of linear combinations of products XY, of which the squares, X^2, are special cases $(Y = X)$—do not belong to $\mathscr{L}$.† We can, however, still give $\mathbf{P}(XY)$ a geometric interpretation by transforming $\mathscr{L}$ geometrically from an affine space into a Euclidean metric space, with a metric defined by the $\mathbf{P}(XY)$, interpreted as the *scalar product* of the vectors X and Y: i.e. by interpreting $\mathbf{P}_Q(X)$ as the *length* of the vector X (limiting ourselves to some $\mathscr{L}^* \subset \mathscr{L}$ if for $X \notin \mathscr{L}^*$ we have $\mathbf{P}_Q(X) = \infty$).

---

† They could all belong to $\mathscr{L}$, if the latter were infinite dimensional; otherwise, a few of them could belong. Anyway, the appearance of X^2, in addition to X, is superfluous (unless one is interested in $\mathbf{P}(X^4)$, $\mathbf{P}(X^2Y)$, etc.).

In fact, $\mathbf{P}(XY)$ satisfies the necessary and sufficient conditions for a scalar product (and therefore generates a Euclidean metric):

it is linear in X and Y, and symmetric

$$(XY = YX, X(Y_1 + Y_2) = XY_1 + XY_2, \mathbf{P} \text{ is linear});$$

it is positive definite $(\mathbf{P}(XX) = \mathbf{P}_Q^2(X) > 0$ if we do not have $X \doteq 0)$.

Remarks. Notice that, for the metric under consideration, it is appropriate to think of coincident random quantities as represented by the same vector (if one wished, one could say that it represents an 'equivalence class' with respect to 'coincidence'). If not, we would have non-zero vectors with zero length.

Under this metric, the length of X would be $\mathbf{P}_Q(X) = \sqrt{(m^2 + \sigma^2)}$, and X and Y would be orthogonal if $\mathbf{P}(XY) = 0$: in general, the cosine of the angle between them would be $\hat{r}$. Fairness implies orthogonality to the axis of certainty. The metric which we use (most often) is not this one but another: it was, however, convenient to begin with this as it is the most natural starting point.†

4.17.2. The metric which serves our purpose is the same as the preceding one (in accordance with the given definition of correlation) but applied to the *separations*, $X - \mathbf{P}(X)$, instead of to the X themselves. The simplest illustration (which is connected with the previous considerations) consists in saying that one takes into consideration only the projections onto the fair plane; i.e. the component orthogonal to the axis of certainty $(X - m$, with $m = \mathbf{P}(X))$, disregarding the parallel component, which is in fact m, or 'mX_0'.

Under this metric, the length of X is $\sigma(X)$; i.e. the length of the projection of X (under the previous metric). The cosine of the angle between X and Y (taking the projections onto the fair plane) is $\mathbf{r}(X, Y)$, and we have therefore: *non-correlation* $(r = 0)$ corresponds to *orthogonality* (of the projections onto the fair hyperplane); *positive correlation* $(0 < r < 1)$ and *negative correlation* $(-1 < r < 0)$ correspond to *acute* and *obtuse* angles, respectively (always between the projections). The extreme cases $(r = \pm 1)$ correspond to parallelism, in the same or opposite direction (again between projections).

In order to avoid constant repetition of the fact that it is the projections that are involved, one could always bear in mind that, in this ambit, if we take as norm (or length, or distance) the standard deviation instead of the quadratic prevision, all random quantities differing by certain constants are identified with one and the same vector of the fair hyperplane, the projection of the original (writing, e.g. $X \doteq Y$). One must be careful not to

---

† In some cases, we shall actually find it necessary to refer to the metric generated by $\mathbf{P}(XY)$: e.g. in connection with *mean-square convergence* (cf. Chapter 6).

become confused, and think in these terms when it is not possible to do so (e.g. in the case of mean-square convergence the norm must be $P_Q(X)$ and not $\sigma(X)$).

4.17.3. The vectorial–geometrical interpretation makes obvious and meaningful all properties relating to previsions of the second order. If we suppose that all the random quantities considered in the following are fair ($P(X) = 0$), we have, for instance:

for the decomposition of X into a component parallel to an arbitrary (non-zero) Y and a component orthogonal, the former will be $\sigma(X)\mathbf{r}(X, Y)$ (the length times the cosine) multiplied by the unit vector in the direction of Y (i.e. $Y/\sigma(Y)$), in other words,

$$(25) \qquad X' = Y \cdot [\mathbf{r}(X, Y)\sigma(X)/\sigma(Y)],$$

and the latter (which is obviously $X'' = X - X'$) has length $\sigma(X)\sqrt{(1 - r^2)}$ (length times sine). It is also characterized by the fact of having the smallest length of all vectors of the form $X - aY$;

in the same way, in order that X' be contained in, and X'' be orthogonal to, a given linear space (for simplicity, we take it to be 2-dimensional—linear combinations of Y and Z), we will have $X' = aY + bZ$ such that

$$X'' = X - X' = X - aY - bZ$$

is orthogonal to Y and Z; hence

$$P(X''Y) = P(XY) - aP(Y^2) - bP(YZ) = 0,$$

$$P(X''Z) = P(XZ) - aP(YZ) - bP(Z^2) = 0,$$

and, if Y and Z are taken to be orthogonal, $P(YZ) = 0$, and unitary, $P(Y^2) = 1 = P(Z^2)$, we have straightaway

$$a = P(XY) = \sigma(X)\mathbf{r}(X, Y),$$

$$b = P(XZ) = \sigma(X)\mathbf{r}(X, Z),$$

$$X' = \sigma(X)[Y\mathbf{r}(X, Y) + Z\mathbf{r}(X, Z)];$$

with a standard procedure (similar to the above), given any linearly independent $X_1, X_2, \ldots, X_n$, one can carry out the orthogonalization by substituting $Y_1, Y_2, \ldots, Y_n$, the Y_i being orthogonal to each other (and, if we wish, unitary). Proceeding in order ($i = 1, 2, \ldots, n$), it suffices to add to X_{i+1} a suitable linear combination of $X_1, \ldots, X_i$ in order to make it orthogonal to these vectors and, if necessary, to normalize (dividing by the length), obtaining Y_{i+1};

and so on.

4.17.4. The standard deviation of the sum of two or more random quantities is particularly important. For two summands, we have

(26)
$$\sigma^2(X + Y) = \mathbf{P}(X + Y)^2 = \mathbf{P}(X^2) + \mathbf{P}(Y^2) + 2\mathbf{P}(XY)$$
$$= \sigma^2(X) + \sigma^2(Y) + 2\mathbf{r}(X, Y)\sigma(X)\sigma(Y),$$

and it is easy to recognize the expression as the length of the sum of two vectors (as it had to be): i.e. the side of a triangle given the other two sides and the (external) angle between them; $c^2 \equiv a^2 + b^2 + 2ab \cos \theta$ (this is Carnot's theorem; if $\cos \theta = 0$, orthogonality, we have Pythagoras' theorem: in the limit cases, $\cos \theta = \pm 1$, i.e. parallelism, $c =$ the sum or difference of a and b). It is important to remember the following: in the case of *orthogonality* (*non-correlation*), the variances are added (*the standard deviations obey Pythagoras' theorem*); in the case of *positive correlation*, the variance and the standard deviation of the sum turn out to be *greater*, and in the case of *negative* correlation *less*, than in the case of non-correlation (the standard deviations of the summands being the same); cf. Figure 4.2.

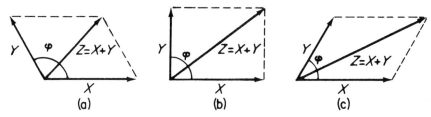

Figure 4.2   (a) Negative correlation. (b) Non-correlation (orthogonality). (c) Positive correlation

The same holds for more than two summands. In this case, of course, one may have correlations which are in part positive, in part negative, and the effect of either the former or the latter may prevail. The general formula is clearly as follows (written directly for a general linear form, always assuming $\mathbf{P}(X_i) = 0$):

(27) $$\sigma^2\left(\sum_i a_i X_i\right) = \mathbf{P}\left(\sum_{ij} a_i a_j X_i X_j\right) = \sum_{ij} a_i a_j \mathbf{P}(X_i X_j) = \sum_{ij} a_i a_j \sigma_i \sigma_j r_{ij};$$

the squared terms ($r_{ii} = 1$) yield $\sum_i a_i^2 \sigma_i^2$; excluding $i = j$ in the general summation, one obtains the contribution of the cross-product terms (zero in the case of orthogonality, positive or negative according to the prevailing correlations *between the summands* $a_i X_i$—not the X_i!—whose signs are those of $a_i a_j r_{ij}$—not of r_{ij}!).

The *covariance matrix*, with entries σ_{ij}, of the random quantities X_i (which we assume to have zero prevision) completely determines the second-order characteristics in the space $\mathscr{L}$ of linear combinations of the X_i (geometrically, in $\mathscr{L}$, it gives the length and angles of the vectors representing the X_i). The *correlation matrix*, with entries r_{ij} ($r_{ij} = \sigma_{ij}/\sigma_i\sigma_j$, $\sigma_i = \sqrt{\sigma_{ii}}$, $r_{ii} = 1$) can be derived from it, giving the angles (r_{ij} is the cosine) but not the lengths. It can still be regarded as a covariance matrix for the standardized X_i; i.e. for the X_i/σ_i (geometrically one is considering the *unit vectors* rather than the vectors).

4.17.5. A fact which is of conceptual and practical importance—and for this reason mentioned already in the Remarks of 4.9.1. for the case of events— is that the size of the *negative* correlation (unlike the positive) must be *bounded*. More precisely, given n random quantities, the arithmetic mean of their $\binom{n}{2}$ correlation coefficients r_{ij} ($i \neq j$) cannot be less than $-1/(n-1)$: in particular, the r_{ij} cannot all be less than $-1/(n-1)$; in the extreme case (as we shall see) they can all be equal to this limit value.

Without loss of generality, we can assume the X_i normalized, $\mathbf{P}(X_i) = 0$ and $\mathbf{P}(X_i^2) = 1$, so that $r_{ij} = \mathbf{P}(X_iX_j)$: we consider their sum, $X = X_1 + X_2 + \cdots + X_n$, and evaluate its variance

$$\sigma^2(X) = \mathbf{P}(X^2) = \mathbf{P}\left(\sum_{ij} X_iX_j\right) = \sum_{ij} \mathbf{P}(X_iX_j) = \sum_i \mathbf{P}(X_i^2) + \sum_{i \neq j} \mathbf{P}(X_iX_j)$$

$$= n + \sum_{i \neq j} r_{ij} = n + n(n-1)\bar{r} = n[1 + (n-1)\bar{r}],$$

where we have set $\bar{r} =$ the arithmetic mean of the r_{ij}, i.e.

$$\bar{r} = \frac{1}{n(n-1)} \sum_{i \neq j} r_{ij}.$$

The variance is non-negative, however, and therefore $\bar{r} \geqslant -1(n-1)$; q.e.d. We note that the extreme value is attained if and only if the sum is identically $= 0$ (or, if we want to be absolutely precise, $\doteqdot 0$, using the notation of 4.17.2): i.e. if the n unit vectors have zero resultant.† In particular, the r_{ij} could have the common value $r = -1/(n-1)$ only if the unit vectors were arranged like the straight lines joining the centre of a regular $(n-1)$-dimensional simplex to the vertices. Figure 4.3 illustrates the case of $n = 3$ (equilateral triangle) and $n = 4$ (regular tetrahedron). We give the basic facts for these cases (and also for $n = 5, 6, 7, 8$):

$n = 3, r = -1/2 = \cos 120°$	$n = 6, r = -1/5 = \cos 101° \, 32'$
$n = 4, r = -1/3 = \cos 108° \, 16'$	$n = 7, r = -1/6 = \cos \; 99° \, 36'$
$n = 5, r = -1/4 = \cos 104° \, 29'$	$n = 8, r = -1/7 = \cos \; 98° \, 12'.$

---

† Observe that they are, therefore, linearly dependent.

Approximately, the angle is a right angle plus $1/(n-1)$ (in radians); in other words, in a possibly more convenient form, plus $3438/(n-1)$ minutes (for $n = 8$ the error is already of the order of $1'$). These numerical examples serve to make clear that one cannot go much beyond orthogonality among random quantities when there are more than just a few of them.

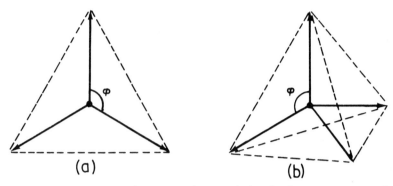

(a)   (b)

Figure 4.3 (a) The maximum negative correlation for 3 vectors: $r = \cos \phi = -\frac{1}{2}$. (b) The maximum negative correlation for 4 vectors: $r = \cos \phi = -\frac{1}{3}$.

4.17.6. All that we have considered so far (4.17.2–4.17.5) has been in terms of the conventional representation of the X_i (and of the X linearly dependent on them) in the abstract space $\mathscr{L}$. If, instead, we wish to consider the meaningful interpretation in terms of the distribution of probability as distribution of mass—an interpretation whose importance was indicated at the end of 4.16.4—we must transfer to the linear ambit $\mathscr{A}$ (the space S_r, with coordinates $x_1, x_2, \ldots, x_r$, where a point represents the outcomes of $X_1, X_2, \ldots, X_r$), since it is over this space that the mass is distributed. The $\mathbf{P}(X_i) = x_i$ identify the barycentres of such distributions (and we again assume the barycentre coincident with the origin, in order to avoid useless petty complications in the notation), and the $\mathbf{P}(X_i X_j) = \sigma_{ij}$ identify the moments of inertia; i.e. *the ellipsoid* (or *kernel*) *of inertia* (and in our case it could be called *of covariance*, like the corresponding matrix).

For our purposes, it is much more meaningful and useful (although the two things are formally equivalent) to consider what we shall call the *ellipsoid of representation*,† which is the reciprocal of the other. With reference to the principal axes (common to the two ellipsoids), the semi-axes measure the corresponding standard deviations, σ_h, in the ellipsoid of representation, whereas, for the ellipsoid of covariance, they give the reciprocals, $1/\sigma_h$ (or K/σ_h; one can take an arbitrary multiplicative constant).

---

† Of course, we speak of 'ellipsoids' in S_r, even if $r > 3$, or $r = 2$ (ellipses), or $r = 1$ (segments). As far as I know, terminology of this kind does not exist in mechanics; statisticians at times refer to the 'ellipsoid of concentration'.

In Mechanics, the latter has been employed (Cauchy–Poinsot), although the former has also been proposed (MacCullach). Part of the reason for preferring this one seems also to hold for Mechanics; in our case, however, there are also rather special and more decisive circumstances (e.g. the fact that we are interested in moments with respect to planes. i.e., in general, to hyperplanes S_{r-1}, rather than moments with respect to straight lines).

The ellipsoid of representation has a concrete meaning: it is the model of a solid having the same moments as the given distribution (assuming it to be homogeneous, and giving it a mass increased in the ratio 1 to $r + 2$— 3 on the line, 4 in the plane, 5 in ordinary space, etc.—or, alternatively, increasing the size in the linear scale 1 to $\sqrt{(r + 2)}$). This is obvious if one thinks of the case of the sphere, to which one can always reduce the problem by imposing a suitable metric on the affine space $\mathscr{A}$ (unless it already has one, either because of an actual geometrical meaning, or because the arbitrariness has already been exploited by reducing to a sphere some ellipsoid previously considered). For the unit sphere (in S_r) the moment about the centre is

$$\int_0^1 \rho^2 \rho^{r-1} \, d\rho \bigg/ \int_0^1 \rho^{r-1} \, d\rho = r/(r + 2),$$

but it is also r times the moment about a diametrical hyperplane, and hence the latter is $1/(r + 2)$. In order to make this equal to 1, it is sufficient to increase either the mass or the radius in the above mentioned way.

In the case of probability and statistics, this reduction to a homogeneous distribution is not the most appropriate procedure: the standard example of the (r-dimensional) normal distribution is much more meaningful (well-known as the 'distribution of errors'). As we shall see when we come to discuss it (Chapter 7, 7.6.7 and Chapter 10, 10.2.4), to each distribution over S_r there corresponds a unique normal distribution having the same second-order characteristics (same covariance matrix), and the ellipsoid of representation characterizes it in the most directly expressive manner.

These brief comments may have led to an appreciation of how many interesting conclusions, although incomplete, of course, can be drawn from incomplete assumptions (even as incomplete and crude as in the case under consideration).

4.17.7. *Inequalities.* We must now establish certain inequalities which are both necessary for the topic in hand, and also serve as simple illustrations of what can be said more generally.†

Tchebychev's inequality gives an upper bound, $1/t^2$, for the probability that $|X|$ is greater than $t\mathbf{P}_Q(X)$; in particular, *for the probability that the*

---

† More general cases than those considered here are developed in the works of E. Volpe (using this geometrical representation); Ernesto Volpe di Prignano, 'Calcolo di limitazioni di probabilità mediante involucri convessi', *Pubbl. n. 16 dell'Ist. Matem. Finanz. Univ. di Trieste* (1966).

standardized deviation is greater than t. For example, the probability that $|X|$ is greater than some multiple of the quadratic prevision is: $<\frac{1}{4}$ for twice; $<\frac{1}{9}$ for three times; $<\frac{1}{25}$ for five times; $<\frac{1}{100}$ for ten times, etc. Without further conditions, this bound is the best possible; however, the bounds are normally crude (the probability is much smaller: we have here placed ourselves in the least favourable position).

The *proof* is obvious if one thinks in terms of mass. If a mass $>1/t^2$ were placed at a distance from the origin $>a$, it would have moment of inertia $>a^2/t^2$; altogether, the moment of inertia is $\mathbf{P}_O^2(X)$, and hence $a < t\mathbf{P}_O(X)$. Placing two masses $1/2t^2$ at $\pm t\mathbf{P}_O(X)$ and the rest at 0, one obtains the limit-case (provided that $t \geq 1$).

Cantelli's inequality is the one-sided analogue of the preceding one: $1/(1 + t^2)$ is the upper bound for the probability that the separation *in a given direction* is greater than $t\sigma$ ($X > m + t\sigma$, or $X < m - t\sigma$, respectively, with $t > 0$). If the mean is not fixed, the question does not arise, the inequality would then be the same as the first one; the improvement is notable only for small t: $t = \frac{1}{2}$, $p = \frac{4}{5}$ instead of 1; $t = \frac{3}{4}$, $p = \frac{64}{100}$ instead of 1; $t = 1$, $p = \frac{1}{2}$ instead of 1; $t = \frac{3}{2}$, $p = \frac{4}{13}$ instead of $\frac{4}{9}$; $t = 2$, $p = \frac{1}{5}$ instead of $\frac{1}{4}$, for $t = 3$ the difference is already hardly noticeable: $p = \frac{1}{10}$ instead of $\frac{1}{9}$.

The proof can be given in a similar way to the above. In order to balance a mass p at $m + t\sigma$, one can place the residual mass $1 - p$ at $m - t\sigma p/(1 - p)$, and this gives a moment of inertia equal to $\sigma^2 t^2[p + (1 - p)p^2/(1 - p)^2]$; $t^2[\dots]$ cannot be greater than 1, $[\dots] = p/(1 - p)$, $t^2 \leq (1 - p)/p = -1 + 1/p$, etc. If the balancing mass is dispersed, the situation can only be made worse.

Although it is outside of our present realm of interest (second-order characteristics), it is worthwhile pointing out how the argument used in proving Tchebychev's inequality can be applied, without any difficulty, to much more general cases. If $\gamma(x)$ is an increasing function ($0 \leq x \leq \infty$), we necessarily have $\mathbf{P}\{|X - m| \geq a\} \leq \mathbf{P}\{\gamma(|X - m|)\}/\gamma(a)$ because a mass $>p$, placed at a distance a from m, alone contributes to $\mathbf{P}\{\gamma(|X - m|)\}$ a quantity $>p\gamma(a)$ (which cannot be greater than the whole thing), and the situation is even worse if the distance is greater.

For example, taking absolute moments of any order r, we have

$$\mathbf{P}(|X| \geq a) \leq \mathbf{P}(|X|^r)/a^r,$$

the Markov inequality: for $r = 2$ this is the Tchebychev case, seen above.

4.18 ON THE COMPARABILITY OF ZERO PROBABILITIES

4.18.1. When we were considering (at the end of Chapter 3) countable additivity and zero probabilities, the question often arose as to whether it makes sense to compare the latter: for example, saying that, if all cases are equally probable, the probability of the union of 12 of them is twice that of the union of 6, and three times that of the union of 4, even if all these probabilities are zero (as in the example of 'an integer N chosen at random'). We assumed this in order to give the statements of a few examples in a more suggestive form; as we indicated then, this is now the time to examine the question.

For the purpose of removing the most radical objection, and as a better means of presenting the sense of the question, a geometrical analogy will suffice. The objection is that zero stands for nothing, and that nothing is simply nothing: this is one of many such vacuous statements on the basis of which certain philosophers pontificate about things of which they understand nothing.†

A set can have measure zero in terms of volume without being empty; it could, for instance, be a part of a surface and have a measure in terms of area (and two areas can be compared). A measure in terms of area could be zero without the set being empty; it could be an arc of a curve and have a measure in terms of length. A linear set might also have measure zero in terms of length (in some sense or other: Jordan–Peano, Borel, Lebesgue) without being empty, but some comparison could also be made in this case (even if it only distinguished sets with single points or 2 or 3, ..., or an infinite number).

> All this would be even more expressive and persuasive if put in terms of more general concepts of measure (with intermediate dimensions also, not just integer) as in Borchardt, Minkowski, Peano, Hausdorff, etc. The example closest to our theme is that in which one defines 'the measure m of dimension α' of a set I to be that for which $V(I_\rho) \sim m\rho^{3-\alpha}$ (I_ρ = the set of points of three-dimensional space with distance $\leqslant \rho$ from I, V = volume, the asymptotic expression to hold as $\rho \to 0$).

4.18.2. In any case, so far as probability is concerned, a direct meaning exists and we have no need of analogies to provide a justification (they may, on occasion, provide encouragement in showing us that our situation is not unique and strange, and may help us by providing visually intuitive models).

Given two events A and B, it is clear that if one has to decide between them—i.e. if one makes the assumption that one of the two is true—a comparison of their probabilities must be made. Expressed mathematically, if we consider their probabilities $\mathbf{P}(A|H)$, $\mathbf{P}(B|H)$ conditional on the 'hypothesis' $H = A \vee B$, their sum is $\geqslant 1$, and their comparison is easy. It could be said that this is the same thing as comparing $\mathbf{P}(A)$ and $\mathbf{P}(B)$: if $\mathbf{P}(H)$, and, *a fortiori*, $\mathbf{P}(A)$ and $\mathbf{P}(B)$, are small, however, the proposed alternative is perhaps psychologically more appropriate as it presumably

---

† *Translators' note.* The author is here referring to what he considers the deleterious influence of Croce's idealism upon Italian culture.

induces one to weigh up the evaluation more accurately by fixing attention on the two cases separately, whereas the reliability of the ratio of two very small numbers—attributed as part of an overall evaluation, in which A and B had no special significance—might well be doubted. When the events A and B (and hence H) have zero probabilities, however, the alternative approach becomes essential. With the direct comparison the ratio of the two probabilities would have the form $0/0$. This does not mean that the ratio is meaningless, but that the method of comparison is not the right one.†

From an axiomatic viewpoint, the extension of the condition of coherence to cover the present case requires a stronger form: we assume tacitly that this has been done (but we will discuss it in the Appendix, Section 16).

Hence, with any event A as reference point, any other event E has a certain ratio of probability with A (a finite positive number, or zero, or infinity): in this way, innumerable 'layers' of events having probabilities 'of the same order' (i.e. with finite ratio) can appear, the 'layers' being ordered in such a way that every event in a higher layer has infinitely greater probability than any event in a lower layer.

4.18.3. An example will suffice as a clarification, both of the general situation, and of the implicit applications mentioned in Chapter 3: this is the example of a 'positive integer N chosen at random'.

We have a partition into an infinite number of events, $E_h = (N = h)$, all with zero probabilities, $P(E_h) = 0$ ($h = 1, 2, \ldots$). This says very little, however: it merely excludes a single case ($\sum_h p_h > 0$) which, from this viewpoint, is 'pathological' (in the sense that, if we think of a function as having been chosen among the entire, unrestricted class of functions of a real variable, to be continuous, even at a single point, is a pathological case). To say that 'all the events E_h are equally probable' is a rather substantial addition: nevertheless, it only suffices to enable us to conclude the following: if A and B are finite unions of the E_h, e.g. of m and n, respectively, then the ratio of their probabilities is m/n; if A is the complement of a finite set we certainly have $P(A) = 1$; if A and its complement are infinite, then $P(A)$ is infinitely greater than any of the $P(E_h)$, but can be any $p \geqslant 0$ (even $p = 1$, or $p = 0$) located somewhere in the scale of the 'layers'.

At first sight, it might seem that one could say something more (perhaps by considering frequencies for the first n numbers and then passing to the limit): e.g. that the probability of obtaining N even is $= \frac{1}{2}$, of obtaining N prime is $= 0$, non-prime $= 1$. In fact, this is not a consequence of the assumption of equiprobability at all; it is sufficient to observe that, by altering the

---

† The knowledge that on a day when a housewife has not bought any sugar she has spent 0, does not allow us to conclude that the price of sugar is meaningless because it is $0/0$; it merely indicates that the information available is not sufficient to determine it.

order, these limits change but the equiprobability does not; on the other hand, the possible evaluations are not only those of the limit-frequency type, up to rearrangements.†

The assumption that $\mathbf{P}(E) = \lim \mathbf{P}(E|N \leqslant n)$ (and possibly, more generally, $\mathbf{P}(A)/\mathbf{P}(B) = \lim [\mathbf{P}(A|N \leqslant n)/\mathbf{P}(B|N \leqslant n)]$; i.e. the limit of the ratio of the numbers of occurrences of A to those of B in the first n integers) is neither compulsory nor ruled-out (for any E, or pairs A, B), where the limit exists. One certainly obtains a coherent evaluation (by continuity; cf. Chapter 3, Section 3.13) in the field where the limit exists, extendable everywhere (Chapter 3, 3.10.7). However, one makes the arbitrary choice from among the infinite possible ones, and automatically satisfying the conditions $\lim \inf \mathbf{P}(E|N \leqslant n) \leqslant \mathbf{P}(E) \leqslant \lim \sup \mathbf{P}(E|N \leqslant n)$.

This choice has no special status from a logical standpoint, but it could be so from a psychological point of view if the order has some significance (e.g. chronological); and indeed it is so if the formulation in terms of an infinite number of possible cases is thought of as, more or less, an idealization of the asymptotic study of the finite problem, with a very large number of cases n.

One can observe, by means of this example, just how rich the 'scale' of 'layers' can be (perhaps more than one would imagine at first sight). For every function $\phi(n)$, tending to zero as $n \to \infty$, we can construct an event (a sequence of integers, $a_1 < a_2 < \ldots < a_n, \ldots$) in such a way that the frequency (n/a_n) tends to zero like $\phi(n)$. It is sufficient to insert into the sequence, as the term a_{n+1}, the number m if otherwise n/m would be less than $\phi(m)$. If we consider $\phi(n) = n^{-\alpha}$, $(\alpha > 0)$, we obtain, for example, an event E_α, and each E_α has infinitely greater probability than those with a larger α (and, as is well-known, the scale is far from being complete: one could insert the $E_{\alpha,\beta}$ corresponding to $\phi(n) = n^{-\alpha}(\log n)^\beta$; and so on).

4.18.4. The method of taking limits, either starting from finite partitions (e.g. $p_h^{(n)} = 1/n$ for $h = 1, 2, \ldots, n$), or countably additive ones (e.g. $p_h^{(n)} = Ka^h$, $a = 1 - 1/n$, $K = n^2/(n - 1)$, $h = 1, 2, \ldots$), with limits which are not countably additive, is, in any case, the most convenient way of constructing distributions which are not countably additive. We must bear in mind, however, that it is a procedure for obtaining *some* coherent distributions in the field in which they are defined by the passage to the limit (since *finite* additivity is preserved), and not necessarily a procedure expressing anything significant.

---

† If while progressively attributing probability to infinite subsets of events (as in Chapter 3, 3.10.7) we always attribute probability = 1 (provided it is not necessarily = 0 by virtue of previous choices), we obtain an *ultrafilter* of events with probability = 1, whereas all the others have probability = 0. Linear combinations of distributions of this 'ultrafilter type' form a much wider class, still disjoint, however, from those of the limit-frequency type.

In particular, one should not think (even inadvertently):

that, assuming the $p_h^{(n)}$ are probabilities conditional on an hypothesis H_n (e.g., $N \leqslant n$, in the first example), the $p_h = \lim p_h^{(n)}$ (and the distribution over infinite subsets which derives from these) give probabilities which are conditional on the hypothesis $H = \lim H_n$ (e.g., referring still to the first example, $H = 1$);

or, even worse, the converse;

or that the events for which probabilities are defined by virtue of the passage to the limit have any special rôle, or that their probabilities have a different meaning from those of the other events (apart from the trivial observation that the former are consequences of the evaluations made by deciding to base oneself, on the passage to the limit, whereas the latter require a separate evaluation: it could have been the other way around if we had started with a different procedure).

4.18.5. Procedures of this kind have often been employed, more or less as a result of interpretations of the type we have here rejected. The most systematic treatments known to me are those by A. Lomnicki (*Fundamenta Mathematicae*, 1923) and by A. Rényi (in many recent works; cf., for example, *Ann. Inst. Poincaré* (1964): prior to this, in German, 1954).

Rényi's approach is constructed with the aim of making considerations of initial probabilities for partitions which are not countably additive fall within the range of the usual formulations, by concealing the non-additivity by means of the passage to the limit. The device consists in accepting that, for the partitions under consideration, countable additivity must be respected, but, in the passage to the limit, the total probability may become *infinite* instead of *one*. The importance of this is mainly in connection with the inductive argument, and so we will return to this topic more explicitly in Chapter 11.

4.19 ON THE VALIDITY OF THE CONGLOMERATIVE PROPERTY

4.19.1. If, conditional on every event H_j of a finite partition, the probability $\mathbf{P}(E|H_j)$ of a given event E is p (or, respectively, lies between p' and p''), then we also have $\mathbf{P}(E) = p$ (or, respectively, $\mathbf{P}(E)$ lies between p' and p''). In fact, we have

$$\mathbf{P}(E) = \mathbf{P}(EH_1 + EH_2 + \ldots + EH_n) = \sum_j \mathbf{P}(E|H_j)\mathbf{P}(H_j) = p \sum_j \mathbf{P}(H_j) = p;$$

(28)

the same holds even if the H_j form an infinite partition, so long as the sum of their probabilities is $= 1$. In fact, if we put

$$H_n^* = 1 - (H_1 + H_2 + \ldots + H_n),$$

we have

(29) $\mathbf{P}(E) = \sum_j \mathbf{P}(E|H_j)\mathbf{P}(H_j)(j \leqslant n) + \mathbf{P}(EH_n^*) = p[1 - \mathbf{P}(H_n^*)] + \mathbf{P}(EH_n^*),$

and hence $\mathbf{P}(E) = p$ because $\mathbf{P}(H_n^*)$ and, *a fortiori*, $\mathbf{P}(EH_n^*)$ tends to 0 as n increases.

4.19.2. Indeed, it would appear natural that this (conglomerative) property should hold for logical reasons, overriding all mathematical demonstrations or justifications, especially if one interprets literally a phrase like 'conditional on each of the possible hypotheses the probability of E is p, and so the fact that $\mathbf{P}(E) = p$ is proved'.

Two counterexamples will demonstrate that this is not so.

Taking an infinite partition of the integers into finite classes (each of three elements) we consider the events $A_h = E_h + E_{2h} + E_{2h+2}$, with $h = 1, 3, 5, \ldots$ odd; conditional on each of the A_h, the probability that N be even is $\frac{2}{3}$; the analogous partition $B_h = E_{h+1} + E_{2h-1} + E_{2h+1}$ would instead give $\frac{1}{3}$ (the asymptotic evaluation gives $\frac{1}{2}$).

Consider an infinite partition of the integers into infinite classes, with A_h (h odd) containing the number h and all multiples of 2^h which are not multiples of 2^{h+2}; conditional on every A_h, the probability that N be even is $= 1$ (independently of any conventions like asymptotic evaluations, there is only one odd number versus an infinite number of even ones and they are all equally probable). Of course, it suffices to change N into $N + 1$ in order to obtain the opposite conclusion: the probability that $N =$ even is 0 conditional on every A_h.

When we are in a position to discuss independence and dependence for general random quantities (Chapter 6, 6.9.5; cf. also Chapter 12, 12.4.3), we shall meet an example which is more meaningful, both from an intuitive and practical point of view (the latitude and longitude of a point of the earth's surface 'chosen at random').

CHAPTER 5

The Evaluation of Probabilities

5.1 HOW SHOULD PROBABILITIES BE EVALUATED?

In order to say something about this subject without running the risk of being misunderstood, it is first of all necessary to rule out the extreme dilemma that a mathematical treatment often poses: that of either saying everything, or of saying nothing. As far as the evaluation of probabilities is concerned, one would be unable to avoid the dilemma of either imposing an unequivocal criterion, or, in the absence of such a criterion, of admitting that nothing really makes sense because everything is completely arbitrary.

Our approach, in what follows, is entirely different. We shall present certain of the kinds of considerations which do often assist people in the evaluation of their probabilities, and might frequently be of use to You as well. On occasion, these lead to evaluations which are generally accepted: You will then be in a position to weigh up the reasons behind this and to decide whether they appear to You as applicable, to a greater or lesser extent, to the cases which You have in mind, and more or less acceptable as bases for your own opinions. On other occasions, they will be vaguer in character, but nonetheless instructive. However, You may want to choose your own evaluations. You are completely *free* in this respect and it is entirely your own *responsibility*; but You should beware of superficiality. The danger is twofold: on the one hand, You may think that the choice, being subjective, and therefore arbitrary, does not require too much of an effort in pinpointing one particular value rather than a different one; on the other hand, it might be thought that no mental effort is required, since it can be avoided by the mechanical application of some standardized procedure.

5.2 BETS AND ODDS

5.2.1. One activity which frequently involves the numerical evaluation of probabilities is that of betting. The motivation behind this latter activity is not usually very serious-minded or praiseworthy, but this is no concern of ours here. We should mention, however, that such motivations (love of gambling, the impulse to bet on the desired outcome, etc.) may to some extent distort the evaluations. On the other hand, motives of a different kind lead to similar effects in the case of insurance, where the first objection does not apply.

However, with all due reservation, it is worthwhile starting off with the case of betting, since it leads to simple and useful insights.

5.2.2. An important aspect of the question (one to which we shall frequently return) is the necessity of 'getting a feeling' for numerical values. Many people if asked how long it takes to get to some given place would either reply 'five minutes' or 'an hour', depending on whether the place is relatively near, or relatively far away: intermediate values are ignored. Another example arises when people are unfamiliar with a given numerical scale: a doctor, although able to judge whether a sick man has a high temperature or not, simply by touching him, would be in trouble if he had to express that temperature on a scale not familiar to him (Fahrenheit when he is used to Centigrade, or vice-versa). Likewise, in probability judgments, there are also those who ignore intermediate possibilities and pronounce 'almost impossible' everything that to them does not appear 'almost certain'. If neither YES nor NO appears sufficiently certain to them, they simply add 'fifty-fifty' or some similar expression. In order to get rid of such gaps in our mental processes it is necessary to be fully aware of this and to get accustomed to an alternative way of thinking.

In this respect, betting certainly provides useful experience. In order to state the conditions for a bet, which have to be precise, it is necessary to have a sufficiently sensitive feeling for the correspondence between a 'numerical evaluation' and 'awareness' of a *degree of belief*. In becoming familar with judging whether it is fair to pay 10, 45, 64 or 97 lire in order to receive 100 lire if a given event occurs, You will acquire a 'feeling' for what 10%, 45%, 64% or 97% probabilities are. Together with this comes an ability to estimate small differences, and a sharpening of that 'feeling for numerical values' which must be improved for the purpose, of course, of analysing actual situations.

5.2.3. These two aspects come together in the particularly delicate question of evaluating very small probabilities (and, complementarily, those very close to 1). Approximations which are adequate (according to the circumstances and purposes involved) in the vicinity of $p = \frac{1}{2}$ (e.g. 50% $\pm$ 5%, $\pm 1\%$, $\pm 0.1\%$) are different from those required in the case of very small probabilities: here, the problem concerns the order of magnitude (whether, for example, a small probability is of the order of 10^{-3}, or 10^{-7}, or $10^{-20}, \ldots$). In this connection, it is convenient to recall Borel's suggestion of calling 'practically impossible', with reference to '*human, earthly, cosmic* and *universal* scales', respectively, events whose probabilities have the orders of magnitude of 10^{-6}, 10^{-15}, 10^{-50} and 10^{-1000}. This is instructive if one wishes to give an idea of how small such numbers (and therefore such probabilities) are, provided that no confusion (in words or, worse, in concepts) with 'impossibility' arises.†

5.2.4. *On the use of 'odds'*. In the jargon used by gamblers, the usual way of expressing numerical evaluations is somewhat different, although, of course, equivalent. Instead of referring to the *probability p*, which (in the sense we have given) is the amount of a bet, we refer to the *odds*,

$$r = p/(1 - p) = p/\tilde{p}.$$

These are usually expressed as a fraction or ratio, $r = h/k = h:k$ (h and k integers, preferably small), by saying that the odds are 'h to k on' the event, or 'k to h against' the event. Of course, given r, i.e. the odds, or, as we shall say, the *probability ratio*, the probability can immediately be obtained by

(1) $\qquad p = r/(r + 1)$, i.e. (if r is written as h/k) $p = h/(h + k)$.

We present a few examples of the correspondence between probabilities and probability ratios, and vice-versa:

p	$p/\tilde{p}$	$= r$	$= h/k$	in words	(check) $h/(h + k) = p$
20%	$20/80$	$= 25\%$	$= 1/4$	'4 to 1 against'	$1/(1 + 4) = 20\%$
$2/7 = 28 \cdot 6\%$	$28 \cdot 6/71 \cdot 4$	$= 40\%$	$= 2/5$	'5 to 2 against'	$2/(2 + 5) = 28 \cdot 6\%$
50%	$50/50$	$= 1$	$= 1/1$	'evens'	$1/(1 + 1) = 50\%$
75%	$75/25$	$= 3$	$= 3/1$	'3 to 1 on'	$3/(3 + 1) = 75\%$.

Observe that to the complementary probability, $\tilde{p} = 1 - p$, there corresponds the reciprocal ratio, $\tilde{p}/p = 1/(p/\tilde{p}) = 1/r$ (i.e. to 'h to k on' there corresponds the symmetrical phrase 'k to h on').‡

---

† Borel himself, and other capable writers, fail to avoid this misrepresentation when they give the status of a principle—'Cournot's principle'—to the confusion (or the attempt at a forced identification) between 'small probabilities', which, by convention, could be termed 'almost impossibility', and 'impossibility' in the true sense. What is overlooked here is that 'prevision' is not 'prediction'. The topic is dealt with in E. Borel, *Valeur pratique et philosophie des probabilités* (cf. p. 4 and note IV), part of the great *Traité du Calcul des Probabilités* which he edited; Gauthier–Villars, Paris (1924) (and subsequent editions).

‡ It would perhaps be better to introduce a notation to indicate that we are passing from probability to 'odds'; similar to that used for 'complementation' ($\tilde{p} = 1 - p$). An analogous approach would be to take $\underline{p} = p/\tilde{p}$ (and if $p = \mathbf{P}(E)$ to use therefore $\underline{\mathbf{P}}(E) = \mathbf{P}(E)/\mathbf{P}(\tilde{E}) = \mathbf{P}(E)/\sim\mathbf{P}(E)$). We prefer merely to draw attention to the possibility without introducing and experimenting with more new ideas than prove to be absolutely necessary. To avoid any difficulties, or risks of confusion in notation, we denote the odds more clearly by writing $O(x) = x/(1 - x)$, $O[\mathbf{P}(E)] = \mathbf{P}(E)/\mathbf{P}(\tilde{E})$.

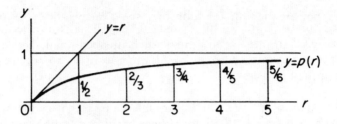

Figure 5.1   The relation between probability (p) and odds (r): $r = p/(1 - p)$

5.2.5. *Extensions.* Probability is preferable by far as a numerical measure (additivity is an invaluable property for any quantity to possess!).† However, there are cases in which it is advisable to employ the probability ratio (especially in cases involving likelihood—Chapter 4—which are often considered in the form of 'Likelihood Ratio'), and it is useful to indicate, at this juncture, the way in which we shall generalize its use (or, in a certain sense, substitute for it) in cases where the need arises.

In accordance with, and in addition to, the conventions introduced in Chapter 3, Section 3.5, concerning the use of the symbol **P**, we can denote that $r = h/k$ by writing

$$
\mathbf{P}(E, \tilde{E}) = (h/(h + k), k/(h + k))
$$

(2)

$$
= (h, k)/(h + k) = K(h, k) = (h:k) = \mathbf{P}(E:\tilde{E}),
$$

where we have successively and implicitly made the following conventions:

a common factor, such as $1/(h + k)$, can be taken outside the parentheses; i.e. $m(a, b) = (ma, mb)$;

such a factor may be taken as understood, denoting it by K, to simply mean that proportionality holds;

the same thing may be indicated by simply using the 'colon' ($:$) as the dividing sign, rather than the comma. This means that two n-tuples of numbers $(a_1, a_2, \ldots, a_n)$ and $(b_1, b_2, \ldots, b_n)$, not all zero, are said to be proportional if $b_i = Ka_i$ where K is a non-zero constant. Proportionality is sometimes denoted by the sign $\propto$ (which is not very good), and can also be expressed by $= K$. We make the convention—once and for all—that K denotes a *generic* coefficient of proportionality, whose value is not necessarily

---

† A newspaper, in considering three candidates for the American presidential election, attributed odds of 2 to 1 on, 3 to 1 against and 5 to 1 against; these are equivalent to probabilities of $\frac{2}{3}$, $\frac{1}{4}$, $\frac{1}{6}$, with sum $(8 + 3 + 2)/12 = 13/12 > 1$. It is difficult for a slip of this nature to pass unnoticed when expressed in terms of probabilities; using percentages especially, it would certainly not escape notice that $67\% + 25\% + 17\% = 109\%$ was inadmissible.

the same, not even for the duration of a given calculation: we can write, for example, $(2, 1, 3) = K(4, 2, 6) = K(6, 3, 9)$. The equals sign is sufficient on its own if the n-tuple with ':' in place of ',' is interpreted as 'up to a coefficient of proportionality' (like homogeneous coordinates); i.e. as a multi-ratio. Hence, for example, $(2:1:3) = (4:2:6) = (6:3:9)$.

Sometimes the omission of the proportionality factor is irrelevant because it is determined by normalization: for example, if it is known that $E_1 \ldots E_n$ constitute a partition, and we write

(3) $$\mathbf{P}(E_1 : E_2 : \ldots : E_n) = (m_1 : m_2 : \ldots : m_n),$$

it is clear that $\mathbf{P}(E_i) = m_i/m$, $m = m_1 + m_2 + \ldots + m_n$, because the sum must equal 1. In other cases (for any E_i whatsoever, even if they are compatible), one can make the common divisor m enter in explicitly, for example by adding in $1 =$ the certain event:

(4) $$\mathbf{P}(E_1 : E_2 : \ldots : E_n : 1) = (m_1 : m_2 : \ldots : m_n : m).$$

The resulting convenience is most obvious when the m_i are small integers. For example, if A, B, C form a partition $(A + B + C = 1)$, by writing $\mathbf{P}(A : B : C) = (1 : 5 : 2)$ (even without the refinement $\mathbf{P}(A : B : C : 1) = (1 : 5 : 2 : 8)$) it becomes obvious that

$$\mathbf{P}(A) = 1/8 = 12 \cdot 5\%, \mathbf{P}(B) = 5/8 = 62 \cdot 5\%, \mathbf{P}(C) = 2/8 = 25\%.$$

At this point we shall also introduce the operation of the *term-by-term* product of multi-ratios, denoting it by $*$:

(5) $$(a_1 : a_2 : \ldots : a_n) * (b_1 : b_2 : \ldots : b_n) = (a_1 b_1 : a_2 b_2 : \ldots : a_n b_n).$$

This frequently provides some advantage in handling small numbers or simple expressions in a long series of calculations, and will turn out to be particularly useful for the applications to likelihood which we mentioned above.

The time has now come to end this digression concerning methods of numerically denoting probabilities and to return to questions of substance.

5.3 HOW TO THINK ABOUT THINGS

5.3.1. In discussing the central features of the analysis which must underlie each evaluation, it will be necessary to go over many things which, although obvious, cannot be left out, and to add a few other points concerning the calculus of probability.

The following recommendations are obvious, but not superfluous:

to think about every aspect of the problem;

to try to imagine how things might go, or, if it is a question of the past, how they might have gone (one must not be content with a single possibility, however plausible and well thought out, since this would involve us in a *prediction*: instead, one should encompass all conceivable possibilities, and also take into account that some might have escaped attention);

to identify those elements which, compared with others, might clarify or obscure certain issues;

to enlarge one's view by comparing a given situation with others, of a more or less similar nature, already encountered;

to attempt to discover the possible reasons lying behind those evaluations of other people with which, to a greater or lesser extent, we are familiar, and then to decide whether or not to take them into account. And so on.

In particular, in those cases where bets are made in public (e.g. horse-races, boxing-matches—in some countries even presidential elections) some sort of 'average public opinion' is known by virtue of the existing odds. More precisely, this 'average opinion' is that which establishes a certain 'marginal balance' in the demand for bets on the various alternatives. This might be taken into consideration in order to judge, after due consideration, whether we wish to adopt it, or to depart from it, and if so in which direction and by how much.

5.3.2. In order to provide something by way of an example, let us consider a tennis match between two champions, *A* and *B*.† You will cast your mind back to previous matches between them (if any); or You will recall matches they had with common opponents (either recently, or a long time ago, under similar or different conditions); You will consider their respective qualities (accuracy, speed, skill, strength, fighting spirit, temper, nerves, style, etc.) and the variation in these since the last occasion of direct or indirect comparison; You will compare their state of health and present form, etc.; You will try to imagine how each quality of the one might affect, favourably or otherwise, his opponent's capacity to settle into the game, to fight back when behind, to avoid losing heart, etc. For instance, you may think that *B*, although on the whole a better player, will lose, because he will soon become demoralized as a result of *A*'s deadly service. However, it would be naïve to stop after this first and lone supposition: it would

---

† This example has already been discussed by Borel and again by Darmois (cf. p. 93 of the Borel work mentioned previously, and again on p. 165 Darmois' note VI). As is clear from this and other examples (like his discussion, again on p. 93, of the evaluation of a weight—similar to our example in Chapter 3, 3.9.7), Borel seems to be inspired—in the greater part of his writings— by the subjectivistic concept of probability: he can thus be regarded as one of the great pioneers, although incompatible statements and interpretations crop up here and there, as was pointed out in the footnote to Section 5.2.3.

mean to aspire to making a prediction rather than a prevision. You will go on next to think of what might happen if this initial difficulty for *B* does not materialize, or is overcome, and little by little You will obtain a summary view—but not a one-sided or unbalanced one—of the situation as a whole. Your ideas about the values to attribute to the winning probabilities for *A* and *B* will in this way become more precise. You may have the opportunity to compare your ideas and previsions with those of other people (in whose competence and information You have a greater or lesser confidence, and whom You may possibly judge to be more or less optimistic about their favourite). In the light of all this, You might think over your own point of view and possibly modify it.

5.3.3. Our additional remarks concerning the calculus of probability consist in pointing out that the conditions of coherence, even if they impose no limits on the freedom of evaluation of any probability, do in practice very much limit the possibility of 'extreme' evaluations. More precisely, an *isolated* eccentric evaluation turns out to be impossible (the same thing happens, for instance, to a liar, who, in order to back up a lie, has to make up a whole series of them; or to a planner, who must modify his entire plan if one element is altered).

It is easy to say 'in my opinion, the probability of *E* is, roughly speaking, twice what the others think it is'. However, if You say this, I might ask 'what then do You consider the probabilities of *A, B, C* to be?', and, after You answer, I may say 'so do You think the probability of *H* is as small as this; $\frac{1}{10}$ of what is generally accepted?', and so on. If You remain secure in your coherent view, You will have a complete and coherent opinion that others may consider 'eccentric' (with as much justification as You would have in calling the common view 'eccentric'), but will not otherwise find defective. However, it will more often happen that as soon as You face the problem squarely, in all its complexity and interconnections, You come to find yourself in disagreement *not only with the others but also with yourself,* by virtue of your eccentric initial evaluation.

We have been talking in terms of bets and the evaluations of probability, and not of previsions of random quantities, although they are the same thing in our approach. This was simply a question of the convenience of fixing ideas in the case where the probabilistic aspect is most easily isolated: however, one should note that the same considerations could in fact be extended to the general situation.

5.4 THE APPROACH THROUGH LOSSES

The betting set-up is related to the 'first criterion' of Chapter 3, Section 3.3; the scheme we are now going to discuss is based on the 'second criterion': it is this latter—as we remarked previously, and will shortly see—which turns out to be the more suitable.

First of all, we shall find it convenient to present this scheme right from the beginning again, referring ourselves now to the case of events. Because this is the simplest case, and because we are treading an already familiar path—which we shall illustrate clearly with diagrams—everything should appear both more straightforward and of wider application.

5.4.1. Instead of some general random quantity X, You must now think in terms of an event E, such that You are free to choose a value x, bearing in mind that You face a loss

(6) $$L = L_x = (E - x)^2.$$

Expanding this (remembering that $E^2 = E$), one obtains the following alternative expressions (in the last one, p is any number whatsoever):

(7)
$$
\begin{aligned}
\text{(a)} \quad L_x &= x^2 + (1 - 2x)E, \\
\text{(b)} \quad &= x^2 \tilde{E} + (1 - x)^2 E, \\
\text{(c)} \quad &= E(1 - p) + (p - x)^2 + (E - p)(p - 2x).
\end{aligned}
$$

They all reveal ((b) most explicitly) that L_x equals x^2 or $(1 - x)^2$ according as $E = 0 = $ false or $E = 1 = $ true.

Since we have already used the criterion as a definition—and hence already know what the probability $p = \mathbf{P}(E)$ of E is—we can, 'being wise after the event', examine how the criterion behaves by looking at $\mathbf{P}(L_x)$, considered as a function of a value x and of a probability p, assumed to be arbitrary (so we adopt the notation $L_x(p)$). Putting $E = p$ in (a), (b), (c) (which are linear in E) we obtain:

(8)
$$
\begin{aligned}
\text{(a)} \quad L_x(p) &= x^2 + (1 - 2x)p \\
\text{(b)} \quad &= x^2 \tilde{p} + \tilde{x}^2 p \\
\text{(c)} \quad &= p(1 - p) + (p - x)^2 = p\tilde{p} + (p - x)^2.
\end{aligned}
$$

5.4.2. We now examine the variation in $L_x(p)$ as p varies, x being an arbitrary fixed value. As might have been expected ((a) shows this up most clearly), L_x varies linearly from $L_x(0) = x^2$ to $L_x(1) = \tilde{x}^2$ (which are the two possible values for L_x, depending on the occurrence of either $\tilde{E}(p = 0)$ or $E(p = 1)$). The straight lines in Figure 5.2, connecting these extreme values, give a visual impression of how they go together: i.e. of how, in order to reduce the penalty resulting in one case, one must increase it in the other.†

---

† This is for $0 \leqslant x \leqslant 1$: we already know, and can also see, that, in every case, $x < 0$ or $x > 1$ is worse than $x = 0$ or $x = 1$, respectively, and is thus automatically ruled out (without need of any convention).

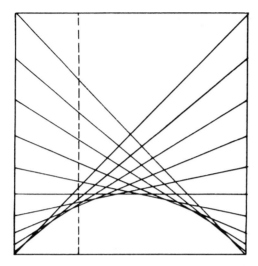

Figure 5.2 The straight lines correspond to the combinations of penalties among which the method allows a choice (the penalty can be reduced in one of the two cases at the expense of increasing it in the other: lowering the ordinate at one end raises it at the other). The ordinate of a particular straight line at the point p is the prevision of the loss for the person who chooses that line and attributes probability p to the event under consideration. In this case, the minimum value that can be attained is given by the ordinate of the parabola (no straight line passes beneath it!), and the optimal choice of straight line is the tangent to the parabola at the point with abscissa p

The figure also shows, in an indirect way, the variation of $L_x(p)$ for varying x, with p fixed. Geometrically, one can see (and (c)) presents it explicitly) that the straight lines are the tangents to the parabola $y = p(1 - p) = p\tilde{p}$, and that none of them can go beneath their envelope (this is within the interval $[0, 1]$: the others would correspond to values $x < 0$ and $x > 1$; cf. footnote). Given p, the best one can do is to take the tangent at p, obtained (as we already know!) by choosing $x = p$: this gives $L_x(p)$ its minimum value (as x varies), $L_x(p) = p\tilde{p}$. Choosing a different x gives rise, in prevision, to an additional loss $(x - p)^2$; i.e. the square of the distance from x to p: (c) shows this explicitly, by splitting the linear function $L_x(p)$ into the sum of $p(1 - p)$ (the parabola) and $(x - p)^2$ (the deviation from the parabola of the tangent at $p = x$). We observe also, and this confirms what has been said already, that this deviation is the same for all the tangents (starting, of course, from their respective points of contact).

The maximum loss is 1, and this is achieved by attributing probability zero to the case that actually occurs: the minimum loss is 0, and is achieved when a probability of 1 (or 100 %) is attributed to this case. For any given x, the loss varies between x^2 and $\tilde{x}^2$ (as we have seen already). For a given p, we already know that the minimum is $p\tilde{p}$ (for $x = p$), and it is readily seen that the minimum is $p \vee \tilde{p}$: more precisely, if $p \leqslant \frac{1}{2}$ it is $1 - p$, obtained by choosing $x = 1$; if $p \geqslant \frac{1}{2}$ it is p, obtained by choosing $x = 0$. If $p = \frac{1}{2}$, we have the maximum of the minimum ($p\tilde{p} = \frac{1}{4}$), and the minimum of the maximum ($p \vee \tilde{p} = \frac{1}{2}$), and hence the largest discrepancy (max $-$ min $= \frac{1}{2} - \frac{1}{4} = \frac{1}{4}$); in general, the discrepancy is $x^2 \vee \tilde{x}^2$, i.e. the maximum of x^2 and $(1 - x)^2$, and attains its maximum ($= 1$) for $x = 0$ and $x = 1$.†

5.4.3. *The case of many alternatives.* We can deal with the case of many alternatives (of a multi-event, of a partition), and the more general case of any number of arbitrary but not incompatible events, by applying the previous scheme to each event separately. In this way, things reduce to the treatment given in Chapter 3, and to the geometric representation which was there illustrated. Here, we simply wish to review the approach in the spirit of the above considerations, and then to look at a few modifications.

It will suffice to consider a partition into three events (such as E_1, E_2 and E_3 of Chapter 3, 3.9.2). We shall call them A, B and C ($A + B + C = 1$) and represent them as points, $A = (1, 0, 0)$, $B = (0, 1, 0)$ and $C = (0, 0, 1)$, in an orthogonal cartesian system. For the time being, we shall distinguish the probabilities, $p = \mathbf{P}(A)$, $q = \mathbf{P}(B)$ and $r = \mathbf{P}(C)$, attributed to them, from the values x, y and z chosen in accordance with the second criterion (we know they must coincide, but we want to investigate what happens if we choose them to be different, either through whim, oversight or ignorance).

We shall denote by P the prevision-point $P = (p, q, r)$; the decision-point will be denoted by P'', $P'' = (x, y, z)$: see Figure 5.3.

The total loss will then be

$$(9) \qquad\qquad L = (A - x)^2 + (B - y)^2 + (C - z)^2$$

and

$$(10) \qquad \mathbf{P}(L) = [p\tilde{p} + q\tilde{q} + r\tilde{r}] + [(p - x)^2 + (q - y)^2 + (r - z)^2];$$

in other words,

$$\mathbf{P}(L) = (\text{first term involving only the prevision-point } P) + (P'' - P)^2,$$

---

† Among other decision criteria which are employed (inspired by points of views which differ from ours) is one which is called the 'minimax' criterion: it consists in taking that decision which minimizes the maximum possible loss. Observe that, in the above situation, this criterion would have us always choose $x = \frac{1}{2}$ (then, in fact, the loss would be $= \frac{1}{4}$, with certainty, whereas every other choice would give a smaller loss in one of the two cases, although greater in the other). Since it is incoherent to attribute probability $\frac{1}{2}$ to all events, such a criterion is absurd (in this kind of application; not so, however, in the theory of games—cf. Chapter 12, 12.7.4—where it provides a solution in situations of a different kind, nor even in this situation under an hypothesis of an extremely convex utility function where it would no longer lead to the choice of $p = \frac{1}{2}$).

the latter being the square of the distance between P'' and P. Hence, in order to avoid an extra loss, whose prevision is equal to the square of the distance between P'' and P, the point P'' must be made to coincide with P.

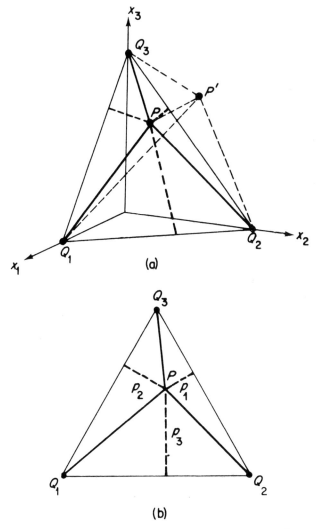

Figure 5.3 The triangles of points such that $x + y + z = 1$ (x, y, z non-negative) seen in space, (a), and in the plane, (b). It is clear from geometrical considerations that the choice of a loss rule corresponding to the point (x, y, z) is inadmissible (in the case of three incompatible events) if it is not within the given triangle. Moreover, if one attributes the probabilities (p, q, r) to the three events, it pays then to choose $x = p$, $y = q$, and $z = r$. In other words, the method rewards truthfulness in expressing one's own evaluations

The argument given previously (Chapter 3, 3.9.2) was saying the same thing, but without reference to a preselected prevision **P**. Given a point $P'' = (x, y, z)$, outside the plane of A, B and C (i.e. with $x + y + z \neq 1$), its orthogonal projection P' onto this plane has distance less than P'' from A, B and C; if P' falls outside the triangle ABC, the above-mentioned distances decrease if one moves from P' to the nearest point P on the boundary. This shows that only the points of the triangles are admissible (in the sense of Pareto optimality; there are no other points giving better results in all cases). The present argument is less fundamental, but more conclusive, because—assuming the notion of probability to be known in some way (e.g. on the basis of the first criterion)—it shows how and why the evaluations x, y, z of the second criterion must be chosen to coincide with the probabilities p, q, r of A, B and C.

5.4.4.   We have here dealt with the most formally immediate case, that of applying to the different events (A, B, C) one and the same scheme with the same maximum loss, namely unity. We know, however (see Chapter 3, 3.3.6), that, so far as the evaluation of probabilities is concerned, and this is what interests us, no modifications would be required were we to use different coefficients: for instance, if one were to take

$$L = a^2(A - x)^2 + b^2(B - y)^2 + c^2(C - z)^2$$

with arbitrary a, b, c. Geometrically, the three orthogonal unit vectors, $A - O$, $B - O$, $C - O$, must now be taken to have lengths a, b and c. This implies—and it is this aspect which may be of interest to us—that the loss, which always equals the square of the distance, is given by $(A - B)^2 = a^2 + b^2$, if in prevision all the probability is concentrated on A, and B actually occurs (and conversely): similarly for $(A - C)^2 = a^2 + c^2$ and $(B - C)^2 = b^2 + c^2$. In the plane of A, B and C, the triangle ABC can be any acute-angled triangle (in the limit, if one of the coefficients is zero, it can be right-angled): in fact, $a^2 = (B - A) \times (C - A) = \overline{AB} . \overline{AC} . \cos \widehat{BAC}$, $\cos \widehat{BAC} > 0$, etc. In any case, the scheme would work in the same way even if ABC were taken to be any triangle whatsoever, although if it were obtuse-angled, we could not obtain it as we just did in orthogonal coordinates (merely by changing the three scales). This is obvious by virtue of the affine properties, a point we have made repeatedly. In the general case, the only condition imposed on the three losses $\overline{AB^2}, \overline{AC^2}, \overline{BC^2}$ is the triangle inequality for $\overline{AB}, \overline{AC}, \overline{BC}$.

5.4.5.   Why are we bothering about the possibility of modifying the shape of the triangle: i.e. the ratios of the losses in the different cases? After all, this is irrelevant from the point of view of evaluating probabilities. Despite this, it may sometimes be appropriate to draw a distinction between the more serious 'mistakes', and the less 'serious' (the former to be punished by greater losses), in those cases in which the losses could also serve as a useful means of comparison when considering how things turn out for different individuals (as we shall see shortly).

A good example, and one to which we shall subsequently return, is that of a football match (or some similar game) in which the following three results are possible: A = victory, B = draw, C = defeat. In the most usual case (triangle ABC equilateral), one considers it 'equally bad' if either a draw or defeat results when one has attributed 100% probability to victory. If, on the other hand, the distance between victory and defeat is considered greater than the distance between each of these and a draw, we could take an isosceles triangle with the angle *B* greater than 60° ; if we take this angle <90°, we have a combination of three losses for the three results, and the loss for victory–defeat will be less than twice the loss for draw–defeat (or for draw–victory). For a right angle, this ratio will be exactly double (the ratio of the sides = $\sqrt{2}$) and the scheme will only be applicable to the events victory and defeat (a draw is only taken into account as complementary to the other two). For angles between 90° and 180°, the interpretation as combinations of losses for the three events no longer holds; for the case of a draw, the loss would have to be *negative* in order for things to proceed smoothly! The 180° case means that we are effectively considering prevision in terms of 'points' (0 for a defeat, 1 for a draw, 2 for a victory), in the sense that previsions like (0, 1, 0) and $(\frac{1}{2}, 0, \frac{1}{2})$—i.e. of being certain of a draw, or of equal probabilities for victory and defeat, excluding a draw—are considered identical.

5.5 APPLICATIONS OF THE LOSS APPROACH

5.5.1. The employment of this method (or something similar) by various people for evaluating probabilities should be given great emphasis and, for many, many reasons, deserves wide publicity.

Sometimes, one is interested in knowing the opinion of a given individual, or of various individuals, concerning the probabilities of certain events under consideration. Sometimes, in order to make some kind of psychological analysis, one is interested in knowing how the various individuals react to information, or other new factors. In certain other cases, it might be interesting to be able to judge, in a more precise fashion, the extent of the 'partial knowledge' of individuals under examination: for instance, one might discourage them from 'guessing'.† And so on.

In all these cases, one should take into account the no less important value of repeated experiences of this kind. They greatly aid one in acquiring the 'feeling for numerical values' with which one expresses 'degrees of belief', and hence they contribute to building up a keen and accurate understanding of the problems of prevision, and of the spirit—not cut-and-dried—in which probability theory must approach them.

5.5.2. With this aim in mind, we must now supply all the details of the method. It must be understood that it is preferable to express one's own evaluations sincerely and accurately, and that otherwise one suffers a loss, equal, *in prevision* (in one's own evaluation), to the square of the distance between one's own true evaluation and the one expressed. In addition, there is a definite advantage in obeying the conditions of coherence (in our example; $x, y, z \geq 0, x + y + z = 1$): to do otherwise is to arrange to suffer

† By 'guessing', we mean 'guessing at random'. This should not be confused with the usage conveyed in Pólya's 'Let us teach guessing', where it means to make useful conjectures (first guess, then prove!).

one part of the loss *with certainty*. If, instead, one wishes to check—in a decision-theoretic sense—the ability of a given individual to do the right thing without having a systematic knowledge of the situation and of the theory, the characteristic features of the method should not be revealed (except for mentioning what losses are). This is a different problem, however; a far cry from those for which we have introduced the method under present discussion (and it seems unlikely, anyway, that anyone could come to sensible decisions without knowing and applying—with great care!—the theory of probability).†

Let us now proceed to some concrete examples of various types of applications.

5.5.3. *The opinions of experts.* It often happens that one turns to the experts for information. This is, in actual fact, nothing other than an evaluation of probability. One is not always in a position to weigh-up for oneself all the probabilities relevant to a given situation; this then is the time to behave like the Prince, who, according to Machiavelli, 'sometimes understands things by himself, sometimes through the understanding of others: while the former is excellent, the latter is also very good'.

An example, one of thousands, is given by the case of a geologist who is asked to give an opinion as to whether it is worth drilling a hole at a particular site during an oil search. This is a useful example to consider, since it has, to some extent, been treated by Grayson,‡ and so the interested reader can delve deeper into those aspects which we shall not discuss. The geologist himself does not have any say in the final decision of whether or not to drill: this decision must be taken (by the 'decision-maker') after consideration of all the various pieces of information, of which that of the geologist is just one. He, for his part, cannot state categorically that oil is present or not present (thus making a prediction rather than a prevision), nor can he sin in the opposite direction and merely list the information about

---

† Experiments of this kind, which are made in order to check the extent to which actual behaviour conforms to the norms derived from the theory of probability, are often considered as 'proving' or 'disproving' the validity of probability theory (or of the related theory of decision-making under conditions of uncertainty). This would be so if such theories were to be regarded as empirical–psychological theories of actual behaviour, but, in fact, it is completely at odds with what we are considering here: a *normative* theory for *coherent* behaviour.

Many criticisms derive from this confusion (or from the refusal to accept that a subjectivistic theory can distinguish incoherent and coherent behaviour, rather than just being an acritical, empirical observation of actual behaviour as it happens to be). This kind of empirical evidence is also of interest from our standpoint, but in the same way as a mathematician might find the mistakes of laymen, students, or even other mathematicians, interesting. He does not modify mathematics by incorporating these 'mistakes', as though, simply because someone has enunciated them, they 'should' be included by virtue of their being part of some psychological truth, or of the indiscriminate collection of mathematical statements made in the course of history.

‡ C. J. Grayson, *Decisions Under Uncertainty: Drilling Decisions by Oil and Gas Operators*, Harvard Business School (1960).

the geology of the area (reliable, but analytical), leaving to others the task of synthesis and drawing some conclusions. The synthesizing and the conclusions about the probable outcome of the drilling—given from a geological standpoint—are precisely what his expertise is called upon to provide.

In actual fact, the geologist's report does provide this answer, but usually couched in extremely vague adjectives or phrases (such as: fairly good prospects, or good, favourable, uncertain, promising, etc.; sometimes preceded by little words like 'very', 'not very', 'quite', 'rather', and followed cautiously by 'unless anything unexpected happens', 'perhaps', 'it's difficult to say', 'in my humble opinion',..., 'God only knows'). The only solution worthy of serious consideration is to have the geologist express the probabilities numerically, and some companies actually do this. The objection could be raised (and often is) that the knowledge of the geologist is too vague to be represented numerically. It would certainly be unwise and over-zealous to assert that the probability of striking oil at a given site is 0·1307594, but to state that the probability is 0·131, or 0·13, or even simply 10–15 %, is always preferable to a string of adjectives whose vagueness depends upon the nature of the opinion itself, on the inadequacy of language, and, perhaps, on a desire to state the conclusions in the least compromising way—i.e. essentially ambiguous, but not appearing to be.†

5.5.4. There remains the problem, however: *how can we interest the expert*—in our case the geologist—*in giving an honest answer; in expressing accurately his deep-felt belief*? This problem was examined by Grayson in the light of the 'first criterion', without any satisfactory solution being obtained. The method we suggest here—that of the 'second criterion'—would seem to give a perfectly satisfactory solution, and is precisely what Grayson requires; 'a system to discourage falsification'. For the practical application at present under consideration, it would be sufficient to agree that some part of the agreed fee (neither insignificant, nor excessive; say, 5–10 %) be held back until the eventual outcome was known, and then the loss deducted (up to a maximum of the amount held back) before payment. In certain cases, however, like those of experts who are consulted regularly, or who hold positions within the firm, one might also add up the losses—expressed as 'scores'—in order to make global comparisons (of the 'goodness' of the previsions of two individuals based on comparisons of the cases examined by both of them). These comparisons could be made separately according

---

† Someone made the acute observation that often the ability to make accurate predictions consists in expressing them in a sufficiently imprecise fashion (this principle is mentioned on p. 213 of Good's anthology—cf. the next footnote but one—and also in a review article of mine; cf. *Civilta delle macchine*, No. 1 (1963), 71–72). On the other hand, the limit-case of Sibylline predictions ('*Ibis, redibis*, ... ') is well-known.

to 'type' of problem, time-period, etc., and could be taken into account when considering the merits of someone in connection with appointments, promotion, etc.

The following discussion is useful both in real-life and as an example.

5.5.5. Forecasting sports results. We consider sports results, football in particular, because they give plenty of scope for experiments of this kind: they can be observed regularly (e.g. every weekend) and sufficiently often; the outcomes are clear-cut (in football, the home team either wins, loses or draws), officially ratified, and the situation is well-known to most people. In addition, there is considerable background information and comment in the newspapers. However, leaving aside the convenience (for the reasons given above) of sports results, we could consider forecasting in any area (e.g. politics, economics, meteorology, everyday affairs, culture, judicial or sanitary matters, personal or business affairs, etc.).

There are, as is well-known, various organized pools for betting on football and horse-racing. These, however, are motivated by the concept of 'prediction', in that they reward those who *guess* all (or almost all) of the *results*. Moreover, the sensitivity of the system is completely distorted by the practice of sharing out the available prize money among the winners. Indeed, the net result of all this is as follows: those who write down ridiculous forecasts, that by chance turn out to be correct, receive fantastic prizes; whereas those who write down forecasts which could reasonably be thought probable receive, if they win, only very small amounts, since the prize, in this case, will presumably have to be shared with many others.† Consequently, the 'most reasonable' way to gamble would not be to bet on the result for which the probability of occurrence is highest, but, instead, to consider the probability multiplied by the prevision of the reciprocal of the number of people betting on it, and to bet on the result for which this is highest.

The betting approach that we discussed previously, illustrating its merits and demerits, is in line with the notion of prevision (as opposed to prediction). The scheme we are now going to present is intended to build on the merits, and eliminate the demerits. It should, therefore, permit us to achieve those goals which we have already mentioned: to develop a feeling for what a prevision (not a prediction) is, and a feeling for the numerical scale on which it is to be expressed; to teach one how to take into account the relevant circumstances, bearing in mind one's own level of competence. Moreover, all this is achieved within the agreeable format of a competition, there being

---

† This brings to mind a rather tragic story: a man died, overwhelmed with joy, on learning that he had guessed correctly all the 13 football results on the Italian football pools. In fact, he was lucky, because otherwise he would have died of disappointment the next day on learning that his winnings were so small (about 3000 Lire), owing to the predictability of the results, which were therefore foreseen by many others besides himself.

the additional opportunity to reflect and to compare, after the event, one's own previsions with those of others, and with the results themselves. It will be necessary to consider rather carefully the latter point; i.e. *'being wise after the event'*. We shall do so in Sections 5.9 and 5.10 of this chapter, and will come back to it on several occasions later in the book.

5.5.6. One could organize a competition more or less along the following lines (this has already been tried, although on a small scale).† The participants have to hand in, each week, previsions for the forthcoming matches, giving, for each match, the probabilities (expressed in percentages) of the three possible outcomes (in the order: win, draw, defeat); writing, for instance, 50–30–20, 82–13–05, 32–36–32, etc. Given the results, one can evaluate, game by game, the losses and the total losses for the day (and, possibly, a prize for the day), as well as the cumulative sum needed for the final classification. This final classification must be seen as the primary objective. If there are prizes, the largest should be reserved for the final placements, and, in order to conform to the spirit of the competition, the prizes must complement losses; i.e. they should depend on them in a *linearly* decreasing fashion. ‡

The lessons of experience tell us much about the necessity of avoiding the mentality of prediction when making previsions. It is true that total success—i.e. no penalty— is achieved if and only if the whole probability, 100 %, is attributed to the case which actually occurs. For this reason, many find it tempting, especially at first, to attempt to get the result spot-on, with evaluations which ignore the possibility of uncertainty (i.e. 100–00–00, 00–100–00 or 00–00–100, which are equivalent, in the notation of the football pools, to the 'predictions' '1', 'X', '2'). However, these participants come to realize very quickly that they have fallen behind—this happens on the individual days, but shows up most in the final classification—relative to those who distribute probability in a sensible way: they soon modify their approach.

We shall come back to this example later.

5.5.7. Replies to multiple-choice questions. One is often required in a 'quiz', or even in an examination (especially in America), to choose from among a few given answers the one which one believes to be correct. The exact details may differ somewhat: one may either have to tick one and only one answer; or be allowed to choose none; or to choose a subset within which the correct

---

† It was tried twice, in 1960–61 and 1961–62, in the Economics Faculty of Rome University. There were about 30 participants (students and a few teachers) on each occasion, and the study centred on the 9 football matches played every week in the first division of the Italian league. Some discussion of this can be found in B. de Finetti, 'Does it make sense to speak of "good probability appraisers"?' in the volume entitled *The Scientist Speculates: An Anthology of Partly-baked Ideas* (edited by I. J. Good) Heinemann, London (1962). The experiment was repeated again in Rome (Faculty of Science) from 1966 on, and experiments of this kind have recently been made in the United States.

‡ If, for example, no prize is to be awarded to those who come last (by whatever ruling is proposed), not only do the tail-enders have no motivation to exercise care in their evaluations, but, on the contrary, they have a vested interest in trying outlandish evaluations, which they presume to be different from those of individuals in a better position. This is their only hope of over-taking them and getting a prize. If the first prize were extremely large, the temptation to behave in this way would be greatest for those in second place on the next to last day. In any case, such a distortion of interest occurs whenever *linearity* is abandoned.

answer is thought to lie (and, in this case, there are two variants, according as one indicates an order of preference or not). In any case, there must be an agreed method of scoring according to the way in which the answers given compare with the correct answers. A problem arises from the necessity of discouraging people from 'guessing': this is often dealt with by estimating statistically what the effect of the assumed presence of 'guessing' would be, in a mass of people.

This latter problem is completely resolved if one applies the method under consideration.† Observe that, in this context, there is no question of events which could be considered 'uncertain' in some 'objective' sense. For example: it is clear that if we ask which of A = Antonio, B = Brutus, C = Caesar said the famous line 'Alea jacta est', we are not asking for any sort of testimony or opinion concerning the fact that some great man uttered the phrase in the course of his life; we simply wish to check whether the examinee knows that the phrase relates to Caesar and the crossing of the Rubicon. In the same way, if we ask whether $\log x + \log y$ equals

$$A = (\log(x + y)), \qquad B = (\log xy) \quad \text{or} \quad C = (\log(e^x + e^y));$$

or whether $\sqrt{26}$ is A = rational, B = algebraic or C = transcendental; or whether, at the battle of Waterloo, Napolean A = won, B = lost or C = drew; or whether the city of Bahia is in A = Argentina, B = Brazil or C = Chile; and so on...; in all these cases, the probability, the doubt to be measured, comes solely from the ignorance, uncertain knowledge, or bad memory of the person questioned.

In every other respect, on the other hand, the situation is identical to that of the football pools: for the person who judges, for the person concerned with his state of doubt, there is no difference. It is sufficient to realize that a person could forecast the football results on a Sunday evening, when the facts are part of the past and are known to everybody (provided they are not known to him), or even a year later, provided that he then recollects them with something less than certainty.

5.5.8. The adoption of the proposed system in the case of multiple-choice questions would turn out to be instructive, in addition to the reasons which hold generally (i.e. learning how to express one's own opinion by translating it into numerical values), for the 'lesson' which would show how it is also *advantageous* (where sensible rather than stupid rules are in use) to strive for the greatest honesty and accuracy in expressing one's own doubts or lacunae. Conversely, stupid rules (like stupid laws) encourage dishonesty and reticence, encourage that complex of underhand and stupid actions

---

† The betting approach, on the other hand, could not be used. Anyone in a state of some doubt would certainly lose against an opponent (e.g. the examiner) who knows the right answer.

which are euphemistically described by the phrase 'trying to be clever'; in our case, they encourage 'guessing'.

For the examiners too, it would be extremely useful to have precise information about those who 'know' (e.g. those who write down Antonio 00%, Brutus 00%, Caesar 100%), with the suspicion of 'guessing' now removed, and even more to be able to make a detailed analysis, on the basis of precise and meaningful data, of the frequency, intensity and nature of the doubts (possibly with a view to investigating their origin and suggesting ways of dealing with inadequacies in the teaching). In addition, they would be able to examine the degree of accuracy with which the evaluations are made (e.g. not simply using 50%–50% if there is uncertainty between two alternatives). In the case under consideration, there could, of course, be any number of alternatives whatsoever; in the examples above, we considered three for convenience, and in order to be able to retain the analogy with football results, and the possibility of imagining the situation as always representable in terms of Figure 5.3.

5.5.9. *Applications in economics.* In the field of economics, the importance of probability is, in certain respects, greater than in any other field. Not only is uncertainty a dominant feature, but the course of events is itself largely dependent on people's behaviour, which is itself determined, in a more or less unconscious and confused fashion, by evaluations and arguments of a probabilistic nature. It is, therefore, probability theory, in the broadest and most natural sense, that best aids understanding in this area (and not those fragments of the theory which never progress beyond the drawing of 'equally likely' balls from an urn, or 'stable' frequencies).

This point of view was presented in a clear and authoritative manner by T. Haavelmo in a celebrated critical speech delivered as president of the Econometric Society,† where he stated that previsions and evaluations of subjective probabilities '*are realities in the minds of people*' and that it was to be hoped that '*ways and means can and will be found to obtain actual measurements of such data*'.

Another point, of particular importance for applications in operational research, is the possibility of making use of those evaluations of probability which represent a decision-maker's own opinions. For example, only the decision-maker himself can say what probabilities he attributes to the different reactions of his most direct competitors to possible decisions of his. How, though, are we to interrogate him? Indirect approaches are necessary; questions about his preferences under some hypothetical sets of conditions, should be posed, in such a way as to provide, in turn, both a complete picture

---

† Trygve Haavelmo, *The rôle of the econometrician in the advancement of economic theory*, Presidential Address, Meeting of the Econometric Society, Philadelphia, 29 Dec. 1957; cf. *Econometrica* **26** (1958), 351–357.

and a check of consistency. These are, however, expedients to make up for the lack of training in expressing oneself in terms of probabilities; the difficulty would not exist if such training became general practice.

Finally (in order not to dwell on too many other aspects†), there are important applications to the more theoretical field of econometric models. As E. Malinvaud says, in his treatise on statistical methods in econometrics,‡ the justification of the introduction of random models into econometrics rests, in his view, on an appeal to subjective probabilities, so that 'l'établissement d'une statistique subjectiviste qui reposerait sur le principe de Bayes' would be desirable (even though, in his opinion, research in this direction is, as yet, not sufficiently advanced to make a systematic application possible: on the other hand, there are those, e.g. A. Zellner,§ who are attempting to do this).

5.6 SUBSIDIARY CRITERIA FOR EVALUATING PROBABILITIES

Having analysed the meaning and the method of evaluating probabilities that a person might be led or compelled to make in order to sort out his ideas about what might occur, and to choose wisely any decision that has to be made, we are now in a position, and are in fact obliged, to return to the essence of the problem of evaluation. We wish to discover whether the task of translating more or less vague impressions and opinions into numerical form could be facilitated by using some suitable subsidiary criterion. Fortunately, this turns out to be the case.

This happy circumstance derives, in general, from the observation that in many cases in the calculus of probability, under restrictions which are often very natural, certain probabilities, which are calculated on the basis of certain others, vary very little as one's evaluations of these other probabilities are varied. Consequently, even if the latter seem, to a given individual, rather vague, the former may very well appear to him capable of being evaluated with sufficient precision and confidence. As a brief aside on the question of interpersonal comparisons, we note that this explains why individuals often make practically identical judgments of prevision, even though they start off with very different opinions.

---

† I have recently provided a wide ranging discussion of these topics (with a fairly mathematical treatment) in 'L'incertezza nell'economia', part I of: B. de Finetti and F. Emanuelli, *Economia delle assicurazioni*, Vol. XVI of *Trattato italiano di economia* (Edited by C. Arena and G. Del Vecchio), Utet, Torino (1967).

‡ Edmond Malinvaud, *Méthodes statistiques dans l'économétrie*, Dunod, Paris (1964).

§ Arnold Zellner, *An Introduction to Bayesian Inference in Econometrics*, Wiley (1970). It should be noted, however, that, although the treatment is Bayesian, the interpretation is not subjectivistic. The choice of the initial distribution does not derive from a case-by-case consideration of the factual circumstances, but from adopting once and for all a mathematically convenient form for each type of problem.

These general considerations will become clearer as we proceed further. For the time being, we restrict ourselves to illustrating the two subsidiary criteria which are of the greatest and most immediate interest : the first one we shall deal with in a reasonably detailed manner ; the second, which, from a logical point of view, is based on material we shall meet much later on, is dealt with in a necessarily superficial way.

5.7 PARTITIONS INTO EQUALLY PROBABLE EVENTS

5.7.1. Every quantitative measurement is made both easier and more precise when it is possible to reduce it to a qualitative comparison. For example; it is much easier to say that A. N. Other has eaten $\frac{2}{9}$ (i.e. about 22·2 %) of a cake knowing that it was divided into 18 pieces, which could be taken as equal, and that he has eaten four of them, than to directly estimate that his portion was 22·2 % of the whole, undivided cake. In precisely the same way, it is obvious that if I judge n events of a partition to be equally likely, I cannot avoid attributing probability $p = 1/n$ to each of them (because the sum of the n terms, each equal to p, must be 1). Judgments of this kind arise rather frequently : it is sufficient that, given the present state of information, one finds oneself in a situation of *symmetry*. This will often, although not necessarily always, reduce to a state of *symmetry* regarding certain physical, or at any rate external, circumstances, which we regard as essential and relevant elements of our state of information.

When tossing a coin, we usually attribute the same probability $\frac{1}{2}$ to both faces, and, similarly, probability $\frac{1}{6}$ to each of the six faces of a die. If we have n balls in an urn, we again, in general, attribute the same probability $1/n$ to any particular one of them being drawn : in this case, if we also know that m of the balls are white, we have no choice but to attribute probability m/n to the drawing of a white ball. This judgment of equiprobability (relative to a single toss, throw or drawing—this is not the place to consider more complicated cases) reflects a symmetric situation which is often made objectively precise by stating that the balls must be identical, the coin and the die perfect (physically symmetric), etc. However, the criterion remains essentially subjective, because the choice, of a more or less arbitrary character, of those more or less objective requirements which are to be included in this concept of 'identical', reflects the subjective distinction drawn by each individual of what is, and what is not, a circumstance that influences his opinion. It was necessary to point this out, in order to avoid giving the impression that in problems of this kind we are dealing with a different kind of probability ; objective rather than subjective. It is true however, that in this context opinions generally do coincide (although the agreement is less strong and unconditional than one would tend to think). Independently of all this, we can always talk about the case of equiprobability, provided we state (or take it as implicit) that this simply means that You (or the individual concerned) attribute the same values to the probabilities in question.

5.7.2. Returning to our examples, we observe that by means of these kinds of set-ups—it might be sufficient just to consider drawings from an urn—we can easily obtain a representation of events of any given probability (to be more precise, any rational m/n). For example, if one wants to get an idea of the magnitude of a probability expressed to two or three decimal places, e.g. expressed in % terms like 13% or 13·2%, it is sufficient to think of an urn with 100 balls, 13 of which are white (or 1000 balls, 130 or 132 of which are white). One can avoid talking about colours, and changing the percentage of white balls, by simply thinking of the balls as numbered consecutively (from 1 to 100 or 1 to 1000): this enables one to say—albeit in less suggestive language—that 13 % is the probability of drawing a number not exceeding 13 (out of 100; or 130 out of 1000); etc.

Using the 'representations' of this 'scale' one can—if the method seems easier—reduce the evaluation of any probability to comparison with cases of this kind, and forget all about both the betting approach and that in terms of losses. In order to translate into figures the probability—according to You—of striking oil by drilling at a given spot, it is sufficient that You decide how many balls, out of 1000, should be white, in order to obtain the same probability of drawing a white ball; if You think the number should be 131, this implies that You think the probability of striking oil is 13·1 %.

It is convenient to express all this formally:

Theorem. *If the n events of a partition are considered as equally probable, the probability of each of them is $1/n$, and the probability of an event which is the sum of m of them is m/n.*

The classical statement is that, under these conditions, *the probability is given by the ratio of the 'number of favourable cases' (m) to the 'total number of possible cases' (n).*

5.7.3. *Criterion of comparison* (or 'third criterion'—following the two in Chapter 3, Section 3.3). Having at one's disposal a model of a partition into n events, which are judged equally probable (e.g. an urn), *the probability of any event E can be evaluated, by comparison with events composed of sums of events of the partition, with an error of less than $1/n$.* In fact, if E_m and E_{m+1} are sums of m and $m + 1$ events, each of probability $1/n$, and if one judges $P(E_m) \leqslant P(E) \leqslant P(E_{m+1})$, then $m/n \leqslant P(E) \leqslant (m + 1)/n$. In order to make the comparison operational, it is sufficient to express it by saying that You would rather receive one lira if E occurs than one lira if E_m occurs, but vice-versa if the comparison is made with E_{m+1}. In this way, its subjective nature is clear; it remains somewhat in the shade when we speak of 'comparison' in the abstract, with no precise meaning.

There are many points, both historical and critical, which one could raise at this juncture, but they would require overlong, and in part untimely, digressions : they will be considered instead at the end of the Appendix.

Let us just say something, however, in order to make the above a little more precise, at least in its essential features. Evaluations made on the grounds of symmetry are generally accepted as a basis for problems concerning games, drawings from an urn, lotteries, dice, etc., and one often regards as 'equally probable cases' certain outcomes which are 'combined' (like the 6^{10} possible sequences obtained by tossing a die 10 times, or the 90!/85! possible sets of five numbers on the Lottery, or the 90! permutations in a drawing of all the 90 numbers at Bingo, etc.), rather than elementary (like the score obtained at the next throw of a given die, or the number 'drawn first' at a given Lottery wheel next Saturday). Recall the remarks made in Chapter 4, 4.10.3, which are relevant to this procedure.

5.7.4. We note, however, that it is not just in examples of games of chance that considerations of symmetry can act as a guide, but, in fact, in any practical problem whatsoever. For example : if we consider the maximum annual temperature (at a given location) in three consecutive years, then it can either :

increase (type 1–2–3, where 1, 2 and 3 schematically denote the three temperatures in increasing order),

decrease (3–2–1),

be maximal in the middle year (types 1–3–2 and 2–3–1), or be minimal in the middle year (types 2–1–3 and 3–1–2).

Now, whatever one's evaluations of the probabilities of more or less high summer temperatures might be, under certain conditions it may very well be natural for us to attribute the same probability $(\frac{1}{6})$ to each of the possible cases.

Examples. A. *Increases and decreases in agricultural production.* This is a (true!) example of a fallacious analysis, based on the observation that, by comparing agricultural production in successive years, the numbers of *inversions of trend* (i.e. the number of times in which an increase was followed by a decrease, or vice-versa) was about twice the number of *permanences* (i.e. repetitions of an increase or of a decrease). An agricultural expert argued that rich and poor crops alternate, and it required a statistician to point out the mistake (the numbers are in agreement with what we have just seen above).

B. *Breaking an existing record.* In connection with temperatures, agricultural production, or even the results in an annual competition, e.g. the winning throw in the national discus championships (assuming the given hypotheses continue to hold : i.e. there exists no reason to expect an improvement due to better training, more participants, etc.), one can pose the following sorts of problems : what is the probability that in the nth year (of the competition, of keeping temperature records, etc.) a new record is set up? (Ans. $1/n$); that the record (set in the first year) be broken for the first time? (Ans. $1/n(n - 1)$); that the previous record had stood for h years ($h = n - 1, n - 2, \ldots, 3, 2, 1$)? (Ans. $1/(n - 1)$ for any h); what is the prevision of the number of times the record was broken in the first n years? (Ans. $\sum (1/h)(1 \leqslant h \leqslant n)$ $\cong \log n$); and what is the prevision of the number of years that the record lasts until the next improvement? (Ans. $+\infty$). As an exercise, verify these answers and pose yourself some further problems (these are easy to find, although not always easy to solve).

5.8   THE PREVISION OF A FREQUENCY

5.8.1   When considering events $E_1, E_2, \ldots, E_n$, it may happen that we know with certainty what the number of successes $Y = E_1 + E_2 + \ldots + E_n$ (or, equivalently, the frequency Y/n) must be: $Y = y$, say; i.e. $Y/n = y/n$. Clearly (cf. Chapter 3, 3.10.3), the sum of the $p_i = \mathbf{P}(E_i)$ must be equal to y (i.e. their arithmetic mean must be equal to y/n); in particular, if the E_i are judged to be equally probable, $p_i = p$, then we must have $p = y/n$ (the probability equal to the known frequency: for $y = 1$ we have the case of a partition, as considered previously). However, even if the frequency is not known with certainty, the relation still holds if we substitute the prevision of the frequency: *the sum of the probabilities must equal the prevision of the number of successes.* In other words, dividing by n, we have

Theorem. *The arithmetic mean of the probabilities must equal the prevision of the frequency:*

(11) $(p_1 + p_2 + \ldots + p_n)/n = \mathbf{P}(Y/n) = \mathbf{P}(Y)/n.$

In particular, *if the E_i are judged equally probable, $p_i = p$, we have* $p = \mathbf{P}(Y/n) = \mathbf{P}(Y)/n$: *the probability (common to all the events) is equal to the prevision of the frequency.*

5.8.2.   In order that correct use be made of this theorem, we must make very clear that it is essentially trivial: otherwise, we run the risk of goodness knows what being read into it. Observe first of all that the E_i can be any events whatsoever, however diverse, so long as the number of successes is given by addition: e.g. success in an examination, a victory for one's favourite football team, finding a traffic-light green, throwing a double six at dice, and anything else, however dissimilar. The 'theorem' is an identity: it imposes no restrictions, apart from informing us that the same thing, written in two ways, remains one and the same thing (rather like the sum of a double-entry table, which can be taken either over rows or over columns).

Well then: *it is in this very thing—and in nothing else—that the value of any theorem in the calculus of probability lies, and it cannot be otherwise. It is to tell us whether, in making the same evaluation in two different ways, we arrive at different conclusions, and, in this case, to invite us to think again and to rectify the situation by modifying one or the other.*

There is no *unique way* of doing this : we do not begin with one side already fixed and the other to be 'deduced'. Instead, we have on both sides evaluations that should agree, and which must be modified if they do not. How should this be done? Generally speaking, one of the evaluations usually seems to be more immediate, and so one is inclined to look for a modification of the other; however, one should be open-minded about it, since appearance might well be only appearance.

5.8.3. Turning now to our particular case, You might find that the probabilities which You have evaluated, when added together give, for instance, a value which is greater than the number of successes, $P(Y)$, which, in prevision, seems to You reasonable. You must then ask yourself: 'have I given the p_i values which are on average too large, or are the values which I thought of for the number of successes Y (or the frequency Y/n) too low?' It is fairly difficult to answer this if the events are rather disparate, but when they are more alike, and especially if we know the frequency of other similar events, which have already been observed, it often happens that one places greater confidence in prevision of the future frequency (under the assumption that it will remain close to that previously observed).

Why is this so? The answer to this cannot be given at present (see Chapter 11), but, even without going into the whys and wherefores, the idea that there is a degree of stability in the frequency of occurrence of events usually grouped together as 'similar' is one which seems quite intuitive to most people. At the present time, this phenomenon may even be somewhat exaggerated as a result of overly simple and rigid formulations current among many statisticians. However, it rests on a very real foundation, since this is how things appear, even to the naïve layman (who, for example, is really surprised if in a given period certain phenomena re-occur with an unusual frequency). Let us accept things as they are.

As a particular case, suppose the events under consideration are so similar that one judges them equally probable: it will turn out that their probability p will be evaluated on the basis of a frequency f, observed among similar events in the past, and that p will be close to f. Notice that in this case the evaluation is based not only on the prevision of a frequency, *but also requires a judgment of equal probabilities.†*

5.8.4. *Some examples.* Statistics show that the percentage (or frequency) of males among live births is always about 51·7 % (hence, a few more males than females); that, according to the Italian tabulations for 1950–53 and

---

† This is often overlooked : if, for example, one speaks of 'the probability of a newly born baby being a boy', it is not made explicit that one is dealing with one unspecified event out of infinitely many ill-defined events, each of which is understood to be equally probable.

1954–57, respectively, the percentages of deaths in the first year were 6·75 %
and 5·49 % for males, 5·88 % and 4·67 % for females; that the overall annual
percentage of deaths in Italy in 1960 attributable to cancer was 1·51 %, but
broken down into age-groups it was

age:	0–5	5–25	25–55	55–75	over 75
	0·013 %	0·009 %	0·078 %	0·524 %	1·131 %,

and into regions (not distinguishing age-groups) it varied from 0·220 % in
Liguria, 0·210 % in Tuscany, etc., to 0·089 % in Puglia and 0·073 % in Basilicata
and Calabria. To change the subject completely, statistics also show that
the results of championship football matches are distributed (in terms of
home fixtures) as 50 % wins, 30 % draws, 20 % defeats.

Thinking of such frequencies as stable, we could adopt them universally
as probabilities for any similar events, or future cases; or, at least, we could
evaluate the probabilities of individual cases in such a way as to make
them compatible, in arithmetic mean, to these frequencies. However....

5.8.5. *The need for realism.* Even though we have expressed our previous
considerations with a certain amount of caution (which itself might appear
overdone and unnecessary to anyone accustomed to a different approach),
it is necessary, in fact, to go still further and provide additional warnings in
emphasis of that caution. We seek to reduce everything to three questions
(and in answering these we shall delve deeper).

5.8.6. *The first question:* are we justified, in real applications, in attributing
the same probability to all the events of a given type? This question is equally
relevant to both of the subsidiary criteria; i.e. symmetric partitions and
frequencies. However, we must first point out that it is meaningless unless
we bear in mind that the probability is not an *external fact* relating to the
event, but, instead, relating to your state of information regarding the
event, and the previsions which You derive from this state of information.
If You know the innate qualities, the past records and the degree of prepared-
ness of every student, your evaluation of the probability of passing an
examination will vary from student to student. Even with all this background
information, however, if You only know the students by sight (i.e. are
ignorant of the name of any given student) and are asked name by name to
give the probabilities, then your evaluations will all be equal (the same
would be true if knowledge by sight or by name were the other way around).
In much the same way, your probabilities for the results of different football
matches on a given day will be different if You know the merits of the
respective teams, and are in a position to express a prevision for each match.
However, if You had to fill in a pools coupon knowing what the matches
were, but not the order in which they were listed, You could only assign the

same probabilities to them all (the averages of those for the individual matches). For example, it might be 40–20–40 if in about half of the matches the away teams are first-rate and favourite to win, or, if You had to fill it in without even knowing what matches were being played that day, You might adopt a standard average probability like 50–30–20. Even in the near legendary case of drawings from an urn, for instance drawing one from among 90 identical balls (numbered from 1 to 90), equality would not necessarily hold if one knew the position of each ball in the urn at the instant before the drawing took place (You might know, or believe, that the person drawing the ball has a habit of drawing from the top, or from the left-hand side, and so on, and taking this into account might lead You to judge the probabilities to be different).

5.8.7. *The second question*: if I wish to make use of a frequency, which one should I base my opinions on? Given an event E in which You are interested, there are usually several classes of events already observed, which are, in different ways, more or less directly similar to yours, but with each class providing a different frequency: the choice is largely arbitrary.

Let us consider, for example, the problem of life insurance for a certain individual (for simplicity, suppose it is a question of a capital sum being provided if he dies within a year). How shall we determine the 'premium'; i.e. the probability of his death within the year (not taking into account any 'extras'—for expenses, etc.). We could check the statistics of the deaths of individuals in the same country (or region, county, city, district, etc.), of the same age (sex, class, etc.), having the same profession (income, degree, etc.), of similar constitution (height, weight, etc.), same name or initial of surname, or house number, or born in the same month, and so on: or we could group together some number (large or small) of these sort of characteristics, or any others. Each grouping will yield a different frequency, and this forces us to adopt a reasoned evaluation rather than a mechanical one; one which takes into account those classifications which it appears most reasonable to assume related to the phenomenon (for instance, age), and not the others (like the person's name). What is 'reasonable' depends not only on whether and *to what extent* this or that circumstance influences the phenomenon, but also on *how* it has an influence. If, for example, it appears reasonable (on general grounds, and on the basis of corroborative evidence) to think that the death-rate increases with age (once we pass childhood), one would be inclined to stick to this when evaluating the probabilities of death in the immediate future, even for those countries for which the most up-to-date statistics would show oscillations from year to year. One would appeal to some sort of *smoothing* procedure, in an attempt to preserve the general outline, which is considered significant, and to eliminate what are thought of as misleading perturbations.

Finally, one is always faced with the aspect we have already spoken of; that of individual differences (which the insurance companies take into account through the results of the medical examination).

This is a general situation, and examples are easy to come by. We shall consider just one other, which shows how meaningful variations in frequency, for appropriately chosen subdivisions, can occur, even in those situations where it appears to be more correct to view the probability as invariant with respect to any of the background circumstances. Given that the frequency of males among new-born babies is almost completely invariant over time, races, or countries, there would seem to be no possibility of differentiating probabilities on the basis of frequency statistics selected according to some factor or other. On the contrary, the research of Gini (using Geissler's data on Saxony, 1876–85) brought to light a differentiation on the basis of families: there were too many families with an excess of either males or females for it to be 'attributed to chance'.† Presumably, one could always find some differentiation if one could succeed in finding appropriate factors on which to base a classification. On the other hand, clearly, as a kind of converse, for those for whom every attempt at picking out significant factors is unsuccessful, every combination of cases automatically appears uniform (even if this is not so for those who do succeed in picking out such factors).

5.8.8. *The third question*: are we justified in expecting frequencies to be stable? The remarks concerning the second question have already led us to consider the differences in frequencies when we refer to subgroups (e.g., in questions concerning people, age-groups, regional groupings, etc.), not to mention individual differences (as discussed in the first question). The stability of all these frequencies is an hypothesis, incompatible with the variability exhibited by the overall composition in terms of subgroupings (e.g. dividing the population according to age, region, etc.). In actual fact, in practice, we can usually assume that the overall composition changes rather slowly, and therefore that the incompatibility is not obvious over a short time period: from a logical point of view, however (and in some cases from a practical one too), the objection is completely valid. On the other hand, even if we leave all this out of consideration, there may be—and there usually are—causes of variation resulting from the evolution of the situation itself. For example, if we consider mortality, there has been great progress in sanitation, medicine, general living standards, etc., as a result of which mortality has progressively declined significantly (this can be seen even from the snippets of data we

---

† C. Gini, *Il sesso dal punto di vista statistico*, Sandron (1908), Ch. X, 'La variabilità individuale nella tendenza a produrre i due sessi' (pp. 371–393). I do not know whether there has been any more recent research confirming these results: in any case, it is the argument which is of interest here rather than the facts.

reported above, relating to very close time periods like 1950–53, 1954–57). It might, therefore, appear reasonable, in evaluating a future probability, to extrapolate the rate of improvement rather than base oneself on the hypothesis of the preservation of the present level.† In any case, the force of the 'stability of frequencies' as a probabilistic or statistical principle is completely illusory, and without solid foundation.

Similar considerations apply, of course, in other fields. We could add the obvious examples of frequencies of car accidents, and similar matters in connection with technical or economic development. In the case of football, the changing character of systems of play, tactics, and many other things, may alter the influence of playing at home or away, and therefore the probabilities of the three results. In addition, even without changes of this kind, frequencies will be altered if the imbalance between 'top' teams and 'bottom' teams is altered.‡

5.9 FREQUENCY AND 'WISDOM AFTER THE EVENT'

5.9.1. Let us repeat an earlier remark, whose function is to prevent a certain confusion; one which we have already warned against, but to which we are particularly vulnerable in the case of previsions of frequencies.

Previsions are not predictions, and so there is no point in comparing the previsions with the results in order to discuss whether the former have been 'confirmed' or 'contradicted', as if it made sense, being 'wise after the event', to ask whether they were 'right' or 'wrong'. For frequencies, as for everything else, it is a question of prevision not prediction. It is a question of previsions made in the light of a given state of information; these cannot be judged in the light of one's 'wisdom after the event', when the state of information is a different one (indeed, for the given prevision, the latter information is complete: the uncertainty, the evaluation of which was the subject under discussion, no longer exists). Only if one came to realize that there were inadequacies in the analysis and use of the original state of information, which one should have been aware of at that time (like errors in calculation, oversights which one noticed soon after, etc.), would it be permissible to talk of 'mistakes' in making a prevision.

Any reluctance one feels in accepting these obvious explanations is possibly accounted for by their seeming to preclude any possibility of taking past experiences into account when thinking about the future. This is not so,

---

† Questions of this nature have been discussed with particular reference to the actuarial field; cf. R. D. Clarke, The concept of probability, *J. Inst. Actuaries*, 1954.
‡ If, for example, one half of the teams were so much stronger than the others that they beat them with certainty, then about half the matches would have the assigned result; if the frequencies 50–30–20 were retained for the other half, we would have, overall, the frequencies 50–15–35 (the averages of 50–00–50 and 50–30–20).

however : the latter is rather different from 'correcting previous evaluations'. One must emphasize that this phrase is wrong, even though it may only be a confused way of expressing an actual need. It is not, however, a harmless inaccuracy : in actual fact, it distorts the basic question, and generates a tangle of confusions and obscurities.

This should be made absolutely clear. If, on the basis of observations, and, in particular, observed frequencies, one formulates new and different previsions for future events, or for events whose outcome is unknown, it *is not* a question of a *correction*. It is simply a question of a new evaluation, *cohering with the previous one*, and making use—by means of Bayes's theorem—of the new results which enrich one's state of information, drawing out of this the evaluations *corresponding to this new state of information*. For the person making them (You, me, some other individual), these evaluations are as correct *now*, as were, and are, the preceding ones, thought of *then*. There is no contradiction in saying that my watch is correct because it now says 10.05 p.m., and that it was also correct four hours ago, although it then said 6.05 p.m.

5.9.2. Discussions and refinements of this kind, which might seem rather pointless when made in the abstract and reduced to mere phrases, are not only of genuine relevance to the conceptual and mathematical construction of the theory of probability, but they also have implications which demand the attention of everyone ; even those not interested in topics of this nature.

The meaning which attaches to statements about 'being wise after the event' does not seem to correspond in a unique way to attitudes either for or against the considerations just made. It is often both different and opposite. This happens when the sentence is uttered as a reproach to someone who belatedly admits that 'he was wrong'—as if to tell him '*tu l'as voulu...*'. It is conceivable that in some situations this reproach is justified : one often makes mistakes through lack of concentration, or because one was unable to resist the temptation, although fully aware of being in the wrong.

However, the reproach is often made when there is no fault—apart from that of failing to be a prophet. Judgment *by results*, the notion that someone's merit should be measured in terms of his successes, is often passed off as 'realism' : to dwell upon the *ifs* and *buts* is considered meaningless. Of course, it is meaningless as far as the facts are concerned ; no-one doubts that these cannot be reversed or modified by any *ifs* and *buts*. The facts themselves are not open to question, but when we turn to *judgments based on those facts, evaluations of personal responsibility, appreciation or criticism of someone's actions*, it is a different matter. In these matters, it is by no means true that the facts provide any definite answers ; in fact, they provide no answers at all. Their only value might be in helping one towards a better understanding of the range of ifs and buts. It is precisely these which allow one to judge someone's actions in the one way that makes any sense: i.e. taking into account, moment by moment, *the context, the situation and the state of information in which the actions took place.*

It would perhaps be overstating the case to suggest, for these reasons, the removal of any distinction between—let us say—being found guilty of murder and of attempted murder. It could happen that 'missing' killing someone was evidence of a lesser intention of doing so; but if everything hinges on a miraculous piece of surgery, how is the offence in any way less serious, or the culprit more deserving of leniency? Anyway, since legal matters are somewhat of a mystery to me, I do not wish to pursue the question.

Something which can be criticized with more certainty is what seems to me the deplorable habit of picking on someone as a scapegoat when something goes wrong. Apart from being unfair, the practice encourages people to avoid taking on responsibility, so that one gets the worst of all worlds. Those who acted loyally, in a sensible manner, cannot be reproached if, by chance, the outcome was unfavourable; those who blundered (in an honest fashion) are advised to learn from the experience and take more care in the future. In contrast, those who had not done everything possible, in terms of organization, control and efficiency, to reduce the risk of unfavourable outcomes are punished—whether or not anyone was responsible.

To set against such stupidity, there is an alternative practice, which can be taken as an example of the beneficial effect of a mode of thinking based on operational research. It was brought to my attention by Pasquale Saraceno,† and is established practice in the industrial group of which he is one of the leading figures. When examining the actions of the various companies, and especially those with unfavourable outcomes, the analysis is based on drawing a distinction between that which could and should have been foreseen, on the basis of the information at hand, and that which could not possibly have been foreseen. This sort of calm criticism and self-examination is undoubtedly what is required in order to encourage a sense of responsibility in a climate of honesty and mutual confidence.

5.9.3. The remarks above were made in order to underline the importance of breaking away from these destructive hangovers of the confusion between prediction and prevision: this is important from a general—one might even say *moral*—point of view. Let us turn to a technical aspect of the problem, which should help to remove such confusion. I say 'should', because I know only too well that such errors (these, it seems, more than most) are difficult to eliminate; like the Hydra with a thousand heads. Were it not for this,

---

† *Translators' note.* Italian economist; former head of the I.R.I. (the state controlled Institute for Industrial Reconstruction).

I would have simply said, as it seems to me, that each objection raised is decisive in itself, and should be sufficient.

In order to combat the idea that the influence of the facts, or, to be more precise, of information regarding the facts, on prevision should be interpreted as a mechanism for refutation and correction (and also to point out the inadequacy and awkwardness of language which gives this impression), we observed that the 'new' opinion, far from being new, was already contained in the 'old', which, far from being refuted, was used when we took over as the 'new' the opinion it had already provided as appropriate for such an eventuality (as for any other possible outcome).

Let us note at this point that such an 'opinion implicitly contained' in the initial one, and already provided for such a contingency, is integrated with it to such an extent that it practically lends itself to being used *without even the occurrence of the facts under consideration.*

5.9.4. The '*device of imaginary observations*', put forward, in particular, by Good (1950), is a method of evaluating probabilities, and, as such, deserves mention in the present chapter. It is a device which is particularly useful for evaluating very small probabilities, and which is more accurate in this context than the direct approach. A simple example will suffice to make the notion clear.

A person claims that he is able to guess in which hand you are concealing a certain object; You do not believe him. If You are invited to be more precise and say what probability p You attach to the possibility that he really can do what he says, You reply 'very small', but are not really in a position to sort out the different implications of saying 10^{-2}, or 10^{-10}, or some other value. Then, according to Good, one can do better by reformulating the question in the following way. Imagine that You put him to the test, and that he guesses correctly three times in a row, or ten times, or fifty times, ... ; after how many consecutive correct guesses would You consider as equally likely (probabilities $\frac{1}{2}$ and $\frac{1}{2}$) the two possibilities that either his claims are justified, or that he has guessed correctly by chance?

It is easily seen that at each trial where he guesses correctly the probability ratio in favour of his claims doubles (likelihood ratio $1:\frac{1}{2} = 2:1$); after n such trials it is 2^n. If, after n trials, the ratio of the probabilities has become $\frac{1}{2}:\frac{1}{2} = 1$, it must mean that initially it was given by $p:\tilde{p} = 2^{-n}$; in other words, we have approximately, $p = (\frac{1}{2})^n = 10^{-n\log 2} \simeq 10^{-(0.3)n}$. For example: if $n = 10$, we have $p = 10^{-3} = 0.1\%$; if $n = 30$, $p = 10^{-9}$; if $n = 50$, $p = 10^{-15}$.

There is no doubt that, with this interpretation available, a comparison between the meaning of answers such as $p = 10^{-3}$, or $p = 10^{-100}$, is no longer unattainable (although a certain vagueness or unfamiliarity is inherent in questions of this kind, and cannot be removed altogether; any method or device of this kind is intended as an *aid*, not a panacea : once one gets to a certain stage, there is nothing to do but try to sharpen the feeling for numerical values of probabilities, including the very small ones).

The conclusion regarding the principle of this method seems to derive further support, psychologically speaking, from a consideration of the paradoxical—I would even say grotesque—position that a contrary point of view leads one into. Its formulation would have to be along the following lines (any attempt at spicing it up in order to increase its paradoxical and mind-bending flavour would only spoil it):

'*My initial evaluation was* $p' = (\frac{1}{2})^n$;

'*it was based on consideration of a hypothetical possibility; that of a succession of n experiments, in which the person claiming to be able to guess obtains successes on every trial, and on my reaction to such a hypothetical result, consisting precisely in the fact that my final evaluation would then have been* $p'' = \frac{1}{2}$;

'*now, the eventuality considered as the hypothesis has actually occurred, and my reaction has been precisely the presupposed one, therefore . . .*

'*the initial evaluation, which was, and still is, a logical consequence of these assumptions (actual or hypothetical) . . . WAS FALSE!*'.

5.10 SOME WARNINGS

5.10.1. It is necessary to point out a number of pitfalls. Although it is premature to talk about the dangers before we understand their causes, some pointers must be given in order to guard against the doubts and distortions which might get mixed up with what we have said concerning the evaluation of probabilities, giving rise to confused and contradictory notions.

The following remarks should, in one sense, be unnecessary. All the dangers have already been mentioned, and the details already given at the appropriate time would be sufficient to render these additional comments superfluous, if—and this is the difficulty—they remained firmly implanted in one's mind, together with all their ramifications, and with such clarity that any dangers reappearing, in whatever disguise, could be dealt with just as effectively as when they were first encountered. This not being the case, it is preferable, and perhaps necessary, to repeat ourselves; to go over the details mentioned above, in their different variants and versions, pointing out the many forms the dangers may assume. (There are such a number of them that perhaps some, even important ones, will be overlooked; hopefully, though, the pattern-book of objections and counter-objections will be sufficiently representative for the reader to be able to answer, by analogy, possible objections not covered, by means of suitable counter-objections.)

5.10.2. It might be argued that the kind of problems we have considered in this treatment, and for which we have discussed the appropriate methods of evaluating probabilities, are outside the 'true' ambit of the calculus of probability, or, at most, they constitute a small and specialized part of it.

The arguments put forward will be, by and large, the standard ones; however, if they are given with reference to Physics, for instance, they may appear novel, or at least more substantial and difficult to refute.

There are cases where the probabilities in Physics are given by combinatorial arguments, in accordance with the 'classical' idea of 'equally likely cases'; that is, they are given by the Maxwell–Boltzmann, Bose–Einstein, Fermi–Dirac 'statistics' (to use the jargon of physicists): for further details, cf. Chapter 10, Section 10.3. Who can argue in this case that we are dealing with a probability whose value is objectively determined by '*a priori*' considerations? It is precisely this example (as Feller observed, vol. I, pp. 5 and 21) which shows how fallacious any *a priori* conclusion would be: nobody could have foreseen that the computation of 'equally likely cases' had to be carried out using completely different methods in problems where different 'statistics' apply (and the explanation only came later, through the distinction between particles with integer or semi-integer spin).

5.10.3. Everyone will probably agree, therefore, that it makes no sense to be willing to deduce properties of phenomena, or previsions regarding their outcomes, basing oneself solely on superficial, preconceived ideas. The confirmation of experience is required, and this, certainly, leads on to an objective conclusion. One might well say that, for the physicist, probability coincides with frequency.

And this statement is, in a certain sense, true. However, this form of expression is completely wrong from a conceptual point of view, even if at first sight it presents no difficulties.

Let us swiftly demolish, one by one, the main arguments put forward with the intention of transforming probability from being subjective to being objective, by means of more or less overt confusion or connection of the notion with that of frequency.

5.10.4. Firstly, we present an objection frequently raised against the notion of the *probability of an individual event*: either this event occurs or it does not, and therefore it either has probability *one* or *zero*; it makes no sense to attribute to it an intermediate probability p. I accept this argument completely, in that it refers to an objective probability p: but I observe that the same argument holds even in cases where my opponent forgets that it does—when he says that in n 'individual cases' there is an objective meaning to p because np of them will occur. This is not true: either *zero*, or *one*, or *two*,..., or *all* n of them occurs, and the objective probability (if one prefers to use this term as a useless and misleading synonym for frequency) is one of the $n + 1$ values 0, $1/n$, $2/n$,..., h/n,...,$(n - 1)/n$, 1, although it is not known which one.

It is only in a subjective sense that it makes sense to speak of p, as the arithmetic mean of these $n + 1$ possible values, *taking as weights the subjective probabilities of the single frequencies (still 'individual cases'!).*

5.10.5. It might be objected that in many cases (those to which an opponent would limit himself) the probability is concentrated near a certain frequency p, which could be defined as objective probability. But here, and in every case in which something 'very probable' is said to be 'practically certain' (or even 'certain', for the sake of brevity) and, symmetrically, something 'very improbable' is said to be 'practically impossible' (or even 'impossible'), an *either–or* must be clearly established. In fact, such sentences can either say something obvious, with which one has no choice but to agree, or, alternatively, they can completely falsify the meaning of things. The field of probability and statistics is then transformed into a Tower of Babel, in which only the most naïve amateur claims to understand what he says and hears, and this because, in a language devoid of convention, the fundamental distinctions between what is certain and what is not, and between what is impossible and what is not, are abolished. Certainty and impossibility then become confused with high or low degrees of a subjective probability, which is itself denied precisely by this falsification of the language.

On the contrary, the preservation of a clear, terse distinction between certainty and uncertainty, impossibility and possibility, is the unique and essential precondition for making *meaningful* statements (which could be either right or wrong), whereas the alternative transforms every sentence into a nonsense.

5.10.6. We have already made abstract reference to this confusion (Section 5.2.3), so let us confine ourselves here to an illustration in the context of Physics (with the warning that we are anticipating things to come for the purpose of preventive therapy; later, Chapter 7, we will be concerned with the true meaning of 'laws of large numbers' and suchlike).

It cannot be denied that two different explanations of the same phenomenon may turn out to be indistinguishable in practice; particularly when one explanation is deterministic and the other probabilistic. One thinks immediately of the diffusion of heat, or any other similar phenomenon, which can either be considered in terms of a differential equation, describing the continuous development of the phenomenon in a manner governed precisely by deterministic laws, or as a random process in which elementary phenomena occur in a non-deterministic way, but such that there is a high probability of the phenomenon developing at a macroscopic level in a manner practically identical to that indicated by the deterministic theory.

However, this in no way implies that the two explanations are similar, and even less that they are the same, or substitutable. On the contrary, they are exact opposites; diametrically opposed and absolutely incompatible. The deterministic explanation makes certain assumptions which preclude any departure from predetermined behaviour. Any similar explanation, albeit less rigid, which laid down that some conclusion was compulsory and certain, would, at the very least, require some sort of self-regulatory mechanism, some sort of *'feed-back'*. The probabilistic explanation makes no assumptions of this kind: it states nothing other than that everything is possible. If it *appears* to state something more, it is only because such a statement, which may seem quite precise, corresponds to a property common to 'almost all' the possible cases.

A probabilistic explanation of the diffusion of heat *must* take into account the fact that heat could accidentally move from a cold body to a warmer one, making the former even colder, the latter even warmer (in Jeans' example: water being frozen rather than boiled when put on the stove). That this is very improbable is merely due to the fact that the 'unordered' possibilities (heat equally diffused) are far more numerous than the 'ordered' possibilities (all the heat in one direction), and not because the former enjoy some special status.

To rule out the possibility of those cases which seem 'exceptional', in no way *improves* the probabilistic explanation, by somehow making it simpler, or more scientific: on the contrary, it negates it. Acceptance of the probabilistic explanation has the following implications: it means that what we state about the phenomenon must not be regarded as necessary, but, instead, must be attributed to 'chance', and, hence, regarded as only approximate and probable. It means that one must regard it as essential to deny the existence of certain and exact laws which are obeyed only apparently; it means that one must consider as necessary the possibility of studying departures from any rigid law, fluctuations, the effects of discontinuities (the *shot effect*), and all that a cursory identification with a different form of explanation would sweep away without a thought.

5.10.7. What we just said is itself open to misinterpretation. It would be a mistake to infer that an explanation based on a 'tendency to disorder' takes care of every application of probabilistic concepts and not merely the particular example given above. 'Chance' (if we can adopt this convenient terminology as a summary of complicated and uncertain factors without its being taken too seriously) plays a no less important rôle in biological and social processes, where the outcome depends on highly ordered and organized structures, like chromosomes, cells and human beings (and also in Physics, in processes like crystallization).

The following needs to be said in order to disprove the thesis which considers a levelling down into a debased chaos (entropic death) to be an inevitable consequence of the validity of this or that 'law' of probability. *The calculus of probability can say absolutely nothing about reality; in the same way as reality, and all sciences concerned with it, can say nothing about the calculus of probability.* The latter is valid whatever use one makes of it, no matter how, no matter where. One can express in terms of it any opinion whatsoever, no matter how 'reasonable' or otherwise, and the consequences will be reasonable, or not, for me, for You, or anyone, according to the reasonableness of the original opinions of the individual using the calculus. As with the logic of certainty, the logic of the probable adds nothing of its own : it merely helps one to see the implications contained in what has gone before (either in terms of having accepted certain facts, or having evaluated degrees of belief in them, respectively).

Physics can make greater or lesser use of the calculus of probability, but the relationship between the two is simply the relationship between a certain field of research, which remains itself, no matter what tools it uses, and a logical tool, unconditionally valid, which remains itself, whatever use is made of it, in whatever field.

5.10.8. Let us return to the necessity of avoiding the dangers implicit in attempts to confuse certainty with 'high probability'. We have to stress this point because these attempts assume many forms and are always dangerous. In one sentence : to make a mistake of this kind leaves one inevitably faced with all sorts of fallacious arguments and contradictions whenever an attempt is made to state, on the basis of probabilistic considerations, that something must occur, or that its occurrence confirms or disproves some probabilistic assumptions.

From such a point of view, the calculus of probability seems to be regarded, more or less explicitly, as a nothingness ; saying nothing when the probabilities in question are intermediate in value, but capable of miraculous transformation into a warrentor of absolute truths when the probabilities are very large or very small, since, in these cases, the difference can be ignored and one can simply say that something, is true or false. One thus has a mechanism which is considered to be useless when it says that which it is capable of saying, and wishes to say, but is blindly trusted when the things one wants to make it say are not the things it does say or could say.

5.10.9. We present three examples of this form of observation.

First example. The statement that '*an event of small probability does not occur*' is sometimes made, under the heading of 'Cournot's principle' (5.2.3).

A kind of corollary or special case of this is referred to as the '*empirical law of chance*' (meaning that frequency and probability actually behave in many cases according to the 'law of large numbers').

Second example. In accordance with the identification of small probability with impossibility, Neyman finds a contradiction in the behaviour of an individual who travels by aeroplane and at the same time takes out insurance. If he considers it possible to have an accident, why does he travel? If he does not, why does he insure himself? The paradox here specifically relates to 'decision theory', which, in the restricted sense to which it is often reduced by 'objectivist' statisticians, considers only the question 'what decision is appropriate given the *accepted hypothesis*' and not 'what decision is appropriate in the *given state of uncertainty*'.

Third example. Again, in this context (of 'objectivist statisticians'), one aspires to '*accept*' or '*reject*' an hypothesis on the basis of an experiment, instead of considering how its outcome modifies the initial probabilities (which *one wants to do without*!) in order to give the final probabilities (which therefore cannot be obtained!). Here the absurdity reaches new heights, because it cannot even be claimed that 'accept' and 'reject' correspond to the minimal requirement of the probabilities being large or small. The use of these two words is a meaningless convention; an apparent attempt to answer a question by disregarding everything that makes it a meaningful question in the first place.

It is as if, in comparing two weights, we were to decide upon which was the heavier by choosing the one which tilted the balance to its side, without taking into account, and, indeed, refusing to consider it legitimate to take into account, any difference between the arms of the balance, even knowing that the difference could be considerable.

5.11   DETERMINISM, INDETERMINISM, AND OTHER 'ISMS'

5.11.1. Continuing with the same theme, there is a clear philosophical point to be made. It derives from the strange fact that precisely the same disposition to accept an objective probability is often justified in two completely opposite ways.

For some people, the ideal instrument for producing an objective probability with value p would be a totally invariable device, working under strictly unchangeable conditions, and for which the tendency to produce successes with frequency p would be a 'built-in property', or, more specifically, a '*dispositional property*' (following, for example, Hacking). Any perturbations would result in a deviation from the desired result; that is, from the realization of a frequency close to p.

For others, whole-hearted determinists, any such device could but yield the same result; always successes or always failures. The fact that both successes and failures occur implies that there exists something causing perturbations. In general, it is assumed that there are a great number of small, accidental, causal factors, which are largely unknown. The fact that the frequency is expected to be around p would be an effect of the combined and random actions of these causal factors (following, for example, Paul Lévy).

So far as the subjectivistic conception is concerned, it has the advantage and, indeed, the preoccupation of remaining outside of disputes of this kind. The thing that really matters, and which justifies, in fact requires, our arguing on the basis of probabilistic logic, is the impossible nature of the situation in which we find ourselves when we attempt to foresee a given outcome with certainty. This is so whatever the reason: whether it be ignorance of certain deterministic laws; or the non-existence of such laws; or an inability to perform the requisite calculations even though we know the laws; or an inability to obtain precise data (or the impossibility of doing so). At any given time, it does not matter. It is only with respect to the prospects for the advance of science in the future that it matters, and, even here, only in a minor way, since reference to such rigid and preconceived positions seems rather unnecessary anyway.

From time to time, as scientific prospects change, this or that particular mental attitude may be useful, in that it facilitates the formulation of theories which—for the moment—give better agreement with this or that point of view. However, nothing remains for ever unchanged; nothing is absolute. The particular mould in which one sets is not so important: what matters far more is not to set too firmly in any one pattern. To set fast is to no longer be alive.

5.11.2. These same remarks need repeating more generally, in connection with all those ways devised to saddle probability with an objective something (meaning, interpretation, justification, definition or whatever). In the first place, it is a fact that these attempts are not successful, and cannot be so, since, having the resolve to express matters relating to uncertainty in terms of the logic of certainty, they force themselves, *ab initio*, into a vicious circle, with no means of escape—'per la contradizion che no'l consente'.† It is as if someone were to wish to hoist himself up by his boot-laces. Logic only permits the exposure of a tautology on the basis of what is taken as known; a prevision, however, is not simply a tautological consequence of what is already known. To be thus, would be to constitute something implicitly

---

† *Translators' note.* Ruled out by the principle of contradiction. (The line is from Dante's Inferno: canto XXVII, line 120.)

known, and would not involve uncertainty and, therefore, would not give rise to prevision.

However, even if we were to consider someone's arguments to constitute an acceptable basis for an objective meaning of probability (and, in general, such arguments will be different and concerned with special and different types of event, according to the different points of view), our thesis consists in believing that these arguments would be irrelevant anyway. All such conceptions, all the 'isms' they reduce to, are rejected here, but *not* in support of yet another 'ism' (as might be thought; e.g. 'subjectivism' or 'solipsism'), which one wants to put forward and contrapose to the others. The latter are rejected because, whatever the explanation of the uncertainty might be (attributing it to 'chance', 'fate', 'hidden laws', 'Providence', or 'statistical regularities', or to something else—or . . . words(?) . . .), the sole concrete fact which is beyond dispute is that someone (me, You, somebody else) feels himself in a state of uncertainty, and has to decide on and adopt some point of view as a basis for previsions and related decisions.

5.11.3. This subjective meaning is an objective and unquestionable fact: all the rest (even if there were no dispute about it) is, in any case, something of an extra, which, at best, serves to help fix one's ideas. It is analogous to a vivid piece of writing which succeeds in forming something like an idea in our minds, although its meaning is not clear, and an analysis of the sentence in fact shows it to be inconsistent.

It is the case, however, that this view of the logic of uncertainty, complete and clear as it is, is far from achieving general acceptance. Why is this? Perhaps it is only the state of being certain which appears to most people as worthy of consideration and fit to be part of the edifice of science (which, according to the prevailing view, appears to express or aspire to omniscience —notwithstanding the fact that all progress, pushing back as it does the frontier of what is known, makes the horizon of what can be seen as unknown even broader). Perhaps the unknown and the uncertain disturb and annoy us, and provoke those who are most upset to attempt to suppress them, or at least to make them disappear. There is not much point in philosophical arguments or speculations of this kind: they do provide, however, a possible explanation of why these different attitudes exist (we had to mention them, or, at least, to give some indication of how a person who finds our point of view natural could try to understand its lack of acceptance).

These different attitudes are, essentially, only variations on the same theme: the attempt to avoid the problems of uncertainty by simply pretending to overcome it; restricting the treatment to cases in which it can be presented in such a watered down way that it looks like something else.

The classical variant limits itself to cases like those of games of chance (where probability should acquire an objective meaning by virtue of the 'definition' based upon 'equally likely cases'). In the view of the most rigid supporters of this position, every application of the theory of probability outside this field would only be a questionable transposition by analogy.

The position which is at present most widely accepted restricts itself to cases of a certain statistical type (where probability should acquire an objective meaning by virtue of the 'definition' based upon 'frequency'). According to its most rigid adherents, the term 'probability', when used outside this context, has no more in common with its 'scientific' meaning than the 'energy' of a team leader has with the same term as applied to physical motion.†

Other approaches, which, having the aim of acting as guides in decision-making, follow less rigid notions, attempt nevertheless to avoid those components of the argument which many find unpalatable (like the 'initial probabilities' required for Bayesian induction).‡

Others adopt an eclectic attitude, accepting that one can base one's thinking on 'that probability which we evaluate for previsions and decisions' (i.e. the one corresponding to the conception of the present author), but, on the other hand, asserting that 'there is also another type of probability, the one with which statistics is concerned' (or, alternatively, 'the type valid in games of chance', or both).§

We should point out, here and now, that the mathematical treatment is unaffected (or, at most, very little affected) by these disagreements. In this sense, we can give a reassurance that everything we shall say mathematically is independent of questions of this kind, and should be acceptable to everyone. However, the interpretation is often different; there are certain nuances which, when looked into closely, completely change the spirit in which a given statement (perhaps expressible, in the same words, in the imprecise manner of everyday language) is to be understood.

So far as our own attitude is concerned, we wish to make clear that it is not utterly opposed to the attitude we have termed 'eclectic', even though it differs from it in a very real sense.

It is not utterly opposed, because we recognize the importance of the problems, concepts and criteria which are the object of the various practical theories, even though we study them within the framework of the general theory. Only by renouncing their alleged autonomy is it possible to compensate for those deficiencies in the foundations of the particularistic theories

---

† The phrases given here, in characterizing the two attitudes, are due to Castelnuovo (transposition by 'analogy') and von Mises ('energy' and energy), respectively.
‡ The followers of 'objectivistic statistics' in its various schools, including that of A. Wald (who we particularly have in mind here, as the nearest in approach to the Bayesian school).
§ The quotations are from V. Castellano. Typical examples of the eclectic attitude are provided by R. Carnap (who differentiates between 'probability₁', *logical*, and 'probability₂', *statistical*) and I. J. Good who admits the possible value of distinguishing many 'kinds of probability' (although in the context of a conception which is essentially subjectivistic).

which render their conclusions meaningless, and the interpretation of them arbitrary.

It differs from it because we do not accept the existence of probabilities of different kinds, nor the autonomous validity of theories which set out to consider them, leaving aside some of the assumptions of the general theory, all of which are at all times essential.

All this has been summed up in an expressive manner by L. J. Savage (in a rather more specialized context): it is as though one wished to make a probabilistic omelette without breaking probabilistic eggs. There are two possible outcomes: either the result is not an omelette; or the eggs have in fact been used, either surreptitiously or inadvertently. All comments which we shall have occasion to make concerning 'other points of view' will essentially be continuations of the above analogy.

CHAPTER 6

Distributions

6.1 INTRODUCTORY REMARKS

6.1.1. Thus far, we have been occupied with the conceptual aspects of the formulation, and the thoroughness of the treatment reflects what the material seemed to us to require. Likewise, we have chosen to deal with the simplest topics and problems, whose meaning was not obscured by the need to involve complicated mathematics (but in fact contributed by appearing in a clear and simple light).

The time has now come, however, to abandon these self-imposed limitations. We must examine whether, and to what extent, we can implement, in any domain whatsoever, the study of probability in terms of the image most often thought of (in an informal manner); that of the 'distribution of mass'. In actual fact, of course, it is well-known that the notion of a probability distribution (the precise mathematical translation of this image) is taken directly as the starting point in many approaches, particularly modern ones. The aim of the present chapter is to introduce this notion and the requisite mathematical tools, tying them in rigorously with our previous formulation and making any necessary modifications or limitations.

There are therefore two different aims to bear in mind in what follows: on the one hand, to provide a knowledge of the mathematical tools required in further study of the calculus of probability; on the other, to give the mathematical and conceptual details which derive from our previously established formulation and point of view.

6.1.2. We shall try to satisfy the first aim as concisely as possible, quoting, with a minimum of explanation, and without proof, those things that can be found in any book on probability, or whose proof can be obtained either with a standard knowledge of analysis or on an intuitive basis. Alternatively, if the reader wishes, the proofs can be taken for granted and this will not affect applications or further reading.

6.1.3. Our second aim, one of a critical nature, will need a more careful treatment, at greater length. Although we do not wish to dwell upon it more than we have to, any omission or incompleteness in what is necessary would certainly cause misunderstanding and incomprehension (especially among those readers who, by interpreting certain sentences in the standard way, would find them, and quite rightly so, either incomprehensible, or, misunderstanding them, wrong). For this reason, we strongly recommend the reader, and especially those who think that they already know enough about the topic of this chapter, not to skip it, and to dwell, in particular, upon the details relating to the differences, slight but important, between this and the standard interpretations.†

6.2 WHAT WE MEAN BY A 'DISTRIBUTION'

6.2.1. An abstract and general explanation would, at this stage, appear rather vague and colourless. It is more appropriate to consider here the simplest and most important special case, that of distributions on the real line, together with their various interpretations. These interpretations should all be kept in mind, in order that the most convenient one can be called upon in any particular instance. This special case will eventually be revealed to have a relationship with that of random quantities in general.

Proceeding in the usual way, we introduce immediately, as a starting point, and as the main mathematical tool for the definition of a distribution, a function $F(x)$, increasing‡ from 0 (as $x \to -\infty$) to 1 (as $x \to +\infty$), and called *a distribution function*.

6.2.2. As a *first interpretation*, the most intuitive one, we have that of a *distribution of mass* on the real line (with the assumption that 'total mass' = 1). $F(x)$ is the mass to the left of a point x, $1 - F(x)$ the mass to the right; the increment $F(x'') - F(x')$ is the mass in the interval $x' \leqslant x \leqslant x''$. If there is a mass, p_h, *concentrated* at the point x_h, F is discontinuous at x_h and p_h is its 'jump', $F(x_h + 0) - F(x_h - 0)$.§ There is at most a fibite or countable number of such jumps, and F is continuous elsewhere.

A distribution which only has concentrated masses ($\sum_h p_h = 1$) is called *discrete*; one without concentrated masses is called *continuous*. The most

---

† Recall the warnings given already in Chapter 1 (1.2.1).

‡ We use 'increasing' to mean 'non-decreasing'; we shall use 'strictly increasing' if the function is not constant in any interval.

§ These two values must be distinguished when considering $F(x)$ if there is a jump at the point x (and we have a choice according as the mass at x is to be considered together with those on the left or those on the right). For various reasons (see 6.5.1), we prefer to avoid those conventions which make $F(x)$ one-to-one at the discontinuity points (by saying that it assumes *all* the values y, $F(x - 0) \leqslant y \leqslant F(x + 0)$. However, when dealing with statistical distributions, where some convention is necessary, we shall take $F(x) = F(x + 0)$ (as is necessary if 'individuals with h children' is to mean 'including those with exactly h children').

We apologize for the awful notation $F(x + 0)$; it is, however, concise and unambiguous.

familiar case of the latter is that of *absolutely continuous* distributions; those admitting a *density* function, $f(x) = F'(x)$, such that

$$F(x) = \int_{-\infty}^{x} f(x)\,dx.$$

In actual fact, when the term 'continuous' is used, it is this special case which is often understood. There is, however, an intermediate case between the *discrete* and *absolutely continuous*; that of *continuous but not absolutely continuous*. In 6.2.3 we shall make this idea concrete by means of an example (and this example will also have an interesting interpretation in a problem in probability). For the time being, we shall limit ourselves to the definition and the basic properties.

6.2.3. To say that $F(x)$ is continuous means, as everyone knows, that for each ε, however small, every interval whose length is less than some suitable δ contains a mass $< \varepsilon$. To say that it is *absolutely continuous* (Vitali) means something more: that the same is true of the mass contained in any arbitrary number of intervals of total length less than δ.†

Every distribution $F(x)$ can be decomposed into partial distributions of masses of the three types. We first of all set

$$(1) \qquad F(x) = a_C F_C(x) + a_B F_B(x) + a_A F_A(x) \quad (a_C + a_B + a_A = 1),‡$$

where:

$a_C = \sum_h p_h$ is the sum of the concentrated masses (masses of type C),
$a_C F_C(x) = \sum_h p_h(x_h \leqslant x)$ is the sum of these masses in $[-\infty, x]$:

we now consider the residual partial distribution,

$$F_{AB}(x) = F(x) - a_C F_C(x),$$

that is, $F(x)$ without the concentrated masses; it follows that:

a_B = 'total mass of type B' = upper limit of the mass of $F_{AB}(x)$ which can be enclosed within intervals of arbitrarily small total length,
$a_B F_B(x)$ = total mass of type B in $[-\infty, x]$ (detailed definition as above):

we are left with $a_A F_A(x) = F(x) - a_C F_C(x) - a_B F_B(x)$, and this is the absolutely continuous part of the distribution (the masses of the first two types, which do not fulfill the condition of absolute continuity, having been removed).

---

† It makes no difference whether we consider the number of intervals as *finite* or infinite (countable: it cannot be uncountable). It is understood that $\varepsilon > 0$ and $\delta > 0$.
‡ Obviously, if $a_i = 0$ (i.e. one of the components is missing) the corresponding F_i is missing. The meanings of the letters are: C = concentrated; A = absolutely continuous; B = intermediate case between C and A.

It is easy to see that, in a linear combination of distributions,

$$F(x) = c_1 F_1(x) + c_2 F_2(x) \qquad (c_1 + c_2 = 1),$$

the various types are preserved. It follows, therefore, that the F_C, F_B, F_A of an arbitrary linear combination are the linear combinations of the corresponding parts of the summands (in particular, a particular type of mass exists in the linear combination if and only if it exists in at least one of the summands). If we say that a distribution is of type A, B, C, AB, AC, BC, ABC, to indicate the pure types involved in it, we can express our conclusion by saying that in a linear combination the letters of the types combine (e.g. from AC and BC we get ABC).

An example of a type B distribution. The following procedure can be used to construct the well-known Cantor set (of measure zero, even in the Jordan–Peano sense) and a distribution on it (which is therefore of type B).

Let us divide the interval $[0, 1]$ into three equal parts. In the middle interval, $[\frac{1}{3}, \frac{2}{3}]$, we set $F(x) = \frac{1}{2}$, so that no mass is placed there, and half the mass is placed in each of the first and third intervals. This operation is then repeated in these latter two intervals. In each of them we will have three subintervals (of length $(\frac{1}{3})^2 = \frac{1}{9}$), and we set $F(x) = \frac{1}{4}$ (respectively, $= \frac{3}{4}$) on the central intervals, thus excluding masses there. The mass is then placed on the 4 residual intervals, $\frac{1}{4}$ on each.

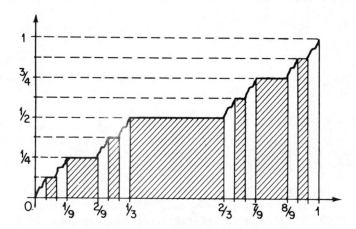

Figure 6.1   The Cantor distribution

Proceeding in this manner (cf. Figure 6.1), after n steps $F(x)$ is defined (with values which are multiples of $(\frac{1}{2})^n$) on the whole interval $[0, 1]$, except for the 2^n residual parts, each of length $(\frac{1}{3})^n$, where all the mass resides ($(\frac{1}{2})^n$ on each residual interval). In the limit, $F(x)$ is defined everywhere and is continuous. It is not, however, absolutely continuous: after n steps the mass is contained within the 2^n intervals each of length $(\frac{1}{3})^n$, and $(\frac{2}{3})^n$ in total. It can therefore be contained within a finite number of intervals of total length less than any given $\varepsilon > 0$.

A probabilistic interpretation. It might be thought that the above construction merely serves to provide a critical comment, giving a pathological example with no practical meaning. On the contrary, we can give a simple practical example of a problem in probability where such a distribution arises.

Suppose we wish to pick a real number in $[0, 1]$ by successively drawing from an urn the digits of its decimal representation:

$$X = 0 \cdot X_1 X_2 X_3 \dots X_n \dots, \quad \text{i.e. } X = \sum X_n / B^n \ (B = \text{base}; \text{e.g. } 10).$$

If a ball representing a figure is missing, all the numbers containing it become impossible (i.e. some intervals are excluded, as in the example given). The above example corresponds to the assumption that $B = 3$, with the figure 1 missing (only the numbers with 0 and 2 are possible, like $0.22020002020022202\dots$).

It is rather surprising to note that this happens even if the balls are all present (unless all of them have the same probability $1/B$).† If one of the figures has probability $p < 1/B$, and we take c between p and $1/B$, and N sufficiently large, the set of numbers X in which that figure appears in the first N places with frequency $\geqslant c$ has measure arbitrarily close to 1 and mass arbitrarily close to 0.‡

6.2.4. Let us observe now how a different interpretation of F permits us to extend considerably its applicability and effectiveness. Given any interval I (with extreme points x' and x''), it suffices to set $F(I) = F(x'') - F(x')$ to obtain F as an additive function for the intervals. If we identify the intervals with their indicator functions ($I(x) = (x' \leqslant x \leqslant x'') = 1$ or 0 according as x belongs to I or not), we obtain F as a linear functional, defined for every $\gamma(x) = \sum_h y_h I_h$ (step functions with values y_h on the disjoint intervals I_h) by $F(\gamma) = \sum_h y_h F(I_h)$. This can be extended to all functions $\gamma(x)$ which can be approximated, in an appropriate way, from above or below, by means of step functions. More precisely, $F(\gamma)$ is determined if, thinking of γ' and γ'' as generic step functions such that

$$\gamma'(x) \leqslant \gamma(x) \leqslant \gamma''(x)$$

everywhere, we have $\sup F(\gamma') = \inf F(\gamma'')$, whence $F(\gamma)$ necessarily has that same value since $\sup F(\gamma') \leqslant F(\gamma) \leqslant \inf F(\gamma'')$.

In actual fact, what we have defined, in a direct and somewhat abstract way, is nothing other than the integral

$$(2) \qquad \phi(\gamma) = \int \gamma(x) \, dF(x) = \int \gamma(x) f(x) \, dx \quad \left(\int \text{ represents } \int_{-\infty}^{+\infty} \right),$$

where the first expression (one which always holds) is the Riemann–Stieltjes integral, and the second (which only holds for absolutely continuous distributions) is the Riemann integral.

---

† This observation is too obvious to be novel; however, I do not remember having seen it before, and I had not thought of it prior to adding it here to the usual example.
‡ This assertion will be seen as obvious as soon as we encounter the basic ideas of 'laws of large numbers' (Chapter 7, Section 5).

As an example, suppose we consider the two functions

$$\gamma(x) = x = \square(x) \quad \text{and} \quad \gamma(x) = x^2 = \square^2(x).$$

In this case, $F(\square) = $ the abscissa of the barycentre, and $F(\square^2)$ the moment of inertia (about the origin) of the mass distribution. In integral form,

$$F(\square) = \int x \, dF(x) = \int xf(x) \, dx, \qquad F(\square^2) = \int x^2 \, dF(x) = \int x^2 f(x) \, dx.$$

As possible interpretations of the function, $\gamma(x)$, one might, for instance, think of it as representing (for the mass at x) the reciprocal of the density, or the percentage by weight of a given component (e.g. of a given metal if we are dealing with an alloy whose composition varies with x), or the (absolute) temperature. In these three cases, the integral, apart from constant terms, will yield the total volume, the weight of the given component, and the quantity of heat, respectively.

6.2.5. A *second interpretation* is the statistical one. It is convenient to mention it here in order to draw attention to the practical importance of the notion of distribution in the field of statistics. This is not only closely connected and related to the probabilistic notion, but also provides it with problems and applications. However, we shall reserve discussion of this until later.

In the final analysis, the image is the same as before : that of a mass distribution. In fact, the distribution of a population of n individuals, on the basis of any quantitative characteristic whatsoever, can be thought of as obtained, in the case of number of children, for example, by placing a mass $1/n$ at the point $x = h$ for each individual with h children ($h = 0, 1, 2, \ldots$), or, in the case of height, at the points $x = x_i$ (distinct if the measurements are sufficiently precise), denoting by $i = 1, 2, \ldots, n$ the n individuals, and by x_i their heights.

In the first example, we have masses $p_h = n_h/n$ concentrated at the points $x = h$ (n_h denotes the number of individuals with h children) and therefore :

$$F(x) = \sum_h (n_h/n) \quad (h \leqslant x)$$

= the percentage of individuals with not more than x children.

In the second example (let us assume that the individuals have been indexed in order of increasing height), we have a jump of $1/n$ at each point x_i (and, if n were large, one could in practice consider the distribution to be continuous —if necessary by 'smoothing'), and the distribution function is given by

$$F(x) = (1/n) \max i \ (x_i \leqslant x) \quad \text{(that is : } F(x) = i/n, \text{ for } x_i \leqslant x \leqslant x_{i+1}).$$

Alternatively, one might be interested in performing some kind of 'weighting' instead of simply 'counting' the individuals (for example : instead of $1/n$ throughout, a 'weight' might be chosen proportional to income, average number of bus journeys per day, cups of coffee consumed, etc.; depending on what was of interest). The 'population' might consist of objects, or events, or anything; but it is customary to retain the terms 'population' and 'individuals'. If a generic and neutral term is required, one can use 'statistical units'. In the general case, units may be counted straightforwardly, or with some appropriate 'weighting'.

This will suffice for the present. We merely recall (cf. Chapter 5, Sections 5.8–5.10) that a statistical distribution *is not* a probability distribution, although it can, in various ways, give rise to one.

6.2.6. In order to clarify, from a different angle, certain aspects of the above (and, more importantly, to mention some further extensions) it is useful at this point to introduce a *third interpretation*. An additive function (non-negative, and with its maximum $= 1$) is also called a *measure*; the change in nomenclature, from mass to measure, is of no importance, but the fact that we have at hand a natural way of looking at such a 'measure'—or, to be precise, the '*F*-measure'—in terms of its own scale (of length) is important.

One works in terms of this scale by looking at $y = F(x)$ instead of at x (as was clear from the definition). We have only to observe that, by drawing the graph of the distribution function (Figure 6.2a), we establish an (ordered) correspondence between the points of the x-axis (all of it) and those of the interval $[0, 1]$. The mass of any arbitrary interval on the x-axis is then measured by the length of its image on the y-axis. Of course, the correspondence is not necessarily one to one (it will be so if $F(x)$ is strictly increasing from $-\infty$ to $+\infty$). To a point of discontinuity on the x-axis there corresponds, on the y-axis, an interval whose length equals the mass which is concentrated at that point; to any interval of the x-axis on which $F(x)$ is constant (no mass) there corresponds a single point of the y-axis. Apart from the interpretation in mechanical terms, this is also clear geometrically. Observe that in both cases the graph $y = F(x)$ (conveniently thought of as containing, for discontinuity points x, all the y's between $F(x \pm 0)$) contains, respectively, vertical or horizontal segments which project to a single point of the orthogonal axis.

In order to concentrate attention on the measure, and to make it easier to visualize developments based upon it, we find it convenient to reverse the rôles of the x- and y-axes, and to look instead at the graph of $x = F^{-1}(y)$ (Figure 6.2b).

Note that the change of variable from x to y transforms, for example, the Stieltjes integrals into ordinary integrals:

$$F(\gamma) = \int \gamma(x) \, dF(x) = \int \gamma(F^{-1}(y)) \, dy = \int \gamma(x) \, dy.$$

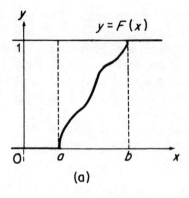
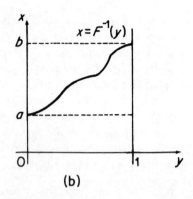

(a)   (b)

Figure 6.2   The graphs of (a) the distribution function, and (b) its inverse: $y = F(x)$ and $x = F^{-1}(y)$. In addition to the present (measure theoretic) interpretation, we have also seen that the statistical interpretation, as *graph of the distribution,* is of interest (and is the most useful from the point of view of applications). In Section 6.4 we shall further consider the probabilistic setting, in which the above admits the following interpretation: one can always construct a random quantity with a preassigned distribution F starting from a Y with a uniform distribution on $[0, 1]$ (or, conversely, $Y = F(X)$ has a uniform distribution on $[0, 1]$ if X has distribution F; some device is necessary in order to make it uniform at the jumps)

6.2.7. We shall see later that this form of representation is also useful for visualizing many problems and situations in the theory or probability and statistics (see, for example, Section 6.6). What is of immediate interest, however, is to exploit *the fact that we have, on the y-axis, the F-measure 'on its natural scale'* in order to look, succinctly, and without formulae, at the question of possible further extensions.

In terms of y, $F(y)$ corresponds to the ordinary (Riemann) integral, and therefore $F(I)$ (where $I = $ set, thought of as identified with its indicator function, $I(x) = (x \in I)$) can be interpreted as the Jordan–Peano measure of the image set of I on the y-axis. The F-measure (apart from the given transformation) is the *J–P measure* (that is, Jordan–Peano), and the *F-measurable* sets are those whose image, on the y-axis, is a *J–P-measurable* set.

6.3   THE PARTING OF THE WAYS

6.3.1. At this point we are faced with a choice.

It is well-known that *there exists a unique extension of the J–P measure to a much larger class of sets.* The methods used are due to Borel and Lebesgue, and the basic idea (put rather crudely) is to argue about a countably infinite collection of sets as if they were a finite collection; in particular, one invokes countable additivity (valid not only for the sum in the ordinary sense, but also for the sum of convergent series). Similar considerations apply to the extension of the notion of integral.

From the viewpoint of the pure mathematician—who is not concerned with the question of how a given definition relates to the exigencies of the application, or to anything outside the mathematics—the choice is merely one of mathematical convenience and elegance. Now there is no doubt at all that the availability of limiting operations under the minimum number of restrictions is the mathematician's ideal. Amongst their other exploits, the great mathematicians of the nineteenth century made wise use of such operations in finding exact results involving sums of divergent series: first-year students often inadvertently assume the legitimacy of such operations and fail the examination when they imitate these exploits. At the beginning of this century it was discovered that there was a large area in which the legitimacy of these limiting operations could be assumed without fear of contradictions, or of failing examinations: it is not surprising therefore that the tide of euphoria is now at its height. Two quotations, chosen at random, will suffice to illustrate this † : 'The definition is therefore *justified ultimately by the elegance and usefulness* of the theory which results from it'; 'Conditions about the continuity of (the integral) are really essential if the operation is to be a useful tool in analysis—*there would not be much of analysis left* if one could not carry out at least sequential limiting operations'.

6.3.2. Are there any reasons for objecting to this from a mathematical standpoint? Rather than 'objections', I think it would be more accurate to speak of 'reservations'; there are, I believe, two reasons for such reservations.

The first concerns what happens *outside* of that special field which results from the above approach. It has been proved (by Vitali, and afterwards, in more general contexts, by Banach, Kuratowski and Ulam) that if one is not content with finite additivity, but insists on countable additivity, then it is no longer possible to extend the 'measure' to *all the sets* (whereas there is nothing to prevent the extension to all sets of a finitely additive function which coincides—when they exist—with the J–P measure, or the L-measure).

Countable additivity cannot, therefore, be conceived of as a general principle which leads us safely around *within* the special field, and allows us to roam *outside*, albeit in an undirected manner, with an infinite number of choices. On the contrary, it is like a good-luck charm which works *inside*

---

† From J. F. C. Kingman and S. J. Taylor, *Introduction to Measure and Probability*, Cambridge University Press (1966), pp. 75 and 101 (the italics are mine).

the field, but which, on stepping outside, becomes an evil geni, leading us into a labyrinth with no way out.†

Never mind, it might be argued: measurable sets will suffice. But from what point of view? Practically speaking, the intervals themselves were perhaps sufficient. From a theoretical standpoint, however, is there any justification for this discrimination between sets of different *status*; the orthodox which we are permitted to consider, and the heretical which must be avoided at all costs? Would it be too far-fetched to suggest an analogy with real numbers, some of which still bear the name irrational because their existence had so scandalized the Pythagoreans?

6.3.3. The second reason for the reservation spoken of earlier concerns what happens *inside* the special field. Here the rules are more restrictive and permit us only to follow a uniquely defined path—like a runway for an automatic landing. This may be a fine thing, but must one be compelled to invoke this aid in all possible cases? Is it too absurd to believe that soap may sometimes have its uses despite the existence of an infinite number of detergents, each of which washes infinitely whiter than any of the others?

We can, happily, provide mathematical analogies in this case, and these will be more illuminating than the whimsical variety (although the latter may help in suggesting in advance the sense of the mathematics).

First a trivial example: if the value of a function is given at a finite number of points (or at a countably infinite number) I can complete it in an infinite number of ways—even under additional conditions (like continuity, etc.). Given n values, I know that the problem has one and only one solution if I add the condition that the function is a polynomial of degree $n - 1$ (if $n = \infty$, and I add the condition that the function be analytic, there is either one solution or no solution): is this a good reason for limiting oneself to this particular solution; or for considering it as 'special'?

A further example seems to me rather relevant. There exist methods for summing series—for example, that of Cesàro—which often give a uniquely determined answer in cases where the usual method of summation leads only to (different) upper and lower limits. Is it right that as a result we should always interpret 'sum of a series' as meaning Cesàro sum, and to

---

† This image of a labyrinth with no way out is an exact description of the situation. In fact, if one wishes to extend the definition of L-measure to non-measureable sets, respecting countable additivity, this can always be done step by step (choosing, for a given set, a value at random between the two extremes of inner and outer measure as determined by the extension so far made). After an infinite number of steps, however, a contradiction can arise, and sooner or later (before exhausting *all* the sets) it certainly arises. (As an analogy; a convergent series remains such if we add 1 onto a finite number of terms, no matter how far we go . . . , but not if we add 1 onto all the terms!)

This observation renders even more artificial the distinction between those sets which are L-measurable and those which are not (none of them has any particular feature which makes it unsuitable).

banish as 'outmoded' the usual notion of convergence? Of course, the compass of the Cesàro procedure (even iterated) is not comparable to that of other innovations, like that of Lebesgue, but, even assuming it to be such, would it then be justified? And would it not be possible that in certain cases there would be interest in ascertaining whether, in fact, the series were convergent according to the old definition (which, although out of date, has not become meaningless)? What if we wanted to know, in that sense, the upper and lower limits? In my opinion, this example is apposite in every respect. In the case of Lebesgue measure, as for Cesàro summability, there is a procedure which (because of additional conditions) often yields a unique answer instead of bounding it inside an interval within which it is not determined. Whether one solution is more useful than the other depends on further analysis, which should be done case by case, motivated by issues of substance, and not—as I confess to having the impression—by a preconceived preference for that which yields a unique and elegant answer *even when the exact answer should instead be 'any value lying between these limits'*.

6.3.4. The above remarks, made from a purely mathematical standpoint, are not designed to prove anything other than that the case for consigning the Riemann integral to the attic now that the Lebesgue integral is available has not itself been proved. The Riemann integral can still be a necessary tool; *not in spite of* its indeterminacy, but *precisely because of* it: this indeterminacy may very well have an essential meaning.

On the other hand, from a mathematical point of view I would by no means presume to discuss topics in analysis which I only know in the context of what I need. In the case of the calculus of probability, however, these questions relate to a fundamental need of the theory. We have already seen (in Chapter 3) the general kinds of reasons which prevent us from accepting countable additivity as an axiom. We shall come across other reasons which, taken with the above considerations, suggest that the use of Lebesgue measure and integration over the special field is not valid. (That is, of course, unless specific conditions are introduced in particular examples in order to meet the conditions which would allow such an application. The distinction, however, is that illustrated by the difference between saying 'I am applying this method because all functions are continuous', and 'I am applying this method because the function that I have chosen is continuous'.)

I do not know whether similar reservations and objections have a sound basis in regard to applications in other fields. In the case of mass, such a degree of detailed analysis is inappropriate (for instance, how could mass at rational points be separated off, or even considered as conceptually distinguishable?). The same thing could be said, in fact even more so, for statistical distributions. Everything leads us, therefore, to the conclusion that, apart from rather indirect issues,† the question is irrelevant in this area.

On the other hand, it is not really surprising if real objections only arise in the field of probability. In fact, we have, in the other fields, empirical assumptions, which are therefore approximate and necessarily lead to some arbitrariness in the mathematical idealization. The probabilistic interpretation, however, must confront logic face to face: this is its sole premise. Logic does not claim that it reaches out to some sort of precision (nor even to a higher level of approximation than is necessary), but neither does it allow the construction of a formally complete structure which does not respect the logical exigencies of a purely logical field of application; nor can it accept one constructed by someone else.

6.4   DISTRIBUTIONS IN PROBABILITY THEORY

6.4.1. Let us now turn to the topic of direct interest to us: that is the application of these mathematical tools within the calculus of probability. Roughly speaking, the application takes the following form: for any random quantity X, one can imagine a distribution of probability over the x-axis by assigning to the distribution function the interpretation $F(x) = \mathbf{P}(X \leqslant x) =$ the probabilistic 'mass' on $[-\infty, x]$. It follows that $F(I) = \mathbf{P}(X \in I)$, and $F(\gamma) = \mathbf{P}(\gamma(X))$, for any set I and function γ for which the notation is applicable.

This formulation, which is deliberately rather vague and neutral, is intended as a curtain-raiser to the questions we shall have to consider later (perhaps it would be more accurate to say that we shall consider them in relation to our particular position). These basically concern the alternatives of either continuing with the Riemann framework, or abandoning it for that of Lebesgue. We shall, however, leave the way open for any further modifications that may be required.

It is worth giving here and now a brief sketch of the two opposed positions. In order to pin a label on them, we might use the term *strong* for those who, as a result of accepting the validity of the Lebesgue procedures in this field, draw stricter and more sophisticated conclusions from the data; and *weak* for those who accept only the conclusions which derive from some smaller number of assumptions, carefully considered, and accepted only after due consideration.

The fact that we do not accept (as an axiom) countable additivity commits us to support of the *weak* position (as we have already mentioned, this is

---

† Like that concerning the precise meaning of a differential equation expressing a physical law; or the definition of the integral on a contour having cusps (this topic has given rise to discussion about the Kutta and Joukowski theorem); and so on.

one of the main planks in our programme; cf. Chapter 1, 1.6.2–1.6.4). The present discussion, apart from giving more insight into the implications of not adopting countable additivity, will consider its relation to other topics, and, although confining itself to the simplest case of distributions on the real line, will, in fact, reveal the general import of the conclusions.

6.4.2. *The strong formulation.* Once we know $F(x)$ we know everything about the probability distribution of a random quantity X. Everything that can be defined in terms of $F(x)$ (and with the Lebesgue extension) has a meaning: nothing else does. The probability that $X \in I$ is either given by $F(I)$, if the set I is F-measurable (Lebesgue–Stieltjes), or has no meaning if I is not F-measurable. The same holds for the prevision of $\gamma(X)$: either the function $\gamma(x)$ is F-measurable, in which case $\mathbf{P}(\gamma(X)) = F(\gamma)$, or the concept has no meaning. The set of possible values for X is also determined by F: it is the set of points for which F is increasing (i.e. the set of points not contained in an interval over which F is constant.†)

In this approach, one operates entirely within the confines of a rigid formulation, prescribed in advance: it was to this type of structure that we applied the description 'Procrustean bed'. Within its confines, '*that which is not compulsory is forbidden*'.

6.4.3. *The weak formulation.* Knowledge of $F(x)$ is only one of the many possible forms of partial knowledge of the probability distribution of a random quantity X (although, in practice, it is one of the most important).

Complete knowledge would demand a 'complete distribution': in other words, a (finitely additive) extension of $F(\gamma)$ to every function γ (and, in particular, to every set I) with no restrictions (on integrability, measurability, or whatever), and such that

$$F(\gamma) = \mathbf{P}(\gamma(X))$$

always holds (in particular $F(I) = \mathbf{P}(X \in I)$). Of course, we are talking of a theoretical abstraction, which can never actually be attained, but we have to make this the starting point, the landmark from which to get our bearings, in order to be in a position to consider all cases of partial knowledge without attributing to any of them some preordained special status.

Knowledge of $F(x)$, which we shall call *distributional* knowledge (or, sometimes, as is more common, knowledge of the distribution, albeit in the

---

† Even from this point of view, there would appear to be no difficulty in allowing something less rigid (e.g. the possibility of excluding a set of measure zero): I do not recall, however, ever having seen this kind of thing done explicitly. Perhaps this is the result of a psychological factor, which causes us to see distributions as prefabricated theoretical schemes, ready for attaching to random quantities, rather than regarding them as deriving from those random quantities, and from the particular circumstances which, depending on the case under consideration, derive from the underlying situation.

restrictive sense explained above), can turn out either to be more than we require, or less than we require, both from the point of view of the possibility of determining it realistically, and in relation to the needs of the situation under study. Sometimes, $\mathbf{P}(X)$, or $\mathbf{P}(X)$ and $\mathbf{P}(X^2)$ together, or some other summary, may be sufficient; in such cases there is no need to look upon the distribution as the basic element from which all else follows. On other occasions, the distribution itself is not enough: this is the case whenever we wish to rid ourselves of the restrictions implicit in the properties of $F(x)$ as commonly accepted; restrictions which are not always appropriate.

In contrast to the strong formulation, the argument in the weak case is always developed with a great deal of freedom of action: there is no obligation to fill in more details of the picture than are strictly necessary, and, on the other hand, there is no limit to the extensions one can choose to make—even up to the (idealized) case of complete knowledge.

6.4.4. *Setting the discussion into motion.* We introduce straightaway some useful notation. Its present purpose is to enable us to distinguish between the various extensions we shall consider in relation to a given F, but it will also enable us to avoid repeated, detailed explanations, whose tendency (despite the intention of avoiding ambiguities) is rather to create confusion.

The general notation is as follows: if $\mathscr{G}$ is a given set of functions γ ($\gamma \in \mathscr{G}$), then F, thought of as defined on $\mathscr{G}$, will be denoted by $F_{\mathscr{G}}$; for every γ not in $\mathscr{G}$, there will be for $F(\gamma)$ (used to denote a generic extension) a bound of the form $F_{\mathscr{G}}^-(\gamma) \leqslant F(\gamma) \leqslant F_{\mathscr{G}}^+(\gamma)$ (we do not dwell here upon the details of this: the interpretation is as set out in Chapter 3, 3.10.1 and 3.10.7, and which will be of use to us later in 6.5.3). We shall adopt, for the time being, as special cases, the following notations for distinguishing the ambit over which F is thought of as defined:

$F_{\mathscr{R}}$: if relative to the Riemann field;
$F_{\mathscr{B}}$: if relative to the Lebesgue field;†
$F_{\mathscr{C}}$: if relative to the complete field; and, finally,
F: if used in a generic sense.

More precisely: $F_{\mathscr{B}}$ always denotes an F which has been extended to mean

$$F_{\mathscr{B}}(\gamma) = \int \gamma(x)\, \mathrm{d}F(x)$$

---

† We shall use $\mathscr{B}$ instead of $\mathscr{L}$ (which was already used, see Chapter 2, for 'linear space'): $\mathscr{B}$, standing for *Borel*, is currently in use with a similar meaning to this (referring to Borel measure, which only differs from Lebesgue measure in so far as the latter extends it wherever it is uniquely defined by the two-sided bound). In our case $\mathscr{B}$ coincides with $\mathscr{L}$ (taken as meaning Lebesgue) because the extension is already implicit in our formulation ($F_{\mathscr{G}}(\gamma)$ is not only defined for $\gamma \in \mathscr{G}$, but for every γ such that $F_{\mathscr{G}}^-(\gamma) = F_{\mathscr{G}}^+(\gamma)$).

(in the Lebesgue–Stieltjes sense) where this makes sense; undefined otherwise. We could, however, denote the upper and lower integrals† by $F_{\mathscr{B}}^{-}$ and $F_{\mathscr{B}}^{+}$, and simply express the bounds $F_{\mathscr{B}}^{-}(\gamma) \leqslant F(\gamma) \leqslant F_{\mathscr{B}}^{-}(\gamma)$. In the above, $F_{\mathscr{B}}$, according to the *strong* formulation, *is all and everything*: the terms in the inequality do not even have a meaning within this framework. In the weak formulation, even if one considers an F which ('by chance', or for some particular reason—any reason—but not by virtue of some postulate) is countably additive over the Lebesgue field, the bounds would still have a meaning.

$F_{\mathscr{C}}$ denotes any F whatsoever, finitely additive, and thought of as defined for all functions γ (the ideal case, thought of in the weak formulation as the basic landmark). In this case, it is clear that bounds on the indeterminacy do not make sense; neither is there any possibility of extension.

When it makes sense (when it does not we consider $F_{\mathscr{R}}^{-}$ and $F_{\mathscr{R}}^{+}$),

$$F_{\mathscr{R}}(\gamma) = \int \gamma(x)\, \mathrm{d}F(x),$$

in the Riemann–Stieltjes sense, expresses *all that one can obtain from F*; that is, *distributional knowledge*, according to the *weak* formulation:

$$F_{\mathscr{R}}^{-}(\gamma) \leqslant F(\gamma) \leqslant F_{\mathscr{R}}^{+}(\gamma).$$

We should make this more precise, but this first requires the following summary.

We summarize briefly the two opposing points of view which present (in terms of the notation introduced above) a choice between:

(*strong*): for a given X, an $F_{\mathscr{B}}$ is to be chosen, and there is nothing more to be said;

(*weak*): for a given X, an $F_{\mathscr{C}}$ should be chosen; in fact, one limits oneself to some partial $F_{\mathscr{G}}$ which serves the purpose; often, one chooses a distribution function $F(x)$, and then it follows that $F_{\mathscr{R}}$ is in $\mathscr{R}$, and that the bound, which lies between $F_{\mathscr{R}}^{-}$ and $F_{\mathscr{R}}^{+}$, is not in $\mathscr{R}$.

6.4.5. Once more a word of warning. When referring to distributions, or distribution functions, F, it is useful to think of them as mathematical entities (e.g. the function $F(x) = 1/2 + (1/\pi)\arctan x$), which are available for representing the probability distribution of any random quantity X, as required. In other words, it is better *not* to think of them as associated with any given X. This distinction is of a psychological nature rather than a point of substance—which explains why the explanation is vague and somewhat confused—but our aim is to warn against misunderstandings that can (and frequently do) arise through some sort of 'identification' of an $F(x)$, an abstract entity, with $P(X \leqslant x)$, which, although equal to it,

---

† For the time being, we are considering only those functions γ which are *bounded* (over the range on which F varies, namely the x for which $0 < F(x) < 1$). The other case will be dealt with specifically in Section 6.5.4.

is a concept dependent on the specific random quantity X which figures in it. A typical example of the misunderstandings to be avoided is the confusion between limit properties of a sequence of distributions and similar behaviour of random quantities which could be associated with those distributions.

6.4.6. Why the 'Procrustean bed'? A preliminary question which it might be useful to discuss (although more for conceptual orientation than as a real question) is the following. Why is it that, at times, some people prefer (as in the *strong* formulation) to adopt a fixed frame of reference, within which one assumes complete knowledge of everything, all the details, no matter how complicated, no matter how delicate, and irrespective of whether they are relevant or not? This, despite the fact that the system is only used to draw particular conclusions, which could have been much more easily obtained by a direct evaluation. All this would appear to be a purely academic exercise; far removed from realism or common-sense.

In seeking the reason for this, one should probably go back to the time when fear was the order of the day, and all manner of paradoxes and doubts resulted. The only hope of salvation was to take refuge within paradox-proof structures—and this was no doubt right, at the time.

We must consider, however, whether it is reasonable, or sensible, to force those who are now strolling across a quiet park to take the same precautions as the pioneers who originally explored the area when it was wild and overgrown, and were ever fearful of poisonous snakes in the grass?

Let us note the following in connection with a specific example:

the use of transfinite induction (Chapter 3, 3.10.7) assures us that we can always proceed in an 'open-ended' way, adding in new events and random entities from outside any prefabricated scheme;

this method of proceeding is the only sensible one; at any moment new problems arise, and the thought of someone having to unscramble the enormous Boolean algebra that he has fixed in his mind, together with the probabilities which are stuck on all over the place, and having to construct a new edifice in order to include each new event, each new piece of information, and to up-date all his probabilities before sticking them back in, this thought is horrifying;

in evaluating probabilities (or a probability distribution), one should also proceed step by step, making them, little by little, more and more precise, for as long as it seems worth continuing. Even Ovid did not record the sudden appearance of a complete Boolean algebra, armed with all its probabilities, and springing from the head of Jove, disguised as Minerva, or rising, like Venus, from the foaming sea.

These remarks have been expressed in a manner which accords with the subjectivistic point of view; they would seem, however, to reflect fully the requirements of any realistic point of view, although perhaps not in such a clear-cut manner.

6.4.7. *The absence of anything having a special status.* We have already said (in 6.4.3) that no partial knowledge was to be accorded special status: not even that provided by $F(x)$. It seems strange to deny special status to probabilities associated with the 'most basic' sets, like intervals (or with continuous functions, as opposed to sets or functions of a 'pathological' nature). Is this objection well founded? Nothing can really be said about this without first considering and analysing the sense in which something has to be 'true', and in what sense, and on what basis, things appear to us as strange or pathological.

With regard to our own enquiry, we must distinguish that which has a *logical* character from that which draws its meaning from *other* sources: this is necessary, because it is only differences of a logical nature which can lead to the possibility of different treatment from a logical point of view. We note, therefore, that, from a logical point of view, in this representation every event corresponds to a set of points, and the only property that is relevant is the fact that one can tell (on the basis of the occurrence of X) whether the 'true' point belongs to the set or not. In this sense, there is nothing that can give rise to special forms of treatment: the above-mentioned property is assumed to be valid everywhere by definition, and other properties do not enter into consideration. From a logical point of view, no other aspects are relevant: for example, topological structures, or some other kind of structures that the space may happen to have for reasons which do not concern us.

Only differences of a logical nature could possibly justify special treatment in a probabilistic context. In general, there is no reason to discriminate between sets, and, in particular, this applies to sets which have, with respect to the outcomes of a random quantity X, the form of intervals, or anything else, however 'pathological'. There is no justification for thinking that some events merit the attributing of a probability to them, and others do not; or that over some particular partitions into events countable additivity holds, but not over others; and so on.

6.4.8. *The argument concerning what happens 'outside $\mathscr{B}$'.* We know that countable additivity cannot hold over the entire field $\mathscr{C}$ (of all events $X \in I$ and random quantities $\gamma(X)$ which can be defined in terms of a random quantity X, in correspondence with all sets I and functions γ). In fact, this was proved by Vitali under the additional assumption of invariance for the measures of *superposable* sets; an assumption which was removed in the extensions mentioned previously.

The above could be taken in itself as a sufficient reason *for rejecting countable additivity as a methodologically absurd condition* (as a general,

axiomatic kind of property) since it sets itself against the absence of any logical distinctions, which alone could justify discrimination between events.†
This would be the case even if we disregarded the reasons we have already put forward (Chapter 3, 3.11, and Chapter 4, 4.18), reasons which, in fact, cannot be disregarded.

6.4.9. *The argument about what happens 'inside $\mathscr{B}$'.* In the particular case of *L*-measurable sets, where we know that countable additivity *can* be assumed without giving rise to any contradictions, there is no reason to assume automatically that countable additivity *must* hold (or that it is entitled to be accepted for some particular reason). Every distinction between measurable and non-measurable sets disappears when we no longer take the topology of the real line into account (imagine reshuffling the points as though they were grains of sand). We present straightaway some counter-examples (they can be disposed of only on the grounds of a prejudice to do so just because they are counterexamples‡).

Here is one of them. Let X be a rational number between 0 and 1, and let us further assume that no rational between 0 and 1 can either, on the basis of our present knowledge, be rejected as impossible, or appear sufficiently probable to merit assigning a non-zero probability to it. In this case, we have a continuous distribution function $F(x)$: we could also limit ourselves to considering the special case of the uniform distribution, $F(x) = x$ $(0 \leqslant x \leqslant 1)$. According to the strong formulation, one would conclude that, with probability 1, the rational number X belongs ... to the set of irrational numbers!

This, and other similar examples (which we shall make use of shortly for other purposes), also show, among other things, that precisely the same distribution function can correspond to random quantities having different ranges of possible values. This will be dealt with in Sections 6.5.2–6.5.3.

6.4.10. *Partial knowledge.* Every piece of partial knowledge will be the knowledge of the complete distribution $F_{\mathscr{C}}(\gamma)$ restricted to some subset or other of the functions γ (it does not matter whether they are functions, sets, or a mixture of the two). For example, one might know $F(x)$ at some particular points (i.e. for a certain partition into intervals) and/or γ for some individual functions. To use the standard examples, these might be the prevision and

---

† More precisely, the discrimination would only be justified if one concentrated the whole probability ($= 1$) on a finite or countable set (of points with positive probabilities, with sum 1). It is absurdly restrictive to pretend this should always be the case; even, e.g., if the 'points' of our field are 'all the possible histories of the universe' (but let us leave aside such extralogical and personal judgements). The fact is, that no continuous measure—in the mild sense of being, like Lebesgue measure, effectively spread over an uncountable set— can satisfy our requirement.
‡ This is the tactic of 'monsterbarring', according to the terminology of Imre Lakatos, in "Proofs and refutations", *Brit. J. Philosophy of Science*, **14** (1963–64), 53–56.

variance (as direct data, and not based on the assumption, either implicit or explicit, of the existence of the distribution of which the prevision is the barycentre, etc., as is usually the case). It would, however, be equally admissible (although, generally speaking, of little interest, and not really practicable) to provide, instead, probabilities for certain pathological sets only (e.g. numbers whose decimal expansions never involve more than n zeroes in the first $2n$ places), or the previsions of some pathological functions (e.g. continuing with the same example, $\gamma(x) = $ sup of the percentage of zeroes in the first n places as n varies).

In short, it is open to us to assume or require that either *everything, a little* or *a great deal* is known about the probabilities and previsions relating to X. Do not lose sight of the fact (even though it is not convenient to repeat it too frequently) that, in using 'known' or 'not known' when thinking in terms of the mathematical formulation (in fact, when thinking of the actual meaning), we mean 'evaluated' or 'not evaluated'.

Of course, it could be, as a special case, that the partial knowledge of the complete distribution is that defined over the intervals : in other words, that given by $F(x)$, known for all x. This is what we have called knowledge of the distribution through the *distribution function*. It is a form of partial knowledge like all the others, but it is of particular interest and we shall wish to, and have to, consider it at greater length, in order to clarify the rôle played (in the present formulation) by $F(x)$.

$F(x)$ remains a standard tool, but re-evaluated (one might say cut down to size) in a manner and for reasons which we shall explain. It does not play any special, privileged rôle *de jure*, but only *de facto* : that is, in relation to the interpretation of X as a magnitude, which is what is of interest in practice, and to the geometric representation on the line, which is what enables it to be visualized. It is for these reasons that it plays a special rôle, by reason of the applications, and from the psychological point of view ; despite the fact that they cannot justify its special *status* from the logical standpoint.

6.4.11. The *re-evaluation* is not solely, however, in this conceptual specification ; nor in the fact that knowledge conveyed by $F(x)$ no longer appears complete in that we require something further ($F(\gamma)$ lying outside the Lebesgue ambit of F), whereas it remains what it is. But it does not remain what it was : it is more restricted. It remains what it was only in the Riemann ambit of F ; outside of this (with no further discrimination between that which is inside or outside the Lebesgue ambit of F) it only provides the bounds we have already encountered

$$F_{\mathscr{R}}^{-}(\gamma) \leqslant F(\gamma) \leqslant F_{\mathscr{R}}^{+}(\gamma).$$

These give the limits for any evaluation of $F(\gamma)$ compatible with knowledge of F in the distributional sense (i.e. knowledge of $F(x)$). We are, of course, dealing with the upper and lower integrals in the Riemann sense; in particular (in the case of sets) we have inner and outer Jordan–Peano measure. This indeterminacy does not imply any fault in the capacity of the concepts to produce a unique answer: on the contrary, as we shall see later in more detail, the indeterminacy turns out to be essential (given our assumptions), in the sense that all and only the values of the interval are in fact admissible (and all equally so). Any of them can be chosen, either by direct evaluation, or by an evaluation which derives from some additional considerations, which must then be set out one by one (and cannot just consist of the assumption of countable additivity, for which one must, case by case, make the choice of the family of partitions on which its validity is to be assumed, and state the choice explicitly).

What we have said so far concerning the rôle of $F(x)$ is more or less the translation and explication in concrete form of the two 'reservations' which we previously put forward in the abstract. But the abandonment of countable additivity implies yet another revision of the meaning of $F(x)$: it is no longer true that a jump at x must correspond to a concentration of probability at the point x (it may only *adhere* to the point, and the point itself might not even belong to the set of possible points). It is also no longer true that $F(x)$ must vary from 0 to 1 (we only require that $0 \leqslant F(-\infty) \leqslant F(+\infty) \leqslant 1$), or that the possible points are those at which $F(x)$ is increasing.

A single observation will suffice. Suppose that the possible points, judged equally likely, form a sequence (e.g. $x_0 - 1, x_0 - \frac{1}{2}, \ldots, x_0 - 1/n, \ldots$) which tends to a given point x_0 from below. In this case $F(x)$ will have a jump of 1 at $x = x_0$, just as if $X = x_0$ with certainty (all the mass concentrated at x_0). In fact, we have $F(x) = (x \geqslant x_0) = 0$ for $x < x_0$, and $= 1$ for $x \geqslant x_0$, because to the left of any point on the left of x_0 there is at most a finite number of possible points, each of which has zero probability; whereas to the left of x_0 (and, *a fortiori*, to the left of any point on the right of x_0) we find all the possible points.

This implies that, in general, if $F(x)$ has a jump p_h at a point x_h, it is always possible (apart from the case when there are no possible points in some left or right neighbourhood of x_h) to decompose p_h, in some way, in the form $p_h = p_h^- + p_h^0 + p_h^+$, where p_h^0 is the mass actually concentrated at x_0, and the other two parts are *adherent* to it on the left and on the right (in the manner illustrated in the example).

This fact alone would seem to provide support for the usefulness of the convention of regarding the value of $F(x)$ to be indeterminate at points of discontinuity (cf. the footnote to 6.2.2). We shall, however, consider this in the next section (6.5.1), where the arguments will be more decisive when put in the context of some further ideas.

The previous example (if we consider sequences tending to $-\infty$ or to $+\infty$) suffices to show that we can, in a similar fashion, have probabilities adherent to $-\infty$ and to $+\infty$. These are given by $F(-\infty)$ and $1 - F(+\infty)$. Those distributions for which (as we have so far assumed, in accordance with the standard formulations) these probabilities are zero we shall call *proper*, and we note that F then actually does vary between 0 and 1; all others will be called *improper* (and we can further specify whether the impropriety is *from below, from above* or *two-sided*).

Our previous remark concerning possible points is also clear, given the possibility of substituting for any point a sequence which converges to it; this topic will be considered further in due course (see 6.5.2).

6.5 AN EQUIVALENT FORMULATION

6.5.1. Knowledge of $F(x)$ (apart from points of discontinuity), in other words, what we are calling distributional knowledge, is equivalent—in the case of a *proper* F†—to knowledge of $F(\gamma)$ for all continuous functions γ, which are bounded over the entire x-axis, from $-\infty$ to $+\infty$. More precisely; these, and only these, functions are F-integrable whatever F might be; conversely, knowledge of $F(\gamma)$ for all continuous γ is sufficient, whatever F might be, to determine $F(x)$ for all x, apart from discontinuity points.

Of course, to say that knowledge of $F(x)$ is equivalent to knowledge of $F(\gamma)$ for all continuous γ does not mean that it has to be known for every such γ. It will be sufficient to know it for a basis in terms of whose linear combinations any continuous function can be approximated. This remark will serve as the foundation for more analytical kinds of treatment (in particular, that for characteristic functions); here it merely serves to assuage possible doubts.

Let us consider the following in more detail, further considering the possibility of 'adherent masses', which we noted above. If $F(x)$ has a jump p_h at the point $x = x_h$, and it were assumed that the mass p_h were concentrated at the point x_h, then (as in the case of the usual assumption of countable additivity) we would take the contribution of this mass to $F(\gamma)$ to be $p_h\gamma(x_h)$. Without the assumption of concentration, however, we can do no more than note that the contribution lies between the maximum and minimum of the five values

$$p_h\gamma(x_h) \quad \text{and} \quad \left.\begin{array}{l}\max \\ \min\end{array}\right\} \lim p_h\gamma(x)$$

as $x \to x_h$ from the left or right, respectively. Proceeding differently (and more simply) it is sufficient to exclude points of discontinuity as subdivision points (this is always possible—there are only a countable number of them).

From this, it is clear that any function $\gamma(x)$ which has even a single discontinuity point is not integrable for all F, since, if we take an F with a

---

† Otherwise one requires in addition the existence of a finite limit for $\gamma(x)$ as $x \to -\infty$, or $x \to +\infty$, or both.

jump at this point, the contribution of this mass to the integral is indeterminate. Conversely, if we know $F(\gamma)$ for the continuous functions γ, we can evaluate $F(x_0)$ from below and above as follows: we take a function $\gamma_1(x)$ which $=1$ from $-\infty$ to $x_0 - \varepsilon$, and $=0$ from x_0 to $+\infty$, and decreases continuously from 1 to 0 within the small interval $x_0 - \varepsilon$ to x_0, and a function $\gamma_2(x) = \gamma_1(x - \varepsilon)$, which is the same as γ_1, except that the decreasing portion is now between x_0 and $x_0 + \varepsilon$. The difference between the two functions is $\leqslant 1$ between $x_0 \pm \varepsilon$ and zero elsewhere; we therefore have that

$$F(\gamma_2) - F(\gamma_1) \leqslant F(x_0 + \varepsilon) - F(x_0 - \varepsilon), \text{ etc.}$$

Everything goes through smoothly, except when we have a discontinuity at $x = x_0$.

The mathematical argument, which seems to me to show conclusively that we should consider $F(x)$ as indeterminate at discontinuity points x, is the following: it is more meaningful to consider the continuous γ, than to consider indicator functions of half-lines or intervals. What seemed to be an *ad hoc* restriction when starting from the intervals, is, instead, rather natural when one considers continuous functions; in this case, one would need an *ad hoc* convention to eliminate it.

On the other hand, this mathematical argument is closely bound up with the point that I consider to be most persuasive both from the point of view of fundamental issues and of applications: the need for some degree of realism when we assume the impossibility of measuring X with absolute certainty. We shall consider in the Appendix (Section 7) limitations imposed on 'possible occurrences' of events due to these kinds of imprecisions; it is clear, however (and we shall confine ourselves to this one observation at present), that to consider $F(x)$ as completely determined, apart from discontinuity points, is equivalent to thinking that X can be measured with as small an error as is desired, but cannot be measured exactly with error $=0$. This suffices to render the case $X = x_0$ with certainty indistinguishable from the case where the mass is adherent to x_0 (e.g. it is certainly at $x_0 - 1/n$, where n is any positive integer whose probability of being less than any preassigned N is equal to zero).†

6.5.2. *The distribution and the possible points.* We have already seen, when examining the special case of a discontinuity point, that there is a lot

---

† Without going into the theoretical justifications (or attempts at justifications), it is a fact that the different conventions reveal practical drawbacks that make their adoption inadvisable. The convention $F(x) = F(x + 0)$ (or, conversely, $F(x) = F(x - 0)$) makes the equation $F_1(x) = 1 - F(x)$ (used in passing from X to $-X$) invalid; writing $F(x) = \frac{1}{2}[F(x + 0) + F(x - 0)]$ avoids this difficulty, but (cf. the end of 6.9.6) one sometimes needs to consider $F_2(x) = [F(x)]^2$, and it is not true that $\{\frac{1}{2}[F(x + 0) + F(x - 0)]\}^2 = \frac{1}{2}[F^2(x + 0) + F^2(x - 0)]$; and so on. In contrast, the convention we are proposing here remains coherent within itself; moreover, it gives a straightforward interpretation of the appropriateness of completing the diagram of Figure 6.2a (Figure 6.2b) with vertical (horizontal) segments.

of arbitrariness concerning the possible points which 'carry' the mass corresponding to the jump; they do not have to enclose the jump-point, they only have to be dense in any neighbourhood of it. Before proceeding any further, we have to examine the general relationship between the set $\mathscr{Q}$ of possible points for a random quantity X—which we shall call the *logical support* of X—and $F(x)$, the distribution function of X; more specifically, the relationship between this set $\mathscr{Q}$ and the set $\mathscr{D}$ of points at which $F(x)$ is increasing—which we shall call the *support of the distribution F* (or the *distributional support* of X). Formally, this is the set of x such that, for any $\varepsilon > 0$, we have

$$F(x + \varepsilon) - F(x - \varepsilon) > 0.$$

Every neighbourhood of x has positive probability; it is therefore possible, and hence contains possible points. It therefore follows that $\mathscr{D}$ is contained in the closure of $\mathscr{Q}$; moreover, this condition is sufficient because, whatever partition one considers (partition, that is, of the line into intervals), no contradiction is possible (every interval with positive mass contains possible points to which it can be attributed).

It is convenient to consider separately the various cases. Let us begin with the intervals on which $F(x)$ is constant (at most a countable collection): these may contain no possible points, but there is nothing that debars them from doing so (they could consist entirely of possible points), so long as the total probability attributed to them is zero. At the other extreme, we have the intervals over which $F(x)$ is strictly increasing. Here, it is necessary and sufficient that the possible points are everywhere dense (it could be that all points are possible). As an example, think of the uniform distribution on $[0, 1]$, with either all points possible, or just the rationals. An isolated point of increase is necessarily a jump-point (but not vice-versa), and we have already discussed this case; either the point itself must be possible, or there must exist an infinite number of possible points adherent to it (of which it is a limit point). Finally, suppose that a point of increase of $F(x)$ is such because each neighbourhood of it contains intervals, or isolated jump-points, where $F(x)$ is increasing. This fact tells us that the given point is an accumulation point of possible points; we can go no further in this case.

We are especially interested in the end-points of the above-mentioned sets. We have adopted (ever since Chapter 3) the notation inf X and sup X for the limits of the logical support; let us now denote by inf F and sup F the limits of the distributional support. These are, respectively, the maximum value of x such that $F(x) = 0$, and minimum value such that $F(x) = 1$ (if F is unbounded—from below, from above, or from both sides—or improper, the values are $\pm\infty$). By virtue of what we said previously, we necessarily have inf $X \leqslant$ inf $F \leqslant$ sup $F \leqslant$ sup X. It is important to note that logical support is a bound for distributional support, but not conversely.

More generally, it is important to realize just how weak the relation between the two forms of support can be. If we are given the distribution, all we can say is that each point of the support is either a possible point, or is arbitrarily close to possible points; in addition to this, possible points (with total probability zero) could exist anywhere and even fill up the whole real line. On the other hand, given the logical support, we can state that the distribution could be anything, so long as it remains constant over intervals not containing any possible points. We are here merely reiterating, in an informal and rather imprecise way, what we have already stated precisely. In this way, however, we may be able to better uncover the intuition lying behind the conclusions. On the one hand, that, corresponding to the concept of being able to take measurements as precisely as one wishes, but not exactly, one is indifferent to the fact that what is regarded as possible can be: either a point or a set of points arbitrarily close to it, respectively; either all the points of an interval or those of a set everywhere dense in it, respectively. On the other hand, that possible points with total probability zero do not affect the distribution, but are not considered as having no importance (and we shall see below that they are important when it comes to considering prevision).

6.5.3. Conclusions reached about sets lead immediately to conclusions regarding their probabilities. In fact, we can see straightaway that $\mathbf{P}(X \in I)$, the probability of a set I, can actually assume any value lying between the inner and outer F-measure (in the Jordan–Peano sense).

Let $\mathscr{D}$ be the set of points for which $F(x)$ is increasing, and partition it into $\mathscr{D}_1$, the intersection of $\mathscr{D}$ with the closure of I (that is, the set of points of $\mathscr{D}$ having points of I in every neighbourhood), and $\mathscr{D}_2$, its complement (points within intervals containing no points of I). Let us assume that in the closure of $\mathscr{D}_1$ only points of I are possible (either all of them, or a subset which is everywhere dense there); only in the intervals containing no points of I do we have recourse to other points in order to obtain the 'possible points' required for $\mathscr{D}_2$. In this way, I turns out to have the maximum possible probability; that is, the outer F-measure (we attribute to I the measure of every interval in which I is dense). By applying the same idea to the complement of I, we obtain the other extreme (the minimum probability for I, given by the inner F-measure; in this case only those intervals containing solely points of I are considered). Clearly, all intermediate cases can be arrived at by mixtures (for example, for a direct interpretation, consider the fact that, without changing the distribution, possible points are taken either to be those of the first version or the second, according as an event E is true or false; by varying the value $p = \mathbf{P}(E)$, $0 \leqslant p \leqslant 1$, we obtain all possible mixtures).

This fact reveals another aspect of the 're-evaluation' of the nature of distributional knowledge: it says very little about what, from a logical viewpoint, is the most important global feature of the distribution; that is, about the logical support.

6.5.4. *The restriction of boundedness.* There remains the question of our restriction to the bounded case: it is an important topic in its own right, and we have rather passed it over (each topic should really come before all the others, and that is just not on). We shall meet a further aspect (the last one!) of the 're-evaluation' of the rôle of the distribution function, and we shall be forced to make (and offer to the reader) some sort of make-shift choice, not entirely satisfactory, in order to be able to draw attention to certain necessary distinctions, without too many annoying notational complications, and without running too many risks of ambiguity.

We have already seen (Chapter 3, 3.12.4–3.12.5) that, without the assumption of countable additivity, there are no upper (lower) bounds for the prevision of a random quantity which is unbounded from above (below). This was seen in the case of discrete random quantities; what happens when we pass from this to the general case?

The question is an extremely deceptive one when looked at in the light of what distributional knowledge is able to tell us. Starting from the knowledge of $F(x)$, the conclusion that we can derive a certain value, $F(\square)$, which 'ought to be' that of $\mathbf{P}(X)$, will be more acceptable if not only the distribution F, but also the logical support of X, is bounded (and knowledge of F gives us no information about this). We shall put this conclusion more precisely, and also examine more closely the value of the partial knowledge that we can obtain in this connection.

First of all, it is convenient to specialize to the case of non-negative random quantities (inf $X \geqslant 0$): given any X, we can, of course, decompose it into the difference of two non-negative random quantities by setting

$$X = X(X \geqslant 0) + X(X \leqslant 0),$$

or, in a different but equivalent form,

$$X = (0 \vee X) + (0 \wedge X).$$

In either case, the first summand has value X if $X \geqslant 0$, and zero otherwise; and the second summand has value X if $X \leqslant 0$ and zero otherwise (and is therefore always non-positive: in order to obtain the difference of non-negative values explicitly, it suffices to write 1st $-$ ($-$2nd) instead of 1st $+$ 2nd).

For X non-negative and bounded, we certainly have

$$\mathbf{P}(X) = F(\square) = \int x \, dF(x).$$

A non-negative X which is unbounded can be turned into a bounded quantity by either 'amputating' or 'truncating' it.† We shall apply the first method, which is simpler. We have

$$\mathbf{P}(X) \geqslant \mathbf{P}[X(X \leqslant K)] = F[\square(\square \leqslant K)] = \int_0^K x \, dF(x);$$

this holds for any K, and hence

$$\mathbf{P}(X) \geqslant \int_0^\infty x \, dF(x) = F(\square),$$

where this defines $F(\square)$ by convention in this case. The integral may be either convergent or divergent: in the latter case, we must have $\mathbf{P}(X) = F(\square) = +\infty$, whereas, in the former, we can only say that all values in the range $F(\square)$ to $+\infty$ are possible for $\mathbf{P}(X)$ (including the two extremes). Note that the case of convergence also includes the case where the distribution is bounded (sup $F < +\infty$), but arbitrarily large possible values of X (with total probability 0) are permitted.

6.5.5. We have adopted as a *convention* the definition $F(\square) = \int x \, dF(x)$; this holds even when the integral is improper (it has to be extended up to $+\infty$), and only makes sense, as a limit, when it converges. This convention can be extended to the general case (to a distribution unbounded either way) with a similar interpretation; that is, with the understanding that

$$\int = \int_{-\infty}^0 + \int_0^{+\infty},$$

if both integrals exist. We have to stress the interpretation we give to our convention, in order to draw a distinction between it and the interpretation it has in the usual formulation (that is, in the strong formulation). In the latter, the convention is taken as *a definition of the prevision* $\mathbf{P}(X)$ *of a random quantity* X *with distribution* $F(X)$: if one of the two integrals diverges, we either have $\mathbf{P}(X) = +\infty$ or $\mathbf{P}(X) = -\infty$; if both diverge, $\mathbf{P}(X)$ has no meaning.

So far as we are concerned, $\mathbf{P}(X)$ will from henceforward have the meaning we have assigned to it; it will not make sense to set up new conventions in

---

† To 'amputate' means to put $Y = X(X \leqslant K)$: to 'truncate' means to set $Z = X \wedge K$; in other words, $Y = Z = X$, so long as $X \leqslant K$, but $Y = 0$ and $Z = K$ otherwise. Clearly we have $Y \leqslant Z \leqslant X$ ($Y = Z = X$ if $X \leqslant K$, and $Y < Z < X$ when $X > K$, since $0 < K < X$).

order to re-define it for this or that special case. Given the knowledge of $F(x)$ one could work out possible bounds for $\mathbf{P}(X)$—always on the basis of the (weak!) conditions of coherence—but one must be careful not to add any further restrictions, and not to interpret the acceptable ones as being in any way more restrictive than they actually are. Not a single one of the values that can be attributed to $\mathbf{P}(X)$ without violating coherence should be ruled out as unacceptable. This would be a mistake; excusable if due to an oversight, but inexcusable if due to carelessness, or an inability to understand the demands of logical rigour.

Our convention should be interpreted entirely differently. It defines $F(\square)$—and, similarly, $F(\gamma)$, for any γ—as information relating to the distribution F (considered as a mathematical entity); in order to avoid any confusion, it would perhaps be better to call $F(\square)$ the *mean value* of the *distribution F*, rather than the *prevision* (a notion concerning a random quantity X). Such a *mean value* is of interest when we are considering the previsions of random quantities X, Y, Z, all having the same distribution F; it is almost never possible, however, to simply state that the previsions must all be equal and coincide with $F(\square)$.

This conventional mean value does, however, play an important rôle for the following three reasons. In the first place, it serves to provide the logical conditions that characterize the set of admissible values for $\mathbf{P}(X)$. Secondly, it always provides a particular admissible evaluation of $\mathbf{P}(X)$, whose acceptance can often be justified by making an additional, meaningful assumption. Thirdly, it turns out that simultaneously accepting this additional assumption for several random quantities cannot lead one into incoherence.

If there is no additional knowledge, there are no logical conclusions to be drawn in passing from $F(x)$ to $\mathbf{P}(X)$. Fortunately, knowledge is available concerning a basic fact of a logical nature: that of the logical support of X (the set of possible values), or simply knowledge of the extremes, inf X and sup X, or, even more simply, knowledge of whether they are finite or infinite. If they are both infinite, nothing more can be said about $\mathbf{P}(X)$—all values $-\infty \leqslant \mathbf{P}(X) \leqslant +\infty$ are admissible. If they are both finite, we must certainly have $\mathbf{P}(X) = F(\square)$. If only one of the extremes is infinite, all values between it and $F(\square)$ are admissible: in other words, if inf $X = -\infty$, we have $-\infty \leqslant \mathbf{P}(X) \leqslant F(\square)$, and, if sup $X = +\infty$, we have $F(\square) \leqslant \mathbf{P}(X) \leqslant +\infty$. In just one special case we also have a uniquely determined value: if $F(\square) = +\infty$ and inf $X > -\infty$, then we certainly have $\mathbf{P}(X) = +\infty$ (and, similarly, if $F(\square) = -\infty$, and sup $X < +\infty$, then $\mathbf{P}(X) = -\infty$).

Turning to the case of arbitrary functions, $\gamma(x)$, there are no essential changes to be made, but there are a couple of details.

In order to remain within the domain of distributional knowledge, we must limit ourselves to considering $F_{\mathscr{R}}$ (integrals in the Riemann–Stieltjes sense, etc.), and, hence, to consideration of γ which are continuous (cf. 6.5.1), or, alternatively, to considering the two values $F_{\mathscr{R}}^-(\gamma) \leqslant F_{\mathscr{R}}^+(\gamma)$ (which are, in general, different). We shall always adopt the latter course, and, consequently, we will omit the $\mathscr{R}$. Extension to unbounded $\gamma(x)$ has to proceed as above; by separating into positive and negative parts, $\gamma(x) = [0 \vee \gamma(x)] + [0 \wedge \gamma(x)]$, and then amputating each of the parts (considering, for example, $[0 \vee \gamma(x)] \cdot [\gamma(x) \leqslant K]$ instead of $0 \vee \gamma(x)$; we shall call this $\gamma_K(x)$): we then take $F^-(\gamma_K)$ and $F^+(\gamma_K)$ relative to these, and obtain $F^-(0 \vee \gamma)$ and $F^+(0 \vee \gamma)$ as limits as $K \to \infty$. Similarly, we deal with $0 \wedge \gamma$, taking $K < 0$ and tending to $- \infty$. Summing, we obtain $F^-(\gamma) = F^-(0 \vee \gamma) + F^-(0 \wedge \gamma)$ (and similarly for F^+). If the sum is of the form $\infty - \infty$, it must obviously be understood as $- \infty$ for $F^-(\gamma)$ and $+ \infty$ for $F^+(\gamma)$.

The second detail (perhaps it would be better to call it a remark) concerns a simplification that can arise in the case of an arbitrary $\gamma(x)$, in comparison with the simplest case, $\gamma(x) = \square(x) = x$, considered above. In fact, if the function γ is bounded ($|\gamma(x)| \leqslant K$ for all x) then $\gamma(X)$ is certainly also bounded (and the same holds for semi-boundedness). If $\gamma(x)$ is not bounded, and all the values of x ($- \infty \leqslant x \leqslant + \infty$) are possible for X, then the random quantity $\gamma(X)$ is also unbounded, in the same manner. It is only in the case of $\gamma(x)$ unbounded and X having a more restricted support that the question of the boundedness of $\gamma(X)$ cannot be settled immediately, but only by examining the values that $\gamma(x)$ assumes on the support of X (it will often, however, be sufficient to check whether it is bounded on the interval $\inf X \leqslant x \leqslant \sup X$; only if it does not turn out to be bounded there will it be necessary to proceed to a more detailed analysis).

6.5.6. This having been said, our previous conclusions, apart from obvious changes, can now be restated, a little more concisely, in the general case. *The admissible values for* $\mathbf{P}[\gamma(X)]$ *are those which satisfy the inequality*

$$F^-(\gamma) \leqslant \mathbf{P}[\gamma(X)] \leqslant F^+(\gamma)$$

when $\gamma(X)$ *is bounded (that is, if* $- \infty < \inf \gamma(X)$, $\sup \gamma(X) < + \infty$*); with* $F^-(\gamma)$ *replaced by* $- \infty$ *if* $\inf \gamma(X) = - \infty$*; with* $F^+(\gamma)$ *replaced by* $+ \infty$ *if* $\sup \gamma(X) = + \infty$.

In other words: in the double inequality, the right-hand side, left-hand side, or both, must be suppressed according as we have unboundedness on the left, right or both.

In particular: we obtain a uniquely determined value for $\mathbf{P}(X)$ only if $F(\gamma)$ exists (that is, $F^-(\gamma) = F^+(\gamma)$). This value is *finite* if $\gamma(X)$ is bounded; *infinite* ($- \infty$ or $+ \infty$) if $\gamma(X)$ is semi-bounded (the direction of the boundedness is obvious).

To see how the present statement contains the previous one as a special case, observe that if both the integrals (from $- \infty$ to 0 and from 0 to $+ \infty$) diverge, then $F^-(\square) = - \infty$ and $F^+(\square) = + \infty$.

6.5.7. *Prevision viewed asymptotically.* If $F(x) = \mathbf{P}(X \leq x)$, the mean value of the distribution F, in addition to its logical interpretation within the confines discussed above, may often have a reasonable claim to be taken as the value of $\mathbf{P}(X)$, even if there are no circumstances compelling one to make this choice.

This is the case when we choose to deal with an unbounded distribution (either one-sided or two-sided), but where the choice might reasonably be seen as an idealized approach to something that, had we been more realistic, should be considered as bounded. To put it more straightforwardly: we think that $F(x)$ represents pretty well our idea of the distribution throughout the range $a \leq x \leq b$, which, practically speaking, includes all the possible values; to also include the 'tail' to infinity is both convenient from a mathematical point of view, and also in practice, since we would not really know just where to set the limits a and b (but this latter point should not be taken too seriously). The most appropriate 'model' is to conceive of using the bounded distribution as 'a limit case of distributions amputated or truncated to intervals, whose limits are so large that an asymptotic expression is appropriate' (that is, for $a \to -\infty$ and $b \to +\infty$, in whatever way).

From among the logically admissible values for $\mathbf{P}(X)$ we shall often select this one when such justifications of asymptotic kind appear to be valid. Sometimes we shall denote this value by $\hat{\mathbf{P}}(X)$: the accent will simply signify that this particular choice has been made (it serves as a shorthand), and will not imply that $\mathbf{P}$ has been thus marked because it is a special value of some sort.

We have stated already that there is no danger of contradiction resulting from the systematic use of $\hat{\mathbf{P}}$; this means that $\hat{\mathbf{P}}$ is additive.

(We observe that in choosing values for $\mathbf{P}(X)$, $\mathbf{P}(Y)$ and $\mathbf{P}(Z)$, it is not enough merely to ensure that each of them is admissible—for example, if we have $Z = X + Y$ with certainty, then our choice must satisfy $\mathbf{P}(Z) = \mathbf{P}(X) + \mathbf{P}(Y)$.)

That this condition is satisfied for $\hat{\mathbf{P}}$ follows from the additivity of the integral. We are, however, dealing with a two-dimensional distribution, and we shall therefore deal with this later (in Sections 6.9.1–6.9.2).

In order to avoid unnecessary complications, we shall, unless otherwise stated, adopt the convention that we shall always take $\mathbf{P} = \hat{\mathbf{P}}$ (exceptions will be made when there is some critical remark worth making). Important points will be made in Section 6.10.3, and in Chapter 7, 7.7.4, concerning the connection with characteristic functions and Khintchin's theorem.

6.5.8. *Probability distributions and distributional knowledge.* We are now in a position to summarize the conclusions we have reached as a result of following through the weak formulation in a coherent fashion, and also the conventions that have proved necessary in order to make the formalism and the language conform to the requirements of the formulation. In fact, we shall not merely provide a summary, but also fill in some more details, mentioning in an integrated manner certain points hitherto made only incidentally: in this way, we shall build up the complete picture.

The distinction, originally presented as if it were a small difference in attitude, between a complete distribution, attached to a random quantity and containing all the information about it, and a distribution function as a mathematical entity, useful for providing a partial indication of the form of a random quantity, is now much more sharply drawn. We have seen, in fact, a number of ways in which the latter form is incomplete and not sufficiently informative; this became clear as we proceeded to 're-evaluate' the notion.

Distributional knowledge, as we introduced it (in a way we considered appropriate to make of it an instrument whose range of application was properly defined), is sufficient to obtain a description of the image of a 'distribution of probability mass' within well-determined 'realistic' limits. One can ask how much mass is contained in an interval (but without being able to state precisely whether the mass adherent to the end-points is inside or outside the interval, and with no possibility of saying anything with respect to a set having a complicated form, or not expressible in terms of intervals). One can ask for the mean value of any continuous function with respect to the mass distribution (but not for functions in general, unless one assumes some further conditions). Nothing, however, can be known precisely concerning which points are possible and, without this knowledge, we cannot even say whether or not the mean value of the distribution is the prevision of an X having that particular distribution function.

To summarize: distributional knowledge is only partial, and has to be made precise before it provides complete knowledge. By making it precise, one can obtain many different probability distributions from it; they all have in common, so to speak, those features that are apparent at first sight, without examining the details more closely under a microscope.

Given this analysis, one can now pick out those properties which the strong formulation obtains from the distribution function by virtue of the assumption of countable additivity. These properties might or might not hold (by chance), and might also hold for non-measurable sets or functions (should these be of interest). Above all, one needs to state precisely what one means by 'possible points'.

> In order to avoid any misunderstandings or ambiguity, and to pay close attention to the distinctions we have drawn, it would be better if we reserved the term *'probability distribution'* for the complete distribution, $F_{\mathscr{C}}$, and always used '*distribution function*' for what, in an abstract sense, should be called 'the equivalence class of all the probability distributions which are the same if we confine ourselves to $F_{\mathscr{R}}$' (to put it briefly, and more intuitively, 'when we look at them with the naked

eye'), and which, in the final analysis, can be said to be $F(x)$. This would be (perhaps?) a little overdone, compared with the standard practice of always saying 'distribution'. At times (when it seems necessary to emphasize the point), we shall be more precise and say 'in the sense of a distribution function'; however, it will generally be left unstated, and clear from the context. What is important is that the reader always bears in mind 'as a matter of principle' that it is necessary to draw a distinction between those things which depend only on $F(x)$, and those which do not.

6.5.9. *A decisive remark.* We have been led, for various reasons, to rule out the assumption of countable additivity. Although it is not directly relevant to our specific purpose, we ought perhaps to give some thought to the reasons why most people are quite happy to accept this assumption as not unreasonable.

Leaving aside the question of analytic 'convenience', seen within the Lebesgue framework (which, in any case, appeared on the scene afterwards), I think the reason lies in our habit of representing everything on the real line (or in finite-dimensional spaces), and in the fact that the line (and these kinds of spaces) does not lend itself to being intuitively divided up into pieces other than those which get included 'by the skin of their teeth'.

To see this, note that the partitions actually made are those which are easiest to make: the 'whole' (length, area, mass, etc.) is divided into a finite number of separate parts, with an epsilonth left over; in order to obtain an infinite partition, one carries on dividing up that epsilonth. If one has to share out a cake among n persons, one could always give $\frac{1}{2}$ to the first one, $\frac{1}{4}$ to the second, $\frac{1}{8}$ to the third,..., $(\frac{1}{2})^{n-1}$ to the last two; if there were a countable infinity of persons, one could cope with them all by this method. But would they be satisfied? Protests would quite likely arise by the time one reached $n = 3$, and, as one proceeded, the number who came to regard this as some kind of practical joke rather than a 'genuine' method of distribution would increase, as would, quite understandably, their anger.

A 'genuine' method, in this sense, for subdividing an interval into a countable partition, is that used by Vitali, in proving the theorem we referred to earlier. The set I_h is formed from points of the form $a + r_h$, where $r_0 = 0$, $r_1, r_2, \ldots, r_n, \ldots$ are the rationals (ordered as a sequence), and the a are the irrational numbers of I_0, chosen so that one and only one representative from each set of irrationals which differ among themselves by rationals is taken. This example has a pathological flavour, however, as a reshuffling of the points, not to mention its evident appeal to the axiom of choice.

In contrast, if we considered a space with a countable number of dimensions, the matter would be obvious. If a point is 'chosen at random' on the sphere $\sum_h x_h^2 = 1$ in the space of elements with countably many coordinates x_h, all zero except—at most—a finite number, then there is equal probability (zero—cf. Chapter 4, and the Appendix, Section 18) that any of the half-lines x_h (positive or negative) will be 'the closest half-line'. Leaving aside the 'random choice', the countably many 'pieces' of the sphere, I'_h and I''_h, defined by 'x_h is the greatest coordinate—in absolute value—and is positive (I') or negative (I'')' are entirely 'symmetric' and 'intuitive' (the number of dimensions is, of course, so much greater than 3).

The essence of the remark can be put, rather more briefly, in another way. By a set of measure zero, the currently fashionable measure theory means a set which is *too empty* to serve as an element of a countable partition. This is a direct consequence of imposing countable additivity as an axiom. This implies, in fact, that a union of a countable number of sets of measure zero (in the Lebesgue sense) is still of measure zero. It is no wonder that in such a docile set-up any kind of process consisting in taking limits is successful, once all the necessary safety devices have been incorporated in the definitions!

6.6   THE PRACTICAL STUDY OF DISTRIBUTION FUNCTIONS

6.6.1. What we are going to say here holds for any kind of distribution: one can, if one wishes to form a particularly meaningful image, think of mass distributions; or (bearing in mind that we are dealing with the 'distribution function') one can think in terms of the probability distribution, which is the thing we are specifically interested in. It will, however, be most useful, particularly for the more practical aspects, to think mainly in terms of the statistical distribution.

In studying a distribution, we may, roughly speaking, distinguish three kinds of ideas and tools:

descriptive properties,

synthetic characteristics,

analytic characteristics.

6.6.2. Many of the properties already mentioned are *descriptive properties*. As examples, we have the following: whether a distribution is bounded or not; proper or improper; whether $F(\square)$ is finite, infinite (negative or positive), or indeterminate ($\infty - \infty$); whether or not there are masses of each type A, B and C (6.2.3), and, in particular, in case A, whether the density is bounded, continuous or analytic; whether this density (or, in case C, the concentrated masses, for example with integer possible values) is increasing, decreasing, or increases to a maximum and then decreases (*unimodal* distribution), or whether the behaviour is different again (for example, *bimodal*; etc.); whether the distribution is symmetric about the origin ($F(-x) + F(x) = 1$) or about some other value $x = \xi$ ($F(\xi - x) + F(\xi + x) = 1$; if the density exists, $f(\xi - x) = f(\xi + x)$, and, in particular, $f(-x) = f(x)$ if $\xi = 0$).

We could continue in this way, but it is sufficient to say that one should note how useful it can be to provide sketches showing these various aspects. Sometimes these alone will be enough for one to draw simple conclusions; more frequently, they provide useful background knowledge to be considered along with quantitative data.

6.6.3. In order to be able to interpret what we shall say later by making use of various graphical devices (and, in this way, to better appreciate both the meanings of the different notions, and the properties and particular advantages of each method), we will mention briefly the principal graphical techniques used.

We shall present them using the language of the statistical distribution (for N 'individuals'), but they are completely general (if we consider the cases of continuous distributions as covered by taking N very large, or, in mathematical terms, mentally taking the 'limit as $N \to \infty$'). For convenience, we shall only deal with bounded distributions over the positive real line ($F(0) = 0$, $F(K) = 1$, $K = \sup F < \infty$). This will be useful for fixing ideas, necessary for some of the points we shall make, and quite sufficient to show how the same things go through in the general case, with appropriate modifications.

The *graph of the distribution function*, $y = F(x)$, is given in Figure 6.2(a); in the statistical case this becomes a step function (which in the limit is a *curve*), called the *cumulative frequency curve*, with a step of $1/N$ at each point x_h, the value, for the hth of the N individuals, taken by the quantity under consideration (for example: age, height, income, etc.). $F(x)$ gives the frequency, that is the percentage,† $n(x)/N$, of the individuals (out of the total of N) for whom the quantity has a value not exceeding x.

As we already pointed out (6.2.5), the 'individuals' must sometimes be counted with different 'weights' p_h (instead of each with $1/N$); it could also happen that several individuals may have the same value x_h (and we then have a mass at that point of $\sum_k p_k(x_k = x_h)$), or, in particular, n/N if the masses are equal and n values coincide). We shall concentrate on the simplest case, however, in order to fix ideas concerning certain aspects of importance, without prejudicing the extension to the more general case.

The graph of the inverse function, $x = F^{-1}(y)$, which we considered already in Section 6.2.6 (see Figure 6.2b), is not widely used. It is, however, a meaningful concept known as the *gradation curve* (Galton); its interpretation is best illustrated in the case of heights—it is the profile obtained by lining up the individuals in increasing order of height (a kind of 'Right dress!').

---

† By 'percentage', we mean the proportion (not the proportion multiplied by 100 as is customary): in other words, $27\% = 0.27$, $27.58\% = 0.2758$, etc. Nothing is altered (we could mention that this way of *writing* it is convenient in that it avoids zeroes on the left, and is more expressive when it comes to *reading* it): the symbol $\%$ is a conventional form of '/100' (divided by 100), as a right operator on any number.

When income is the quantity of interest, one could think, for instance, of a pile of equal coins rather than of the individuals. This image is useful for clarifying the concept required in cases like the present one, where an obvious meaning attaches to the *sum* of the x_h values of the various individuals; here, the total income of a certain group of individuals. The area under the curve, and relative to a given interval $y' \leqslant y \leqslant y''$, represents the total income (reduced, on that scale, from 1 to $1/N$) of the individuals belonging to the group of those for whom the percentage point of 'the least rich among them' lies between y' and y''. In any case, dividing by the length of the segment, one always obtains the mean value (arithmetic mean) of that group of individuals, and this also makes some sense in the case of age and height, etc., although the meaning is rather one of convention, since the sum does not have a straightforward interpretation. In any interval (and, in particular, for the whole interval $[0, 1]$) the mean value is therefore the height of the rectangle of equivalent area (in other words, in more visual terms, leaving equal areas above and below).

In those cases where the sum has an obvious meaning (as in the case of income), a third graphical device is also useful and meaningful. It is known as the 'concentration curve', and is the cumulative version of the previous one (with the total area taken to be unity by convention: e.g. total income $= 1$). Figure 6.3 shows the *concentration curve* $z = G(y)$ (Lorentz), and the *gradation curve* $x = F^{-1}(y)$ displayed together, with total income and average income, respectively, taken as the units of measure. By definition, $G(y)$ represents the fraction of the total income owned by the fraction y of least wealthy individuals. In the case of a uniform distribution (all incomes equal) the curve would be the diagonal of the square $G(y) = y$; in general, the area between the curve and this diagonal—called the area of concentration—when divided by the maximum possible area, $\frac{1}{2}$ (corresponding to all income in the hands of one of the N individuals, N large) is called the *concentration ratio*, and gives an idea of the inequality of distribution (Gini). At each point, the slope of $z = G(y)$ is given by $x = F^{-1}(y)$; the mean corresponds to the point of maximum distance from the diagonal (where $G'(y) = 1$; we have a tangent parallel to the diagonal).

6.6.4. The representation by means of the *density curve* is widely used; in the statistical case this is called the *frequency* curve. It is this representation which best shows up the features of behaviour that we were discussing earlier.

We must point out, however, that the density is often (and, strictly speaking, in the statistical interpretation always) a fiction, or a mathematical idealization. Any actual statistical distribution (with a finite number of individuals, N) must be discrete: we either have N masses p_h (possibly equal—$p_h = 1/N$—possibly not) with $\sum_h p_h = 1$, or fewer than N if several individual values

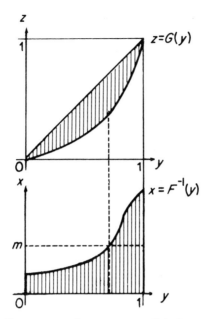

Figure 6.3 The concentration curve $z = G(y)$; for example, in the case of incomes, to the fraction y of the least wealthy, there corresponds the fraction $G(y)$ of total income, which is represented on the graph below (the gradation curve: cf. Figure 6.2(b) and the discussion in 6.2.6) by the fraction of the total area to the left of y; that is, including all incomes $\leqslant x = m\,dz/dy$ (m = average income). Observe, in particular, that $x = m$ at the point where the curve $z = G(y)$ has slope $= 1$ (the tangent is parallel to the diagonal; it is therefore the point of maximum distance from the diagonal). The diagonal, $z = y$, corresponds to the case of equal distribution; in all other cases, we must have $z < y$, z increasing and concave

are equal. Even in the actual case of a distribution of mass, we would find similar discontinuities once we descended to the atomic scale, or even indeterminacy because of thermo-agitation, etc., which would prevent us localizing the masses precisely.

In actual fact, even in Physics, the density is acknowledged to be a sensible tool if we consider the ratios of mass/volume for neighbourhoods of a point which are not too large, so that macroscopic inhomogeneity has little effect, and not too small, so that the effects of structural discontinuity are avoided. In any case, if we make the transition from step function to distribution function without attributing to the latter any unnecessary irregularities of slope, then $f(x) = F'(x)$ can, to a large extent, be considered as determined. On the other hand, the curve is sometimes *smoothed*; that is, modified in order to simplify it, possibly into a more tractable analytic form, more or less of a standard type.

It is sometimes stated, in this context, that one is attempting to remove '*accidental* irregularities'. This, however, can only be done from a probabilistic angle and in the necessary depth. For this reason, we shall not go into the question here. Anything we might say would only tend to give rise to superficial and misleading ideas, which can come about easily enough, even without our saying anything (we shall come back to this in Chapters 11 and 12; we hinted at the underlying idea in Chapter 5, 5.8.7).

The most elementary and, at the same time, the best way of introducing the density in practice (and of constructing the density curve) consists in considering the *average density* over intervals of some appropriate subdivision (neither too coarse nor too fine, for reasons stated already). Unless there is any reason to do otherwise, we usually take equal subintervals (for convenience). The average density in the general interval $[\xi_i, \xi_{i+1}]$, is the incremental ratio of $F(x)$, $[F(\xi_{i+1}) - F(\xi_i)]/(\xi_{i+1} - \xi_i)$. Figure 6.4, formed by rectangles whose bases are the subintervals, and whose height is the average density, is called the *histogram*† (sometimes called a column diagram). Here also, by smoothing, one can pass to a continuous *curve*.

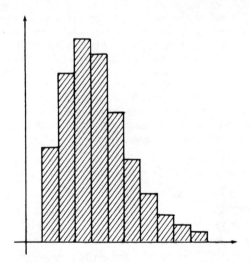

Figure 6.4   An example of a histogram. (It represents the distribution of families in Italy in 1951, according to the number in each)

---

† Note that it is essential to indicate the subdivisions between the rectangles (and that it is not sufficient merely to provide the upper contour). In fact, it is essential to distinguish the case of two (or more) consecutive rectangles of equal height from the case of a single rectangle given by their union. In the first instance there is more information, since we know that the average density is the same in the different subintervals.

6.6.5. The *synthetic characteristics* are the quantitative aspects, which often provide useful information, enabling us to find out all we need to know about the distribution in so far as it relates to a particular problem. It is sufficient to recall Chisini's definition of a *mean* (Chapter 2, Section 2.9), in order to understand how the knowledge of a 'mean' of a distribution can meet our needs. Often, this will be the mean value (arithmetic mean), given by $F(\square)$, or some other *associative* mean, $\gamma^{-1}F(\gamma)$, with γ increasing, corresponding, in the probabilistic interpretation, to the *prevision*, $\mathbf{P}(X) = F(\square)$, or, more generally, to the γ-*prevision*:

$$(3) \qquad \mathbf{P}_\gamma(X) = \gamma^{-1}[\mathbf{P}(\gamma(X))].$$

Sometimes, in addition to the mean (or prevision), one requires the *separation*, $X - \xi$, or the *deviation*, $|X - \xi|$ (the absolute value of the separation), of X from a given point ξ (which may be anything). On occasions, it will be particular choices of ξ which are important, as we have already seen in the case of the standard deviation—the quadratic prevision of $|X - \xi|$ with $\xi = \mathbf{P}(X)$—because it is with this choice of ξ that it assumes its minimum value and maximum significance. Leaving aside the probabilistic interpretation, to consider the separation is simply to consider shifting (from 0 to $-\xi$) the origin of the distribution; to consider the deviation is to turn over that part of the distribution on the negative axis and superimpose it on the positive axis.

Finally, we note that there are other synthetic characteristics which cannot be viewed as means (at least, not without distorting their meanings).

6.6.6. According to the purpose in hand, one can distinguish between measures of *location* and measures of *dispersion* (or spread), which are useful in giving some idea of 'whereabouts' the distribution tends to be concentrated, and 'to what extent' it is concentrated (these are often the two features of greatest interest). Other characteristics which one occasionally attempts to measure by some kind of indices are, for example, the *asymmetry*, the '*kurtosis*', and so on. A brief remark or two will suffice.†

The most meaningful measures of *location* are, generally speaking, the means (in which the Chisini sense; precisely because of the property expressed by his definition). Most often, however, one is interested in measures which behave sensibly under *translation* (and we implicitly mean *homogeneous*: in other words, if X transforms to $aX + b$, the measure is multiplied by a and increased by b). In general, this property does not hold: for example, among associative means only the arithmetic mean has the property.‡

---

† For a more extensive treatment, see M. G. Kendall, and A. Stuart, *The Advanced Theory of Statistics* (3rd. Ed.), vol. I, Griffin, London (1969), pp. 32–93.
‡ It holds for the others if the scale is transformed by $y = \gamma(x)$.

Examples of measures of location which do have the required property are the commonly used *median* (or median value) and *mode* (or modal value) of a distribution.

The *mode* is the value for which the density is a maximum. It is clearly defined and meaningful in the case of distributions whose densities have regular behaviour, and which are unimodal (that is, have a unique maximum), especially when defined in terms of simple functions. The more we depart from such well-behaved situations, the less clearly defined and meaningful it becomes.

The *median* is the central value of the distribution, the value which splits it in half; that is, such that $F(x) = \frac{1}{2}$ (or, more explicitly, $x = F^{-1}(\frac{1}{2})$). It is the value which has the property of minimizing $\mathbf{P}[|X - \xi|]$, the prevision of the deviation.†

The median is a special case—the most important—of a *positional value*, or *quantile*, of a distribution. The definition of the *p-quantile* $(0 \leqslant p \leqslant 1)$ follows along the same lines; $x_p = F^{-1}(p)$, that is, the value which divides up the distribution into a mass p on the left, and $1 - p$ on the right. For $p = 0$ and $p = 1$, we have inf X and sup X (making the natural convention of choosing one of these values rather than any value $<$ inf X or $>$ sup X). These values have the translation property, but are not suitable (for $p \neq \frac{1}{2}$) as really meaningful measures of location; they are useful as 'milestones', well-suited to describing the distribution in terms of intuitive subdivisions, especially when considering *quartiles* $(p = \frac{1}{4}$ or $p = \frac{3}{4})$, *deciles* and *centiles* (p multiples of $\frac{1}{10}$ or $\frac{1}{100}$), or for furnishing measures of dispersion (as we shall see).

In the case of measures of *dispersion* (or, if looked at in the opposite sense, measures of *concentration*), it will also prove important to consider a *homogeneity* property (similar to the translation property considered above). For the most important measures, when we consider $aX + b$ the measure is multiplied by a (and b has no effect).

Let us consider the special case of a distribution transformed into its '*normalized*' (or standardized) form, by taking the mean value as the origin, and the standard deviation as the unit ($m = 0$, $\sigma = 1$). If we denote by α^* the index for the normalized distribution, then, after transformation, the translation property would lead to $\alpha = m + \sigma\alpha^*$, and the homogeneity property to $\alpha = \sigma\alpha^*$. If $\alpha = \alpha^*$ (in other words, invariance under translation and change of scale) the index could be called *morphological*, because it

---

† This is obvious if one thinks about it. Shifting ξ to $\xi + d\xi (d\xi > 0)$ increases by $d\xi$ the deviation for all masses to the left of ξ, and decreases by the same amount the deviation for those on the right. It is therefore sensible to move towards the median, at which point the masses on the left and right are equal. This property (with an appropriate modification) allows us to eliminate the indeterminacy which occurs in $F(\xi) = \frac{1}{2}$ throughout some interval. One can define (D. Jackson, 1921) the median as the limit as $\varepsilon \to 0$ of $\xi(\varepsilon)$ = the value at which the prevision of the deviation to the power $1 + \varepsilon$ ($\varepsilon > 0$) is minimal.

expresses a characteristic of the form of the distribution, that is, of the *kind* of distribution (this terminology is often useful for denoting all those distributions which differ from each other only by changes of origin and scale; in other words, the $F(ax + b)$ for given F and any a and b; sometimes, we are limited to $a > 0$ and/or $b = 0$). Observe that we carried out the normalization using m and σ, but this is by no means the only possibility, nor is it even always possible (σ may be infinite, or m indeterminate); we used this method because it is the most common, and the most useful from several points of view. As an example of the other possibilities, we mention the possibility of taking the *median* and the *interquartile range*, in place of m and σ (this has the advantage that it is always meaningful, and avoids the over-sensitivity of σ to the 'tails' of the distribution; its disadvantage is that it is rather crude).

Examples of morphological properties are provided by *asymmetry* and *kurtosis*, for which one can take as indices the cubic and quartic means of the separation—$\mathbf{P}[(X - m)^n]^{1/n}$ for $n = 3$ and $n = 4$, respectively, divided by σ.† The first index is equal to 0 in the case of symmetry (or of deviations from symmetry which cancel each other out),‡ and is positive or negative according as the left-hand or right-hand tail is more pronounced. Kurtosis, measured by the second index, is the property of whether the density is sharp or flat around its maximum, and its main use is in discovering whether a density which appears to be *normal* (cf. 6.11.3) is, instead, *leptokurtic* or *platykurtic*; that is, more peaked or more flat than it should be around the maximum. The index given distinguishes between the three cases according as it is $=, >, < \sqrt[4]{3}$.

Let us now go back to the case of dispersion, and mention, in addition to the mean deviations (from m or any other value), the means of the differences, $\mathbf{P}[|X - Y|]$ or $\mathbf{P}_y[|X - Y|]$, where X and Y are independent random quantities having the distribution under consideration. The *mean difference*,§ $\mathbf{P}[|X - Y|]$, is expressible (for distributions on the positive axis) in terms of the area of concentration (cf. 6.6.3, Figure 6.3); the *quadratic mean difference*, $\mathbf{P}_Q[|X - Y|]$, does not give us anything new, it is clearly equal to

---

† More usually, powers are used: it seems preferable and more meaningful to take ratios of means of dimensionality 1 with respect to the variable.

‡ Observe how this cancelling out depends on the particular choice of the index. In general, any index which translates an essentially qualitative property into a quantitative measure introduces a degree of arbitrariness. One should take account of this both by exercising caution in interpreting the conclusions, and also by avoiding abstract verbal discussions concerning the 'preferability' of various indices; this question should, if at all, be examined in relationship to the concrete needs of the problem.

§ In the case of the statistical distribution (with N individuals) one considers mean differences *with* and *without repetition*. The latter implies that one excludes X and Y referring to the same individual (excluding the fact that it can be drawn twice) and the index is then multiplied by $N/(N - 1)$. In fact, the probability of a repeat drawing is $1/N$; hence, we have '*index with*' = $(1 - 1/N)$. "*index without*" + $(1/N) \cdot 0$ (0 being the difference between X and Y when they coincide).

$\sqrt{2}\sigma$ ($\sqrt{(\sigma^2 + \sigma^2)}$). Other indices can be set up in terms of quantiles: the *interquartile range* and the *interdecile range* are, respectively, the differences between the quantiles with $p = \frac{1}{4}$ and $p = \frac{3}{4}$, and with $p = \frac{1}{10}$ and $p = \frac{9}{10}$; the limits, $p = 0$ and $p = 1$, give the range of the distribution; sup $-$ inf.

A somewhat different concept of dispersion lies behind the function $l(p)$ $(0 \leqslant p \leqslant 1)$ defined by $l(p) =$ 'the minimum length of a segment containing mass (probability) p' $= \inf \{\lambda \sup_x [F(x + \lambda) - F(x)] \geqslant p\}$. Clearly, $l(p) = 0$ for $p \leqslant$ 'the maximum jump' (the maximum probability concentrated at a point; in particular, if there are no concentrated masses then $l(p) = 0$ only when p $= 0$); $l(p)$ is increasing, and tends to the range of the distribution as $p \to 1$. If $l'(0) = c > 0$, the distribution has a bounded density, and its maximum is $1/c$ (and conversely).

6.7   LIMITS OF DISTRIBUTIONS

6.7.1. We have had occasion to note that certain properties and synthetic characteristics of the distribution function are rather insensitive to 'small changes in the form of the distribution', while others are very sensitive. To make this more precise, we must first say what we mean by a 'small change'; at the very least, this implies saying what we mean by a sequence of distributions, $F_n(x)$, tending to a given distribution $F(x)$ as $n \to \infty$. Better still, when this is possible, it means defining a notion of 'distance' between two distributions, allowing us to recast $F_n \to F$ in the form dist $(F_n, F) \to 0$.

Fortunately, there is little doubt about what form of convergence is appropriate in the case of proper distributions (and we shall limit ourselves to this case). To say that $F_n \to F$ will always mean convergence of $F_n(x)$ to $F(x)$ at all continuity points of F (or, alternatively, convergence of $F_n(\gamma)$ to $F(\gamma)$ for every bounded and continuous γ). An equivalent formulation is expressed by the condition:

given any $\varepsilon > 0$, the inequalities

(4) $F(x - \varepsilon) - \varepsilon \leqslant F_n(x) \leqslant F(x + \varepsilon) + \varepsilon$ $(-\infty \leqslant x \leqslant \infty)$

are satisfied for all n greater than some N.

A condition of this form makes it evident that the smallest value of ε for which it holds can be defined as the *distance*, dist (F_n, F), between F_n and F (geometrically, this is the greatest distance between the curves $y = F_n(x)$ and $y = F(x)$ in the direction of the bisector $y = -x$). We shall not prove this; we merely observe that this corresponds to the idea that a given imprecision is tolerated not only in the ordinates (a small change in the mass, in the probability), but also in the abscissae (small changes in the position of the mass, even the concentrated mass).

It often happens that a sequence F_n does not tend to a particular distribution F, but only to a distribution of *the same kind* as F (as defined in 6.6.6).

In other words, $F_n(a_n x + b_n)$ tends to F if we *choose* the constants a_n and b_n in an *appropriate manner*. The most common case is that of the normalized distribution $F_n([x - m_n]/\sigma_n)$ (with $a_n = 1/\sigma_n$ and $b_n = -m_n/\sigma_n$), but this is not the only one, and is not always applicable, even when all the variances (of the F_n and of F) are finite and convergence to F occurs (by choosing the constants differently).†

6.7.2. We can straightaway make some important points.

Every distribution can be approximated to any desired degree by means of discrete distributions, or by means of absolutely continuous distributions.

It suffices to observe that this follows, for example, if we set

(5) $F_n(x)$ = the largest multiple of $1/n$ which is less than $F(x) + 1/2n$,

or, respectively,

(6) $$F_n(x) = \int_0^1 F(x + u/n)\, du,$$

from which it follows that

(6') $$f_n(x) = F_n'(x) = n[F(x + 1/n) - F(x)] \leqslant n.$$

As a result:

A property which has been established only for discrete distributions (or only in the absolutely continuous case, or simply for cases with bounded density) holds for all distributions if that property is continuous (a property is *continuous* if it holds for F whenever it holds for the F_n such that $F_n \to F$).

It is easy to show that continuity usually holds for most of the properties which are required. It is much less long-winded to write out the proof (even if it follows the same lines) in one or other of the special cases, whichever is convenient for our purpose.

It is useful to bear in mind that in order for a sequence F_n to *be convergent* (assuming that the F_n tend to a proper limit F) *it is necessary that the F_n be equally proper* (in the sense that $F_n(x) - F_n(-x)$ tends to 1 as $x \to \infty$, uniformly with respect to n); and, conversely, that this condition is sufficient to ensure *the sequence F_n, or at least a subsequence, tends to a proper limit distribution*. (Ascoli's theorem).

---

† The masses which move away (as n increases) and which die away (as $n \to \infty$) without changing the limit of the distributions may, for example, change the σ_n.

Example. Let F_n have masses $\frac{1}{2}(1 - 1/n)$ at ± 1 and masses $\frac{1}{2}n$ at $\pm n$; we have $\sigma_n \sim \sqrt{n} \to \infty$; the normalized F_n would have two masses $\sim \frac{1}{2}$ at $\pm x_n$, $x_n \sim 1/\sqrt{n} \to 0$ (and two which become negligible) and would tend to a distribution concentrated at 0; the F_n (unnormalized) tend, on the other hand, to F, with masses $\frac{1}{2}$ at ± 1.

6.8 VARIOUS NOTIONS OF CONVERGENCE
FOR RANDOM QUANTITIES

6.8.1. In the most natural interpretation, the notion of convergence deals with *sequences* of random quantities. However, although for the sake of simplicity we shall deal with sequences $X_1, X_2, \ldots, X_n, \ldots (n \to \infty)$, nothing would be altered were we to deal with X_t, with $t \to t_0$ (real parameter), or, similarly, with X_t associated with elements t of any space whatsoever (in which $t \to t_0$ makes sense). Instead of a sequence, we might be dealing with a series (but this amounts to the same thing when we consider the sequence of partial sums); instead of a random quantity, we might be dealing with random points in general (for example, 'vectors' or n-tuples of random quantities), provided that in these spaces the concepts involved also make sense.

Here we are merely concerned with setting out the basic ideas, and noting, in particular, the numerous points at which the *weak* conception, to which we adhere, leads to formulations and conclusions different from those usually obtained as a result of following the *strong* conception.†

6.8.2. In the first place, it is possible to have definite convergence, uniform or non-uniform, either with a definite limit or not; by *definite* we mean independent of the evaluation of the probabilities; in other words, something that can be decided purely on the basis of what is known to be *possible* or *impossible*.

As an example of definite, uniform convergence to a definite limit, consider the total gain in a sequence of coin tosses (Heads and Tails). A 'success' is defined by the occurrence of a Head, or by 100 consecutive Tails following the last success; the gain is $(\frac{1}{2})^n$ for the nth success, and 0 for a failure. The total possible gain is 1, and it is certain that after at most $100n$ tosses the first n terms will have been summed.

Definite, uniform convergence, but to an uncertain (random) limit, occurs in a sequence of coin tosses if the successive gains are $\pm\frac{1}{2}$, $\pm(\frac{1}{2})^2$, $\pm(\frac{1}{2})^3, \ldots, \pm(\frac{1}{2})^n, \ldots (+$ for Heads, $-$ for Tails); the remaining gain after n tosses is (in absolute value) certainly $\leqslant(\frac{1}{2})^n$, but the limit could be any number between -1 and $+1$.

In the following example, convergence is definite, non-uniform, and may be either to a definite or to an uncertain (random) limit. We have an urn containing $2N$ balls, a finite number, but for which no upper bound is known. There are $N + X$ white balls and $N - X$ black balls, where $X = x$ may be known (certain; e.g. $x = 0$), or may be unknown (e.g. any number between ± 100).

---

† It is not a question, of course, of declaring a preference for weak convergence or strong convergence (although the identity of the terminology does reflect a relationship between the concepts). In both the weak and strong formulations these and other notions of convergence exist, and each might present some difficulties of interpretation in one or the other formulation.

The balls are drawn without replacement, and the gains are ± 1 ($+$ for white, $-$ for black). After all drawings, the gain will be $2X$ and will remain so thereafter (we assume, to avoid nuances of language, that when the urn is empty some other fictitious drawings, all of gain 0, are made). The limit is $2X$, either known or unknown, but objectively determined right from the very beginning.

So far, probabilities have not entered onto the scene (nor, therefore, have probabilistic kinds of properties, like stochastic independence). One might ask, however, whether knowing the limit X (as a certain value, x), or attributing to it some probability distribution $F(x)$ (if it is uncertain), imposes some constraints on the evaluations of the probability distributions $F_n(x)$ of the X_n (or conversely: it amounts to the same thing).†

In the case of uniform convergence the answer is yes: if we are to have $|X_n - X| < \varepsilon_n$ with certainty, then F_n and F must be 'close to each other' in the sense that $F_n(x - \varepsilon_n) < F(x) < F_n(x + \varepsilon_n)$ (and conversely: $F(x - \varepsilon_n) < F_n(x) < F(x + \varepsilon_n)$). In particular, if $X = x_0$ with certainty, we must have $F_n(x_0 - \varepsilon_n) = 0$, $F_n(x_0 + \varepsilon_n) = 1$. When we are dealing with non-uniform convergence, this does not hold in general (unless we accept countable additivity). In the example of the urn, if $2N$ has an improper distribution (for example, equal probabilities (zero) for each N) then the probabilities of the behaviour of the gain in the first n tosses (however large n is) are the same as for the game of Heads and Tails (whether the difference between the number of white and black balls is known, e.g. $= 0$, or bounded, e.g. between ± 100 with certainty). Whatever happens, until the urn is emptied (and we know that there is no forewarning that this is about to happen) nothing can be said about the limit (if it is not already known), and knowledge of this limit (if we have it) does not modify the F_n.

6.8.3. Notions of convergence in the probabilistic sense carry a meaning very different from just saying that (with greater or lesser probability) $X_n \to X$ (in the analytic sense of being numbers),‡ and from saying that

---

† In general, one should consider the joint probability distribution for $X_1, X_2, \ldots, X_n$, for every n; the mention of this fact will suffice here.

‡ In connection with the terminological distinction between *stochastic* and *random* (Chapter 1.1.10.2), we offer here a remark which seems to clarify the various considerations about the X_n (concerning their 'convergence' in various senses), and at the same time to clarify the terminological question. The fact of the numbers X_n, when they are known, tending or not tending to a limit (in some sense or another; convergence pure and simple, Cesàro, Hölder, etc.) can either be *certain* (true or false with certainty), or *uncertain*, given the present state of information: the convergence is then said to be *random*.

Convergence in the probabilistic sense (either the variants we are going to consider, or others) is called *stochastic convergence* because it is not concerned with the values of the X_n, but with circumstances which relate to the evaluation of probabilities (concerning the X_n and possibly an X, which may or may not be their limit in some sense) made by someone in his present state of information. This is something relating not to the facts, but to an opinion about them based on a certain state of information.

$F_n \to F$ (this can be true for the distributions of X_n and X, without the latter having anything in common).†

We give straightaway the three most important types of convergence.

Convergence in quadratic mean. X_n is said to converge to X in quadratic mean, and we write $X_n \overset{.}{\to} X$, if $\mathbf{P}_Q(X_n - X) \to 0$ as $n \to \infty$ (or, equivalently, if $\mathbf{P}(X_n - X)^2 \to 0$). This notion is the simplest, and the most useful in practice; it is related to what we have already said concerning second-order previsions.

Weak convergence (or *convergence in probability*). X_n is said to converge weakly to X, and we write $X_n \overset{\cdot}{\to} X$, if, for any $\varepsilon > 0$,

$$\mathbf{P}(|X_n - X| > \varepsilon) \to 0 \quad \text{as} \quad n \to \infty.$$

More explicitly (in order to make a more clear-cut comparison with the case to be considered next) we can state it in the form: for any given $\varepsilon > 0$ and $\theta > 0$, and for all n greater than some appropriately chosen N, all the probabilities $\mathbf{P}(|X_n - X| > \varepsilon)$ are $< \theta$, or (alternatively) all probabilities $\mathbf{P}(|X_n - X| < \varepsilon)$ are $> 1 - \theta$.

Strong convergence (or *almost sure convergence*).‡ X_n is said to converge strongly to X, and we write $X_n \overset{..}{\to} X$, if for any $\varepsilon > 0$, $\theta > 0$, and for all n greater than some appropriately chosen N, we not only have all the probabilities $\mathbf{P}(|X_n - X| > \varepsilon)$ that each deviation *separately* is greater than ε being $< \theta$, but we also have the same holding for the probability of even a single one out of an arbitrarily large finite number of deviations from N onwards $(n, n + 1, n + 2, \ldots, n + k, \ldots, n + K; n \geqslant N, K$ arbitrary$)$ being $> \varepsilon$. Expressed mathematically,

$$\mathbf{P}\left[\bigvee_{k=0}^{K} |X_{n+k} - X| > \varepsilon \right] < \theta \quad \left(\bigvee_{k=0}^{K} = \text{max for } k = 0, 1, \ldots, K \right),$$

or

$$\mathbf{P}\left[\prod_{k=0}^{K} (|X_{n+k} - X| < \varepsilon) \right] > 1 - \theta$$

$(\prod = \text{product (arith.} = \text{logical) of the events } (|X_{n+k} - X| < \varepsilon).)$

Put briefly: the probability of any of the deviations being greater than ε must be $< \theta$; in other words, the probability that they are *all* less than ε must be $> 1 - \theta$.

---

† A warning against confusing these two notions is necessary, not because in themselves they are open to confusion, but because of the dangers of using inappropriate terminology (such as 'random *variable*': cf. Chapter 1, 1.7.2 and 1.10.2).

‡ A form of terminology which is inaccurate in the weak formulation; cf. the remark to follow, and the last but one footnote.

Remark. In the strong formulation the definition can be more simply stated by talking of '*all* the deviations from N on', rather than of a finite number (K), however large. From a conceptual viewpoint, the question becomes a rather delicate one because an infinite number of events are involved. As usual, this modification is only admissible if countable additivity is assumed.

6.8.4. *The Borel–Cantelli Lemmas.* For a sequence of events E_i, it is required to provide bounds for the probabilities of having at least one success, or no successes, or at least h successes (that is, if Y denotes the number of successes, $Y \geqslant 1$, $Y = 0$, $Y \geqslant h$); all that can be assumed is knowledge of the $p_i = \mathbf{P}(E_i)$. In the *weak* version, this will only make sense if we limit ourselves to finite subsets (with, of course, the possibility of considering asymptotic results when these subsets cover the whole infinite range). In the *strong* version (as originally considered by Cantelli and Borel, and still standard) the asymptotic results should be interpreted as conclusions about the total number of successes out of the infinite number of events which form the sequence.

For a finite number of events, with probabilities $p_1, p_2, \ldots, p_n$, if we put $\bar{y} = \mathbf{P}(Y) = \sum p_i =$ prevision of the number of successes, we have (unconditionally) an upper bound on the probability of the number of successes:

$$\mathbf{P}(Y \geqslant 1) = \mathbf{P}(\text{event-sum of the } E_i) \leqslant \bar{y}, \quad \mathbf{P}(Y \geqslant h) \leqslant \bar{y}/h.$$

(In fact, $h\mathbf{P}(Y \geqslant h) = \mathbf{P}[h(Y \geqslant h)]$ and $h(Y \geqslant h)$, which is $= 0$ if $0 \leqslant Y < h$ and is $= h$ if $Y \geqslant h$, is always $\leqslant Y: \vdash h(Y \geqslant h) \leqslant Y.$)

We therefore have that if for the sequence E_i the sum of the p_i converges, let us say $\sum p_i = a < \infty$, then $\bar{y} \leqslant a$ for any finite subset, and the previous bounds are valid *a fortiori* (with a in place of $\bar{y}$). One can now say that for any $\varepsilon > 0$, and for $h \geqslant a/\varepsilon$, we have a probability $< \varepsilon$ of obtaining more than h successes among the first K events of the sequence (it does not matter how large K is). In addition, if we only use the bound for $h = 1$, and we start with an n sufficiently large for the rest of the series to be $< \varepsilon (\sum_{i>n} p_i < \varepsilon)$, we can say that the probability of finding even a single success out of K events (K arbitrarily large, but finite) from E_n on is always $< \varepsilon$.

In the strong version we have the following: *if the series of probabilities converges, it is practically certain* (the probability $= 1$) *that the number of successes is finite.*

This is the Cantelli lemma; the Borel lemma states the converse, but with the additional condition of stochastic independence.† In the *strong* version,

---

† It is obvious that this would not hold without any extra condition: think of the case in which the E_i are all incompatible with some E having $P(E) \geqslant a > 0$, such that E implies no successes; i.e. $Y = 0$ (and, in particular $Y_n = 0$ out of the first n of the E_i), so that $P(Y = 0)$ and $P(Y_n = 0)$ are both $\geqslant a > 0$ (instead of $= 0$ and $\to 0$, respectively). If, however, the series of the $p_i = P(E_i)$ diverges, the E_i cannot then be independent (cf. the following inequality for $P(Y_n = 0)$).

the divergence of $\sum_i p_i$ implies that the number of successes is infinite; the *weak* version is much the same in this case, because Y, if not infinite, must be a completely improper random quantity (with distribution adherent to $+\infty$).

The bound which is required can be established immediately using the elementary inequality $e^x \geqslant 1 + x$; the probability of no successes in n *independent* events is

$$\mathbf{P}(Y = 0) = (1 - p_1)(1 - p_2)\ldots(1 - p_n) \leqslant e^{-p_1}e^{-p_2}\ldots e^{-p_n}$$

$$= e^{-(p_1 + p_2 + \ldots + p_n)} = e^{-\bar{y}};$$

stated explicitly,

$$\mathbf{P}(Y = 0) \leqslant e^{-\bar{y}}, \quad \mathbf{P}(Y \geqslant 1) \geqslant 1 - e^{-\bar{y}},$$

and, more generally, we have the similar result

$$\mathbf{P}(Y \leqslant h) \leqslant e^{-\bar{y}}[1 + (\alpha\bar{y}) + \tfrac{1}{2}(\alpha\bar{y})^2 + \ldots + 1/h!(\alpha\bar{y})^h], \quad \alpha = e^{\max p_i}.$$

If the series $\sum_i p_i$ diverges, $\bar{y}$, relative to the first K events, tends to $+\infty$ as K increases, and this is also true if we start from the nth event. The conclusion is that there is a probability $\to 1$ of finding at least one success starting from any arbitrary n, and, hence, a number exceeding any bound. Alternatively, this can be established directly from the fact that $\mathbf{P}(Y \leqslant h)$ also tends to 0, for any h.

6.8.5. *A corollary for strong convergence.* In order that strong convergence holds, it is sufficient that the $\mathbf{P}(|X_n - X| > \varepsilon)$ constitute the terms of a convergent series† (and do not merely tend to 0, as required for weak convergence). This condition is also necessary if the $|X_n - X|$ are stochastically independent (or if the events $|X_n - X| > \varepsilon$ are). This is seldom so in cases of interest, but one can often obtain the negative result by finding a subsequence of terms, which are sufficiently far apart to be 'practically independent', for which the series of probabilities diverges (when we consider something being 'sufficiently independent', we are thinking of some condition or other to be translated into a rigorous form as appropriate for the case in question).

6.8.6. *Relationships between the different types of convergence. Weak* convergence is implied both by *strong* convergence (as is obvious from the definition), and by convergence in *quadratic mean* (by virtue of Tchebychev's inequality, Chapter 4, 4.17.7. Neither of the latter two implies the other.

In addition to convergence in quadratic mean (also known as convergence in *2nd-order mean*, or in *mean-square*), one also considers, though less

---

† *A fortiori*, it is sufficient that the series $\sum \mathbf{P}(X_n - X)^2$ converges.

frequently, convergence *in pth-order mean* (where p is any positive number), defined by $\mathbf{P}(|X_n - X|^p > \varepsilon) \to 0$; the condition becomes more restrictive as p increases, and always implies weak convergence.

Definite uniform convergence implies all the above.

Convergence of distributions is implied by weak convergence (and so, *a fortiori*, by all the others).

It is sufficient to note that if the random quantities X and Y are 'sufficiently close to each other' in the sense that $\mathbf{P}(|X - Y| > \varepsilon) < \theta$ (for given $\varepsilon, \theta > 0$), then their distributions F and G are 'sufficiently close to one another'† in the sense that (for all x) $F(x - \varepsilon) - \theta \leqslant G(x) \leqslant F(x + \varepsilon) + \theta$. In fact, in order that $X \leqslant x - \varepsilon$, it suffices that either $Y \leqslant x$ or $|X - Y| \geqslant \varepsilon$. Expressed mathematically,

$$(X \leqslant x - \varepsilon) \leqslant (Y \leqslant x) \vee (|X - Y| \geqslant \varepsilon) \leqslant (Y \leqslant x) + (|X - Y| \geqslant \varepsilon);$$

taking probabilities, it follows that $F(x - \varepsilon) \leqslant G(x) + \mathbf{P}(|X - Y| \geqslant \varepsilon)$, and the final term is $< \theta$, by assumption. This proves the first half of the inequality; the other half follows by symmetry.

In the case of weak convergence, however we take ε and θ, the inequalities hold for X_n and X from some $n = N$ on, and hence $F_n \to F$.

6.8.7. *Mutual convergence* (or *Cauchy convergence*). Suppose that for a given sequence X_n we know that $X_n - X_m \to 0$ (in some sense) as $m, n \to \infty$: what can be said about the convergence (in the same sense) of X_n to some random quantity X? If we adopt the strong formulation, we can say that such an X exists. For all the types of convergence that we have considered, '*il n'y a pas lieu de distinguer la convergence mutuelle et la convergence vers une limite*' (to quote P. Lévy, *Addition*, p. 58, Th. 18).

The answer is even more conclusively yes if we are dealing with a random quantity which is a measureable function $X(\omega)$ of the points of a space Ω (and, in this case, we should just mention that the various probabilistic notions, and in particular the notions of convergence, reduce to concepts in analysis—apart from changes in terminology: for example, convergence in probability instead of *in measure*; almost certain convergence instead of *almost everywhere*).

Without the assumption of countable additivity, and with no reference to a 'space of points' (cf. the quotations from von Neumann and Ulam, Chapter 2,

---

† It is clear that we could define a *distance* between random quantities conforming to this idea (completely analogous to what we did for distributions in 6.7.1): dist $(X, Y) =$ 'the minimum value that can be given to ε and θ for which the given condition remains satisfied'. Note that there is a difficulty with regard to the dimensionality (θ is a probability, a pure number, and ε is in general a length): however (as in many such cases, for example the one given in 6.7.1, where this fact was disguised by denoting both θ and ε in the same way, by ε) this difficulty is irrelevant, because changes in 'distance' due to expressing ε in different units, does not alter the thing which interests us; that is, the topology based on 'dist $\to 0$'.

2.4.3), we might well say that an X_n for which, for example, $\mathbf{P}(X_n - X_m)^2 < \varepsilon$ for all but a finite number of X_m, 'represents the limit to within ε'. There is no possibility, however, of thinking of defining X *by the given passage to the limit.*

In order to be able to talk about X, it is necessary that it be a well-defined *quantity*, independently of the incidental fact of whether it is known or not (and then, in this sense, a random quantity). There are various possi-bilities (which we distinguish for the purpose of giving examples, not because of fundamental differences): X could be random on account of circumstances *logically independent* of the X_n (and therefore, in principle, capable of being measured or known through relevant procedures or information); it could be definable as some function of a finite number of the X_n (as an example, to underline the absence of any restriction on the possibilities, rather than because it makes any sense, one could think of

$$X = \tfrac{1}{2}(X_{1577} + X_{7814}) + \pi^{X_{62}}(e^{X_{54}} - e^{X_{737296}})$$

or anything else that comes to mind), and these also might depend on some further random factors (e.g. on a random quantity Y which may or may not have any connection with the problem); finally, it might depend on all the X_n (and possibly on other things as well; for instance a Y such as we just mentioned).

In particular, it could in this case be

$$X = \begin{cases} \lim X_n & \text{(if the sequence of the } values \text{ of the } X_n \text{ turn out to be convergent)} \\ 0 & \text{(otherwise)} \end{cases}$$

(and, if one wished, convergence could be taken in the Cesàro sense, or some other). Here too, X is in fact a well-defined quantity (although it can actually only be known after we know the values of all the X_n).

The sentence concerning convergence would only make sense, however, if for such an X, *actually defined independently of the incidental circumstance of what is at present known or unknown*, it were possible to show that, *in the condition of ignorance deriving from these given circumstances*, our present evaluation of probabilities for the X_n and X are such as to imply $X_n \to X$ in some probabilistic sense (quadratic mean, weak, strong, ...). On the contrary, we know that this is not the case in general, not even when $\lim X_n = X$, and still less can it be assumed for an undefinable X which has to appear, phantom-like, from the Cauchy property, and then miracu-lously materialize.

However, mutual convergence (in the weak sense, and *a fortiori* in other, more restrictive, cases) does determine, if not a random quantity X, the limit distribution F. The discussion given above (at the end of 6.8.6) estab-lishes, in fact, that the distributions F_n and F_m, of X_n and X_m, become arbitrarily 'close', and therefore close to one and the same well-defined F, for n and m sufficiently large. In order to be able to state that there exists a limit distribution F such that $F_n \to F$, it is sufficient, for example, to prove that $\mathbf{P}(X_n - X_m)^2 \to 0$ as m and n tend to ∞.

6.8.8. Zero–One law (*Kolmogorov*). We must at least give a mention of a phenomenon which was present in the Borel lemma, and is of a general character, constantly cropping up. In order to be brief (since we only want to deal with it in passing), we shall express ourselves in terms of the strong formulation.

Given an infinite number of independent events E_i, the probability that only a finite number of them occur ($Y < \infty$) is always 1 if the sum of the probabilities converges, and is always 0 if the sum diverges; intermediate probabilities are not possible.

We shall not give a proof, but the main idea is contained in the following: suppose that an event A (such as $Y < \infty$ in the above) is independent of any property A_n which depends only on the first n trials (for example, whether Y is finite cannot be altered by considering a finite number of trials), but is defined, in the limit as $n \to \infty$, by the A_n. Because of independence, $\mathbf{P}(A_n A) = \mathbf{P}(A_n)\mathbf{P}(A)$; taking the limit $A_n \to A$, we have

$$\mathbf{P}(AA) = \mathbf{P}(A) = \mathbf{P}(A)\mathbf{P}(A) = [\mathbf{P}(A)]^2$$

which implies $\mathbf{P}(A) = [\mathbf{P}(A)]^2$, and hence the only possible values are 0 and 1.

6.9 DISTRIBUTIONS IN TWO (OR MORE) DIMENSIONS

6.9.1. Everything we have said in the one-dimensional case extends straightforwardly to two dimensions (or more: in general, we shall present the extension for $n = 2$, and indicate how to proceed to $n = 3$, etc.). The extension has to be considered now because, even if we only wished to deal with random quantities, as soon as we consider two of them we have to deal with the distribution of the pair (X, Y) as a random point in the plane (x, y). This will not, however, be the only kind of application.

A distribution (always to be interpreted as distribution function) over the (x, y)-plane will always be defined by a *joint distribution function*.

$F(x, y) =$ 'the mass contained in the quadrant SW of the point (x, y)';† the mass in the rectangle $x' \leqslant x \leqslant x''$, $y' \leqslant y \leqslant y''$ is then given by

(7) $$F(x'', y'') - F(x'', y') - F(x', y'') + F(x', y');$$

---

† Adopting the practical terminology favoured by economists, we label the 1st, 2nd, 3rd and 4th quadrants as NE, NW, SW and SE (and use these also in referring to directions, etc.; the intuitive reference is to a map with N oriented upwards, as usual). Here we implicitly consider F as undefined where it is discontinuous, and so on. Let us simply remark that all the same conceptual details, which we have discussed at length in the one-dimensional case, can be filled in: we shall only do so when some new feature arises, which is something other than a more or less obvious extension of what has gone before.

cf. Figure 6.5: rectangle = whole quadrant − hatched quadrants + double-hatched quadrant (since this was taken away twice). The relation can be interpreted as an operation involving masses, or probabilities, or, more basically, a linear combination of the 4 events 'belonging to the various quadrants under consideration'.

It may be that the masses are concentrated at points, or distributed in an absolutely continuous manner; there are, however, a great variety of intermediate cases (think, for example, of a mass distributed continuously along a line!).

The density (if and when it exists) is given by

$$(8) \qquad f(x, y) = \frac{\partial^2 F}{\partial x \, \partial y}$$

(the limit of the probability given above, with $x'' = x' + h$ and $y'' = y' + k$, divided by the area hk as h and $k \to 0$).

We can define $F(\gamma)$ for functions $\gamma(x, y)$ of two variables, always in the Riemann–Stieltjes sense (and, if γ is not integrable, we have $F^-(\gamma) < F^+(\gamma)$; the probabilistic interpretation is as the bound for $\mathbf{P}[\gamma(X, Y)]$, and, in particular, if $F(\gamma)$ exists, as its evaluation: throughout, the boundedness conditions for the possible values are to be understood, or, if not, the choice of $\hat{\mathbf{P}}$ is understood, etc.).

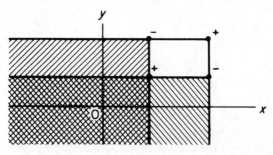

Figure 6.5 Quadrants of the (x, y)-plane, in terms of which the joint distribution function $F(x, y)$ is defined (SW quadrants), and a method of indicating the rectangles with their linear combinations (and, hence, their probabilities in terms of linear combinations of the values $F(x, y)$ at the vertices)

In particular, if $\gamma(x, y)$ represents a set I ($\gamma = 1$ on I and $\gamma = 0$ outside), $F(\gamma) = \mathbf{P}(I)$.

Important examples. If $Z = X + Y$, the distribution function of Z is given by

$$(9) \qquad \mathbf{P}(Z \leqslant z) = F(x + y \leqslant z)$$
$$= F \text{ (the half-plane to the SW of the line } x + y = z)$$

(in other words, 'the mass contained there'). If $Z = XY$, we have

(10)
$$\mathbf{P}(Z \leqslant z) = F(xy \leqslant z)$$
$$= F\text{(the region bounded}\dagger \text{ by the hyperbola } xy = z)$$

(in other words, 'the mass contained there'). If $Z = Y/X$, we have

(11) $\quad \mathbf{P}(Z \leqslant z) = F(y/x \leqslant z) = F[(y \leqslant zx)(x > 0) + (y \geqslant zx)(x < 0)]$
$\quad = F$ (the NW and SE corner regions between the y-axis and the line $y = zx$)

(in other words, 'the mass contained there'). If $Z = \sqrt{(X^2 + Y^2)}$, we have

(12)
$$\mathbf{P}(Z \leqslant z) = F(x^2 + y^2 \leqslant z^2)$$
$$= F\text{(the disc centred at 0 with radius } z)$$

(in other words, 'the mass contained there').
And it would be easy to continue in this manner.

6.9.2. Let us now see how to obtain these results more explicitly. The standard method—integration, using cartesian coordinates—requires us to make the inequality explicit in terms of one of the variables, y, say. In the examples given we have:

$$\begin{aligned}
&sum, && y \leqslant z - x; \\
&product, && (y \leqslant z/x)(x > 0) + (y \geqslant z/x)(x < 0); \\
&quotient, && (y \leqslant zx)(x > 0) + (y \geqslant zx)(x < 0); \\
&distance, && |y| \leqslant \sqrt{(z^2 - x^2)}.
\end{aligned}$$

In these four cases, the integrals (always either $\int \mathrm{d}F$ or $\int f(x, y)\,\mathrm{d}x\,\mathrm{d}y$) will be

(9′)
$$\int_{-\infty}^{\infty} \mathrm{d}x \int_{-\infty}^{z-x} \mathrm{d}y \dots ;$$

(10′)
$$\int_{-\infty}^{0} \mathrm{d}x \int_{z/x}^{+\infty} \mathrm{d}y \dots + \int_{0}^{+\infty} \mathrm{d}x \int_{-\infty}^{z/x} \mathrm{d}y \dots ;$$

(11′)
$$\int_{-\infty}^{0} \mathrm{d}x \int_{zx}^{\infty} \mathrm{d}y \dots + \int_{0}^{\infty} \mathrm{d}x \int_{-\infty}^{zx} \mathrm{d}y \dots ;$$

(12′)
$$\int_{-z}^{z} \mathrm{d}x \int_{-\sqrt{(z^2-x^2)}}^{+\sqrt{(z^2-x^2)}} \mathrm{d}y \dots .$$

In general, if $Z = \gamma(X, Y)$ we have $\mathbf{P}(Z \leqslant z) = F(\gamma(x, y) \leqslant z) = F_\gamma(z)$ (say), and if the inequality can easily be made explicit with respect to y, obtaining, in the simplest case, $y \leqslant g(x, z)$ (or, possibly, $g_1(x, z) \leqslant y \leqslant g_2(x, z)$), we

---

† 'Interior' or 'exterior' region, according as $z > 0$ or < 0.

shall have

$$F_y(z) = \int_{-\infty}^{\infty} dx \int_{g_1(x,z)}^{g_2(x,z)} dy \ldots.$$

Clearly, it may sometimes be more convenient to adopt other coordinate systems (e.g. polar coordinates); remembering, of course, to multiply by the Jacobian.

Let us indicate also how one obtains directly the *density* $f_y(z) = dF_y(z)/dz$ (in those cases where everthing goes through smoothly). From the expression for $F_y(z)$, assuming that $F(x, y)$ has a density $f(x, y)$, we obtain

$$f_y(z) = \frac{d}{dz} \int_{-\infty}^{\infty} dx \int_{g_1(x,z)}^{g_2(x,z)} f(x, y) \, dy$$

$$= \int_{-\infty}^{\infty} dx \left[f(x, g_2(x, z)) \frac{\partial}{\partial z} g_2(x, z) - \text{the same thing for } g_1 \right].$$

For the examples we have considered, this gives

(9″)   *sum*: $g_1 = -\infty, g_2 = z - x; f_s(z) = \int_{-\infty}^{\infty} f(x, z - x) \, dx$;

(10″)   *product*: $x < 0: g_1 = z/x, g_1' = 1/x, g_2 = +\infty$;

$x > 0: g_1 = -\infty, g_2 = z/x, g_2' = 1/x$;

$f_p(z) = \int_{-\infty}^{\infty} \frac{1}{|x|} f(x, z/x) \, dx$;

(11″)   *quotient*:   (as above, with x in place of $1/x$):

$f_q(z) = \int_{-\infty}^{\infty} |x| f(x, zx) \, dx$;

(12″)   *distance*: $-g_1 = g_2 = \sqrt{(z^2 - x^2)}$;

$-g_1' = g_2' = z/\sqrt{(z^2 - x^2)}$;

$$f_d(z) = \int_{-z}^{z} \frac{z}{\sqrt{(z^2 - x^2)}} \{f(x, \sqrt{(z^2 - x^2)}) + f(x, -\sqrt{(z^2 - x^2)})\} \, dx.$$

The first example, the simplest, should be noted well, since the case of the sum is basic for most theoretical developments and applications.

We add one last example, where the answer comes out directly: for the *maximum*, $Z = X \vee Y$, the distribution function is given by

(13) $F(z) = F(z, z)$ (in fact, $(Z \le z) = [(X \vee Y) \le z] = (X \le z)(Y \le z)$);

similarly, for the *minimum*, $Z = X \wedge Y$, the distribution function is given by

(14) $F(z) = F(z, +\infty) + F(+\infty, z) - F(z, z).$

By means of $F(\gamma)$, we can also, in this case, express various 'synthetic characteristics' of distributions of two variables. For example, for the moments we take $\gamma(x, y) = x^r y^s$ and obtain $M_{r,s} = P(X^r Y^s) = \int x^r y^s \, dF = \int x^r y^s f(x, y) \, dx \, dy$. We have already seen the first- and second-order moments with respect to the origin: $P(X)$ and $P(Y)$, the coordinates of the barycentres; $P(X^2)$, $P(Y^2)$ and $P(XY)$, the second-order terms (the moments with respect to the barycentres are

$$P(X^2) - [P(X)]^2, \quad P(Y^2) - [P(Y)]^2 \quad \text{and} \quad P(XY) - P(X)P(Y),$$

the variances and the covariance). We already know that, in terms of second-order properties, these moments completely characterize the distribution: in particular, we have seen that the cancelling out of the mixed barycentric moment $(P(XY) - P(X)P(Y) = 0$, i.e. $P(XY) = P(X)P(Y)$, the property referred to as non-correlation) is a necessary condition for X and Y to be stochastically independent.

6.9.3. *Stochastic independence of random quantities.* The time has come for us to consider the notion of stochastic independence in the context of random quantities (and, essentially, in the most general case, since the delicate issues have a unique character). Up until now, the concept has only been defined (in Chapter 4) for events (4.9.2) and for random quantities with only a finite number of possible values (4.10.1). The extension to the general case is essentially intuitive; we mentioned this (in 4.16.2), where we also pointed out that a detailed and critical approach was required.

The meaning of stochastic independence was: 'that whatever one learns or assumes about X does not modify one's opinion about Y'; put more 'technically', 'every event concerning Y is stochastically independent of every event concerning X'.

Naturally, when it comes to considering n random quantities, these (like events) will *not* be called independent if the independence is merely *pairwise*, but only if each of them is independent of anything one knows or assumes *concerning all the others* simultaneously (that is, of each event concerning all these other random quantities).

Once again we are faced with the question: *which events* do we include in this definition? We might be tempted to say '*all of them*' (and so refer ourselves to $F_{\mathscr{C}}$; but we know that this is a rather unimaginable abstraction); we might say (along with the supporters of the 'strong' formulation) 'all those of the Lebesgue field, or at least the Borel field' (thus referring ourselves to $F_{\mathscr{B}}$; but this runs counter to the objections we have made against countable additivity and the strong formulation); we might limit ourselves to the intervals (and things expressible in terms of them; this leaves us in the field $F_{\mathscr{R}}$). Note, however, that the question does not require a discussion and a decision as to which answer provides the *correct* definition: the best solution would probably be to consider all three definitions (or perhaps none of these), drawing a distinction between 'complete', 'strong' and 'weak' independence. We shall limit ourselves, however, to the weak definition since it is the only one which does not make too unrealistic assumptions about our knowledge. In fact, it is the usual definition, apart from the fact that this notion has a completed appearance when the unique extension to the Lebesgue field is assumed, along with non-existence outside it.

The assumption that events of the form $X \leqslant x$ are independent of those of the form $Y \leqslant y$ (for any x and y) is sufficient to imply that $F(x, y) = F_1(x)F_2(y)$, where $F_1(x) = F(x, +\infty)$ and $F_2(y) = F(+\infty, y)$ are the distribution functions of X and Y (with the usual qualification of indeterminacy at jump points). It follows immediately that there is also independence for the intervals:

$$\mathbf{P}[(x' \leqslant X \leqslant x'')(y' \leqslant Y \leqslant y'')]$$
$$= F_1(x'')F_2(y'') - F_1(x'')F_2(y') - F_1(x')F_2(y'') + F_1(x')F_2(y')$$
$$= [F_1(x'') - F_1(x')] \cdot [F_2(y'') - F_2(y')].$$

This implies independence for step functions of the single variables x or y, and hence for continuous functions. We conclude that the condition defined by

(15) $F(x, y) =$ products of functions involving x only and y only,

is also equivalent to the following condition:

for any product of continuous functions, $\gamma(x, y) = \gamma_1(x)\gamma_2(y)$, we have

(15′) $F(\gamma) = F(\gamma_1)F(\gamma_2),$

in other words,

(15″) $\mathbf{P}\{\gamma_1(X)\gamma_2(Y)\} = \mathbf{P}\{\gamma_1(X)\}\mathbf{P}\{\gamma_2(Y)\}.$

6.9.4. Observe, however, how far removed this condition is from the intuitive notion of stochastic independence. We can always assume that the possible points are those of the set of $A_{r,s}$, with coordinates $x_{r,s} = r + s\sqrt{2}$, $y_{r,s} = r + s\sqrt{3}$ (a countable set, since the points are defined in terms of two rationals r and s).

This set is in fact everywhere dense in the plane and can be the logical support of any distribution function; in particular, of a distribution which makes X and Y stochastically independent. But, on the other hand, to each possible value for X there corresponds a unique possible value for Y, and conversely (because, given x, there exists at most one pair of rational values r and s giving $x = r + s\sqrt{2}$; if there were another pair, so that $x = r' + s'\sqrt{2}$, we would have $\sqrt{2} = (r - r')/(s - s')$, an absurdity).† We can thus have logical dependence (even one-to-one and onto) at the same time as (distributional) stochastic independence. We must bear in mind just how unsatisfactory this definition is from a logical viewpoint, even if it seems difficult to improve on it within the ambit of realistic possibilities.

Remark. Observe that such 'paradoxes' can *also occur in the discrete case*, if the probabilities are thought of not as being concentrated at the points (x_h, y_k), but as *adherent* to them (which is excluded, as in Chapter 4, 4.10.1, if we talk of a 'finite number of possible values'—but this subtlety might be overlooked). Therefore, the decision (here in 6.9.3) 'not to give a precise value to F at jump points' is essential.

If a point (x_h, y_k) is not a possible point, but instead (or also) a limit point of a sequence of possible points, each having zero probability, but with positive total probability, a great number of different cases of distributional independence $(p_{hk} = p'_h p''_k)$ are possible, but other kinds are not (not even logical independence).

6.9.5. On the other hand, we ought to point out that paradoxes (of *non-conglomerability*; cf. Chapter 4, 4.19.2) arise in connection with 'stochastic independence' without any need to look at pathological examples (or, as some would say, to make them up). The following is a well-known example: if we choose a point 'at random' on the surface of a sphere, equal areas have equal probabilities; and if we happen to know which great circle the point has landed on, then equal arcs will have equal probabilities; if, in addition, we have a system of geographical coordinates (latitude and longitude, say, as on the earth) these coordinates are independent.

In fact, distributional independence holds; the surface element whose latitude lies between ϕ and $\phi + d\phi$, and whose longitude lies between λ and $\lambda + d\lambda$, has area $\cos \phi \, d\phi \, d\lambda$, and (apart from the normalization constant) this is also its probability. Longitude has a uniform distribution $(1/(2\pi)$ between $\pm \pi$), and latitude has a distribution whose density is given by $f(\phi) = \frac{1}{2} \cos \phi$ (between $\pm \pi/2$); the density for the area (in the λ, ϕ-plane) is the product

$$\frac{1}{2\pi} \cdot \frac{1}{2} \cos \phi = \frac{1}{4\pi} \cos \phi.$$

But then, because of the other assumptions, even if we know the longitude precisely—in other words, the meridian to which the point belongs—the probability distribution of the latitude should always have density $\frac{1}{2} \cos \phi$; on the other hand, since we are dealing with (half) a great circle, the density should be uniform $(= 1/\pi)$.

---

† From this it also follows that $y = f(x)$ is additive, where it is defined: $f(x' + x'') = f(x') + f(x'')$ but not linear (for $s = 0$, $f(x) = x$; for $r = 0$, $f(x) = \sqrt{(3/2)}x$); and the graph of such a function is dense in the whole plane (cf., for example, B. de Finetti, *Mathematica logico-intuitiva*, No. 40, 'Sulla proprietà distributiva', in particular, Figure 30, pp. 91–92 in 3rd Ed., Cremonese, Rome (1959)).

The paradox is easily resolved if we argue in terms of 'imprecision'. If, instead of thinking of the point lying exactly on that curve, one thinks in terms of the fact having been ascertained to within some margin of error, however small, one sees that the two answers are coherent. We give two different versions: if the imprecision concerns λ, then, instead of a meridian curve, we have a zone which narrows from the equator to the poles as $\cos \phi$; if, instead, we think in terms of having measured the distance from a plane passing through the centre of the earth (that is, the distance from the great circle) then (finding the distance to be 0) we have a zone of constant width.

It is easy to avoid paradoxes by avoiding any reference to limit-cases, except when considering these explicitly as such (never speak of 'the probability of something conditional on $X = x_0$', but 'conditional on $X = x_0 + \varepsilon$', perhaps giving the limit as $\varepsilon \to 0$). Many authors (the first of them being, I think, Kolmogorov in 1933) explicitly state that the problem only makes sense under this restriction (since, otherwise, conditional probability would formally be given by expressions of the form $0/0$). From a theoretical point of view, viewed from the standpoint to which we adhere, such a conclusion seems rather drastic (although it avoids some difficulties, others take their place). Theoretically, it does not seem possible to avoid the necessary comparisons among the zero probabilities which would yield an actual probability for the 'precise' fact, rather than the zero probability usually attributed (cf. Chapter 4, 4.18); practically speaking, it is convenient to attempt to use the Kolmogorov limit argument, by considering it in conjunction with what is empirically known about the imprecision (when actually present) and not merely as a convention or a dogma. We shall mention this again later (Chapter 12, 12.4.3).

6.9.6. *Operations on stochastically independent random quantities: Convolutions.* Let us now return to consideration of a random quantity $Z = \gamma(X, Y)$, a function of two other random quantities (as in 6.9.2), in the rather special case where X and Y are stochastically independent. This implies that $F(x, y) = F_1(x)F_2(y)$ and $f(x, y) = f_1(x)f_2(y)$ (if these exist), and we have $dF(x, y) = dF_1(x) dF_2(y) = f_1(x)f_2(y) dx dy$.

The fundamental case, which we shall encounter and make use of over and over again, is that of the *sum*, $Z = X + Y$, for which $F(z)$ and $f(z) = F'(z)$ are given by

(16)
$$F(z) = \int_{-\infty}^{+\infty} dF_1(x) \int_{-\infty}^{z-x} dF_2(y) = \int_{-\infty}^{+\infty} F_2(z - x) \, dF_1(x)$$

$$= \int_{-\infty}^{+\infty} F_2(z - x) f_1(x) \, dx,$$

(17)
$$f(z) = \int_{-\infty}^{\infty} f_1(x) f_2(z - x) \, dx.$$

The rôles of F_1 and F_2 can, of course be interchanged (choose the simplest way!) and, as usual, we make the qualification that the expressions in terms of densities only hold when the latter exist.

The operation on the distributions which gives F in terms of F_1 and F_2, and f in terms of f_1 and f_2, is called *convolution*. It is usually denoted by the symbols ✴ and ∗, and we write $F = F_1 ✴ F_2$, $f = f_1 ∗ f_2$.

The operation can clearly be repeated to give the distribution of the sum of three independent random quantities (and so on for any finite sum). It follows from the definition that convolution is associative, commutative and even distributive. In the special case where all the summands are identically distributed (that is, have the same distribution function F), the convolution is denoted by $F^{✴n}$ (and $f^{✴n}$).

The following is a brief summary of the other cases we considered:

(18) *product:* $$F(z) = \int_0^{+\infty} F_2(z/x)\,dF_1(x),†$$

$$f(z) = \int_{-\infty}^{+\infty} \frac{1}{|x|} f_1(x) f_2(z/x)\,dx\,;$$

(19) *quotient:* $$F(z) = \int_0^{+\infty} F_2(zx)\,dF_1(x),†$$

$$f(z) = \int_{-\infty}^{+\infty} |x| f_1(x) f_2(zx)\,dx\,;$$

(20) *distance:* $$F(z) = \int_{-z}^{z} [F_2(\sqrt{(z^2 - x^2)}) - F_2(-\sqrt{(z^2 - x^2)})]\,dF_1(x),$$

$$f(z) = \int_{-z}^{z} \frac{z}{\sqrt{(z^2 - x^2)}} f_1(x)[f_2(\sqrt{(z^2 - x^2)})$$
$$+ f_2(-\sqrt{(z^2 - x^2)})]\,dx\,;$$

(21) *maximum:* $F(z) = F_1(z)F_2(z), \quad f(z) = F_1(z)f_2(z) + F_2(z)f_1(z).$

6.9.7. *Synthetic characteristics for sums of independent random quantities.* Let Z be the sum of two or more independent random quantities; we shall include both $Z = X + Y$ and $Z = X_1 + X_2 + \ldots + X_n$ in order to draw attention both to the notationally simplest case and to the general one.

We shall consider now some of the points that can be made concerning their synthetic characteristics. We shall use the indices $i = 1, 2, \ldots, n$ for aspects concerning the summands, and $\overline{n}$ for what concerns the sum of

---

† For the sake of brevity, the term $\int_{-x}^{0}$ (anti-symmetric) is omitted; if X is not certainly positive, it must be included.

n terms; when the summands are identically distributed, we shall drop the indices.

In the case of the prevision, $m = \mathbf{P}(X)$, we have additivity (in all circumstances); for the variance, $\sigma^2 = \mathbf{P}(X - m)^2$, additivity holds when the summands are uncorrelated (and, *a fortiori*, when they are independent):

$$(22) \qquad m_{\overline{n}} = m_1 + m_2 + \ldots + m_n \quad (= n \cdot m)$$

$$(23) \qquad \sigma_{\overline{n}}^2 = \sigma_1^2 + \sigma_2^2 + \ldots + \sigma_n^2 \quad (= n\sigma^2; \sigma_{\overline{n}} = \sqrt{n}\sigma).$$

For the third-order moments, we have

$$\mathbf{P}(Z^3) = \mathbf{P}(X + Y)^3 = \mathbf{P}(X^3) + 3\mathbf{P}(X^2 Y) + 3\mathbf{P}(X Y^2) + \mathbf{P}(Y^3),$$

and, in the case of independence,

$$\mathbf{P}(Z^3) = \mathbf{P}(X^3) + 3\mathbf{P}(X^2)\mathbf{P}(Y) + 3\mathbf{P}(X)\mathbf{P}(Y^2) + \mathbf{P}(Y^3).$$

For $Z = \sum X_i$, with the summands independently and identically distributed, if we denote by

$$M_1 = m = \mathbf{P}(X), \quad M_2 = m^2 + \sigma^2 = \mathbf{P}(X^2), \quad M_3 = \mathbf{P}(X^3)$$

the moments (of 1st, 2nd and 3rd orders, respectively) of the summands, and by $(M_3)_{\overline{n}}$ that of the sum, we have similarly

$$(M_3)_{\overline{n}} = \sum_{ijh} \mathbf{P}(X_i X_j X_h) = nM_3 + 3n(n - 1)M_1 M_2 + n(n - 1)(n - 2)M_1^3.$$
$$(24)$$

On the basis of this formula, the reader can see how things proceed in the general case by noting the following simple points (and these will not apply only to M_3, with summands not identically distributed, but to moments of any order, whether the summands are identically distributed or not):

the 3rd power (or the general rth power) of a sum of n terms is the sum of the n^3 (or n^r) products (including repetitions) of the summands 3 at a time (or r at a time);

the prevision of each product is $(M_3)_i$ if it contains precisely the same factor X_i three times; $(M_2)_i(M_1)_j$ if the product is $X_i X_i X_j$; $(M_1)_i(M_1)_j(M_1)_k$ if the product is $X_i X_j X_k$ (with distinct factors); for r summands, things become more complicated, but the idea is the same;

in the case of identically distributed summands, it is sufficient to suppress the indices i, j and k, and count up the number of the three kinds of term M_3, $M_2 M_1$, M_1^3 (and there are n choices for i; $3n(n - 1)$ ways of putting a j in one of the three positions and an $i \neq j$ in the remaining two; $n(n - 1)(n - 2)$ ways of arranging the n elements 3 at a time); for a general r, we have products of the form $M_1^a M_2^b \ldots M_n^m$, with $a + 2b + 3c + \ldots + mn = n$, if the product contains a single factors, b which appear twice, c which appear three times, ... , and m (either 0 or 1) n-tuples.

As far as the extreme values, inf Z and sup Z, are concerned, in the case of independence we can definitely say that inf $Z = \sum$ inf X_i and sup $Z = \sum$ sup X_i (in general one can only note the obvious inequalities, $\geqslant$ and $\leqslant$, respectively).

6.9.8. One obvious additional result is that for the sum of independent random quantities (i.e. the convolution of distributions) the range of variation of the distribution must increase: if $F = F_1 * F_2$,

$$\sup F - \inf F > \sup F_1 - \inf F_1 ;$$

(with equality only in the trivial case of F_2 concentrated at a single point). The same conclusion holds, however, in a much more general context: the *dispersion* $l(p)$ must also increase (for all $0 < p \leqslant 1$; the above corresponds to the extreme case $p = 1$). Suppose that in the distribution F there is an interval of length l enclosing a mass $\geqslant p$; let the interval be $a, a + l$: if we assume that in F_1 every interval of length l contains a mass $< p$ (cf. 6.6.6) we are led to the following absurd conclusion:

$$p \leqslant F(a + l) - F(a)$$

$$= \int_{-\infty}^{\infty} \{F_1(a + l - x) - F_1(a - x)\} \, dF_2(x) < p \int_{-\infty}^{\infty} dF_2(x) = p.$$

It follows, as an important corollary, that, for the convolution, 'regularity' must increase: the resulting distribution enjoys all those regularity properties enjoyed by at least one of the component distributions. For example: the property of not having any masses greater than some given p; the property of continuity; or of being absolutely continuous; or of having a density never greater than some given bound; properties of existence or bounds for successive derivatives; or the property of being analytic.

It can easily be seen, for instance, that the mathematics used in 6.7.2 to construct a continuous distribution 'close' to some given one, was essentially an application of the following: given any random quantity, in order to obtain a distribution with density $\leqslant 1/\varepsilon$, it is sufficient to add to it a random quantity with a uniform distribution in the interval $[0, \varepsilon]$ (for example, a 'rounding error'). An 'accidental' error with a *normal* distribution—which we shall meet soon—is sufficient to make the distribution analytic.

In addition to the moments, $\gamma = \square^r$, which we have already considered, there is another class of previsions $F(\gamma)$ of great importance: that of the exponential functions $\gamma = a^{\square}$. The basic property of these functions yields, for $Z = X + Y$ (or $Z = \sum X_i$),

$$a^Z = a^{X+Y} = a^X a^Y, \quad a^Z = a^{\sum_i X_i} = a^{X_1} a^{X_2} \ldots a^{X_n},$$

so that, in the case of independence,

(25) $\qquad \mathbf{P}(a^Z) = \mathbf{P}(a^X)\mathbf{P}(a^Y), \quad \mathbf{P}(a^Z) = \mathbf{P}(a^{X_1})\mathbf{P}(a^{X_2}) \ldots \mathbf{P}(a^{X_n}).$

We shall see shortly how this property can be exploited.

6.10   THE METHOD OF CHARACTERISTIC FUNCTIONS

6.10.1. The synthetic characteristics provide partial information of varying usefulness and interest; we have examined some of the most important kinds. One could ask, however, whether it is possible for a sufficiently rich set of 'synthetic characteristics' to be sufficient to completely characterize a distribution?

In terms of the $F(\gamma)$, the answer (in a general form) has already been given (in 6.4.4), since, in order to determine $F(\gamma)$, we said that it was sufficient to know $F(\gamma)$ for all continuous γ (it is also sufficient to know it for a subset which permits approximation to any desired degree of accuracy from above and below). It is known that in certain cases (for example, for bounded distributions) this can even be obtained by means of polynomials, and hence knowledge of (all) the moments, $F(\square^r)$, $r = 1, 2, \ldots, n, \ldots$, turns out to be sufficient (and, in fact, the researches of Tchebychev and others have dealt with this topic; Castelnuovo's treatise gives a masterly account of the research in this field). On the other hand, this method of moments also appears in the approach that we shall adopt.

This is the approach based on the property of the exponential function that we noted above. It consists in considering the prevision for the exponential function as the base varies in an appropriately chosen set (the reals, or, better still, complex values with absolute value $= 1$). The method is called that of *generating functions*, or *characteristic functions* (according to the variant adopted). In order to avoid using more than one term (which is often misleading, since it prevents one seeing the essential identity of things expressed in slightly different forms) we shall always use the name 'characteristic function'.

This powerful technique has a rather curious history :[†] it has entered into consistent and systematic usage only recently (especially following the brilliant applications of it made by P. Lévy in about 1925), after having been

---

[†] A clear, concise and essentially complete account can be found in H.L. Seal, 'The historical development of the use of generating functions in probability theory', *Bull. Ass. Actuaires Suisses*, **49** (1949), 209–228.

discovered, applied, abandoned and then rediscovered in a variety of applications and circumstances (from De Moivre to Lagrange, from Laplace to Poisson).

6.10.2. In the simplest case (the original application of De Moivre) the method consists in noting that if X is a random *integer*, and t any real (or complex), then $\mathbf{P}(t^X) = \sum_h p_h t^h$ is a polynomial in which the coefficient of t^h is the probability of obtaining the value $X = h$ (h an integer, often—but not necessarily—positive). One also notes—and this is the *fundamental property* that we mentioned—that if X and Y are *stochastically independent* random quantities, so are t^X and t^Y, and hence

(26) $$\mathbf{P}(t^{X+Y}) = \mathbf{P}(t^X t^Y) = \mathbf{P}(t^X)\mathbf{P}(t^Y).$$

If $\mathbf{P}(t^X) = \sum_h p_h t^h$ and $\mathbf{P}(t^Y) = \sum_k q_k t^k$, and we take the product

(27) $$\sum_{hk} p_h q_k t^{h+k} = \sum_i t^i \sum_h p_h q_{i-h},$$

we have an 'automatic' way of computing the probabilities

(28) $$r_i = \mathbf{P}(X + Y = i) = \sum_h p_h q_{i-h};$$

that is, of obtaining *the distribution of the sum, $Z = X + Y$.*

This fundamental property (that is, that the product $\mathbf{P}(t^X)\mathbf{P}(t^Y)$ corresponds to the sum $(X + Y)$ clearly holds even if X and Y are not integer, so long as t^X and t^Y continue to make sense. In order that this be so, one could limit oneself to t on the positive real axis, or, alternatively, write $t = e^z$, with the convention that in place of $t^X = (e^z)^X$ one considers $e^{zX}(= e^{(zX)})$, which always makes sense.†

Instead of $\mathbf{P}(t^X)$ we therefore consider $\mathbf{P}(e^{zX})$ (which is equivalent when t is real and positive and z is real, and more general in that it allows the removal of these restrictions). If X has an unbounded distribution, $\mathbf{P}(e^{zX})$ could diverge; this could never happen if z were purely imaginary (since then $|e^{zX}| = 1$). In order to map the imaginary axis (which has this nice property we have just mentioned) onto the real axis (which is more convenient as the standard support for representing functions of a real variable) we set $z = iu$, and then $t = e^z = e^{iu}$; in this way $\mathbf{P}(e^{iuX})$ becomes a function of u, which is certainly defined for all u on the real axis (where, however, it will in general assume complex values), and possibly outside it as well.

But, in the general case, will knowledge of $\mathbf{P}(e^{iuX})$ be sufficient to determine the probability distribution? We shall see that the answer to this is yes.

---

† To the infinity of values $z = z_0 + 2ki\pi$, having values e^z which coincide for a given t, there correspond different values for the non-existent 't^x', i.e. $e^{(z_0 + 2k\pi i)x}$.

The answer is unconditional if we know $\mathbf{P}(e^{iuX})$ for all real u (or if we know $\mathbf{P}(e^{zX})$ for all purely imaginary z); under suitable conditions, it also holds for $\mathbf{P}(t^X)$ and $\mathbf{P}(e^{zX})$ and for $t > 0$ and z real.

This justifies the name *characteristic function* given to

$$(29) \qquad \phi(u) = \mathbf{P}(e^{iuX})$$

(and sometimes also to $\mathbf{P}(e^{zX})$); and the name *generating function* given to $g(t) = \mathbf{P}(t^X)$. We shall always use $\phi(u) = \mathbf{P}(e^{iuX})$, permitting ourselves to write (when X is an integer, and it is convenient to do so) $\phi(u) = $ (an expression in t), implying that $t \equiv e^{iu}$ (and we shall not speak of generating functions: one of the two terms is superfluous).

In the case of discrete distributions (masses p_h at the points x_h), or of distributions admitting a density function $f(x)$, the characteristic function can be expressed in the form

$$(30') \qquad \phi(u) = \sum_h p_h e^{iux_h}$$

or

$$(30'') \qquad \phi(u) = \int e^{iux} f(x)\, dx,$$

respectively: in the general case, we have (using the Riemann–Stieltjes integral)

$$(30) \qquad \phi(u) = \int e^{iux}\, dF(x) = F(e^{iu\square}) \quad \left(\int = \int_{-\infty}^{+\infty} \right).$$

Of course, if one prefers to avoid the imaginary number under the prevision and integral signs, or if one wishes rather to give the real and imaginary parts separately, it suffices to recall that $e^{ix} = \cos x + i \sin x$ and to write

$$(29') \quad \phi(u) = \mathbf{P}(\cos uX) + i\mathbf{P}(\sin uX) = \int \cos ux\, dF(x) + i \int \sin ux\, dF(x).$$

6.10.3. There is a one-to-one, onto and continuous correspondence between characteristic functions $\phi(u)$ and proper distributions $F(x)$. The inverse relation (in the simplest case, where

$$\int_{-\infty}^{\infty} |\phi(u)|\, du < \infty$$

and $f(x)$ is then continuous and bounded) is given by

$$(31) \qquad f(x) = \frac{1}{2\pi} \int_{-\infty}^{*+\infty} e^{-iux}\phi(u)\, du \dagger$$

and has a symmetric relationship with (30″); this remarkable fact will be important in applications. By continuity we mean that *the convergence of $\phi_n(u) \to \phi(u)$, uniformly in any bounded interval, is equivalent to the convergence of $F_n(x) \to F(x)$ for all x (except for the discontinuity points of F).*

The fundamental property that we began with states that: *to the convolution of distributions, $F = F_1 * F_2$* (or of densities $f = f_1 * f_2$) *there corresponds the product, $\phi = \phi_1\phi_2$, of characteristic functions.*

Moreover, to any linear combination, $F = \sum_h c_h F_h$, there corresponds the same linear combination, $\phi = \sum_h c_h \phi_h$. These properties in themselves are sufficient to solve many problems; they are also useful for deriving new distributions and for modifying distributions in order to make formulae like (31) applicable (by means of approximations) in cases where they are not directly applicable.

It is useful to bear in mind the following properties (for proofs and details see, for example, Feller, II, pp. 472 ff.): $\phi(u)$ is *continuous*; $\phi(0) = 1$ and $|\phi(u)| \leqslant 1$; the real part of $\phi(u)$ is *even* and the imaginary part *odd*; $\phi(u)$ is *real* if and only if the distribution is *symmetric*; changing X into aX, changes $F(x)$ into $F(x/a)$, and $\phi(u)$ into $\phi(au)$.

For the moments $\mathbf{P}(X^h) = M_h$ (where $\mathbf{P} = \hat{\mathbf{P}}$!) which exist, the following expansion is valid

$$\phi(u) = 1 + iuM_1 - u^2M_2/2! - iu^3M_3/3! + \ldots + (iu)^h M_h/h! + \ldots$$

(and corresponds, formally, to $\mathbf{P}(e^{iuX}) = \mathbf{P}(1 + iuX - u^2X^2/2! - \ldots)$). If all the moments exist, the series has a non-zero radius of convergence ρ, and $\phi(u)$, and therefore the distribution, is completely determined by the sequence of moments.‡

These and other properties reveal a relationship to be borne in mind in the following qualitative sense: the smaller the 'tails' of the distribution at

---

† The asterisk at the upper limit of the integral sign means that the principal value (in the Cauchy sense) is to be understood: i.e.

$$\lim \int_{-a}^{a} \quad \text{as} \quad a \to \infty.$$

Formula (31) is the classical Fourier *inversion theorem*.

‡ Since $1/\rho = \lim \sup (e/n)\sqrt[n]{|M_n|}$, a necessary and sufficient condition for the function to be analytic is that $\sqrt[n]{|M_n|}$, the mean of order n, does not increase faster than n (i.e. remains $\leqslant Kn$ with K finite). The necessary and sufficient condition for the distribution to be determined by the moments is that the sum of the reciprocals, $1/\sqrt[n]{|M_n|}$, diverges (Carleman). This is a little less restrictive than the above, which implies that the sum of the reciprocals, $\geqslant 1/Kn = K^{-1}(1/n)$, diverges almost as rapidly as the harmonic series.

infinity (i.e. the faster $F(x)$ tends to 0 or 1), the more regular the behaviour of $\phi(u)$ near the origin; the smoother (in terms of differentiability, etc.) the distribution is, the more regular is the behaviour of $\phi(u)$ at infinity.

6.10.4. *Geometrical representation of the mathematical nature of the problems.* We note that the functions of u, z and t that we have been considering are the transformations of the distribution function known in analysis as the Fourier, Laplace and Mellin transforms, respectively. As we have already indicated (but reiterate for the sake of anyone who has come across these transforms separately and has not noticed the fact), we are always dealing with precisely the same transform, apart from a change of variable. Figure 6.6 indicates, schematically, the planes of the (complex)

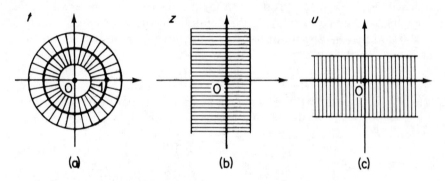

Figure 6.6 The planes of the three variables t, z and u, in terms of which the characteristic function can be expressed, together with the lines or regions where it is defined. Usually we operate in terms of u (the Fourier transform); $z = iu$ and $t = e^z = e^{iu}$ are occasionally to be preferred (the Laplace and Mellin transforms, respectively)

variables u, $z = iu$ and $t = e^z$; the line on which the function is always defined is marked in heavily (the real axis in the case of u, the imaginary axis for z, and the unit circle for t), and the striped region indicates where it is defined in the analytic case:

$$-\alpha' < \mathscr{I}(u) = \mathscr{R}(z) < \alpha'', \quad |t'| < |t| < |t''|†$$

(where $0 \leqslant \alpha', \alpha'' \leqslant +\infty$; $0 \leqslant |t'| = e^{\alpha'} \leqslant 1 \leqslant e^{\alpha''} = |t''| \leqslant \infty$).

We have so far seen illustrations of the complex planes of the three variables (t, z, u). In order to 'visualize' the meaning and the properties of the characteristic function $\phi(u)$ (for u real) in the complex plane of $w = \phi(u)$, we draw

---

† The annulus of convergence for the Laurent series (Figure 6.6a); the strips for the Dirichlet series (Figure 6.6b and, changing axes, Figure 6.6c). $\mathscr{R}$ and $\mathscr{I}$ denote the real and imaginary parts.

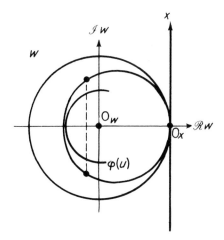

Figure 6.7 The plane of $w = \phi(u)$, and the interpretation of $\phi(u)$ as the barycentre of the distribution 'wrapped around the circumference $|w| = 1$'

it (Figure 6.7), indicating the unit circle, $|w| = 1$, and the tangent at the point $w = 1$ (the straight line $\mathscr{R}(w) = 1$). This point is denoted by 0_x, because it is the origin of the x coordinate, thought of both as the abscissa on this tangent line and as parameter (angle or arc length) on the circumference. In order to avoid confusion, the origin $w = 0$ has been denoted by 0_w.

If we think of the distribution of X as located on the x-axis, then e^{iX} has the same distribution 'wrapped around the unit circle', and similarly for u^X and e^{iuX} (with only a modification of scale from 1 to u, reflected if u is negative). The characteristic function $\phi(u) = \mathbf{P}(e^{iuX})$ is the barycentre (necessarily inside the circle, unless the distribution is concentrated at a single point), and it follows therefore that $|\phi(u)| \leqslant 1$. If $u = 0$, we always have, of course, $\phi(u) = 1$; in general, however, we have $|\phi(u)| < 1$, the only other exceptional cases being the following. Firstly, a trivial case consisting of a single mass concentrated at $x = a$; in this case we always have $\phi(u) = e^{iua}$, and, hence, $|\phi(u)| = 1$. The second exception is that of a distribution concentrated at the points of an arithmetic progression, $x = c \pm 2k\pi/u_0$; clearly $|\phi(u_0)| = 1$, and the same will hold for all multiples of u_0.

If we think in terms of the graph of $w = \phi(u)$, many properties (those we have already mentioned and others) become obvious. As an example, the change from X to $-X$ implies that the distribution (on the line, or wrapped around the circle) is reflected in the real axis; the same is also true for the barycentre, so that the characteristic function of $-X$ is the conjugate of the $\phi(u)$ corresponding to X; $\phi(-u) = \phi^*(u)$. An important corollary follows: given any $\phi(u)$, we can obtain a symmetric characteristic function,

$|\phi(u)|^2 = \phi(u)\phi^*(u)$. The corresponding distribution is called the *symmetrized*† version of the $F(x)$ we started with, and is obtained from the convolution of $F(x)$ and $1 - F(-x)$; it is the distribution of the difference $X_1 - X_2$, where X_1 and X_2 are independent, both with distribution $F(x)$.‡

For u purely imaginary (and we shall write $u = iv$, with real v, so that $v = iu = z$), we have, separating the contribution of the probability distribution on the negative semi-axis from that on the positive axis, and from that concentrated at the origin, if any ($p_0 = F(+0) - F(-0)$),

$$(32) \quad \phi(-iv) = \int_{-\infty}^{\infty} e^{vx} \, dF(x) = \int_{-\infty}^{0} e^{vx} \, dF(x) + \int_{0}^{\infty} e^{vx} \, dF(x) + p_0.$$

The contribution in $[-\infty, 0]$ is clearly finite for $v \geqslant 0$, and possibly for negative v between 0 and some $-\alpha'$ (everywhere if $\alpha' = \infty$); by symmetry, the contribution in $[0, \infty]$ is finite for $v \leqslant 0$, and possibly for positive v between 0 and some α'' (everywhere if $\alpha'' = \infty$). If it exists in the interval $[-\alpha', \alpha'']$ of the imaginary axis, ϕ is positive, real and concave (upwards), like each of the e^{vx} of which it is a mixture. The meaning of the bounds, $-\alpha'$ and α'', and some other aspects, becomes clear if we introduce the notion of *twinned*§ distributions, a notion which is of interest in its own right.

The twins of $F(x)$ (and the relationship is mutual) are defined to be those $F_v(x)$ for which

$$(33) \quad dF_v(x) = Ke^{vx} \, dF(x), \quad \text{with } 1/K = \phi(-iv);$$

this defines distributions whenever $\phi(-iv)$ makes sense.

When the densities exist, we have

$$(34) \quad f_v(x) = Ke^{vx} f(x),$$

and the meaning may be clearer (because the notation is more familiar). We see immediately that the characteristic function of $F_v(x)$ is given by $\phi_v(u) = K\phi(u + iv)$ (where $\phi = \phi_0$ is the characteristic function of $F(x)$), and it follows that $\phi(u)$ is defined throughout the strip $-\alpha' < \mathscr{I}(u) < \alpha''$ (in other words, there is no further restriction due to singularities outside

---

† Another form of symmetric distribution is obtained by taking the average of the given distribution $F(x)$ and its reflection $1 - F(-x)$; this gives a distribution function $\frac{1}{2}[1 + F(x) - F(-x)]$ with characteristic function $\frac{1}{2}[\phi(u) + \phi(-u)]$. It is the distribution we obtain when we toss a coin before deciding whether to take $+|X|$ or $-|X|$.

‡ Symmetrized distributions are also considered in the statistical context. The prevision of $X_1 - X_2$ is zero, but the quadratic prevision and that of $|X_1 - X_2|$ constitute 'indices of variability' (the first one is clearly simply $\sigma(X)$ multiplied by $\sqrt{2}$); $\mathbf{P}(|X_1 - X_2|)$ turns out to be the concentration ratio multiplied by $2\mathbf{P}(X)$, which, for a given $\mathbf{P}(X)$, is the maximum possible value: cf. 6.6.3.

§ The term *conjugate* (cf. Keilson, 1965) is used in other contexts (cf., for example Chapter 12, 12.4.2). I therefore suggest the term given in the text. Feller (II, p. 410) refers to the property in question as the *translation principle* (but, as far as I know, does not give a name to such distributions).

the imaginary axis for u; in particular, if α' and α'' are both positive, $\phi(u)$ is analytic, and the minimum of the two bounds is the radius of convergence).

Expressed in a non-mathematical way, the conclusion is that $\phi(iv)$ exists (and hence so does $\phi(u)$ over the entire line $\mathcal{I}(u) = v$) if the twin distribution $F_v(x)$ exists, and that this happens if the tail of $F(x)$ on the positive semi-axis (for positive v; conversely for negative v) is thinner than the tail of the exponential distribution $f(x) = Ke^{-vx}$; α' and α'' are zero, infinite or finite, according as the tail (on the left or on the right) is fatter or thinner than every exponential, or comparable with an exponential, respectively.

6.11   SOME EXAMPLES OF CHARACTERISTIC FUNCTIONS

6.11.1. This is a convenient point at which to note and calculate explicitly the characteristic functions of some common distributions. In part, these will be cases of importance for applications; in part, they will be examples whose main purpose is to show how one can often avoid direct calculation with shrewd use of the properties of characteristic functions, keeping an eye on their interpretations. Until we actually illustrate these ideas with reference to the applications, the sense of this must inevitably remain some-what unclear, but just a brief mention of the nature of the applications will suffice to give the basic idea.

6.11.2. In the case of an event E (with probability $p = \mathbf{P}(E)$), or for a bet $s(E - p^*)$ on E (with gain s if E occurs, loss p^*s if it does not—the bet is fair if $p^* = p$), we have, respectively,

$$(35) \qquad \phi(u) = \mathbf{P}(e^{iuE}) = \tilde{p}e^{iu0} + pe^{iu1} = 1 + p(e^{iu} - 1),$$

$$(36) \qquad \phi(u) = \mathbf{P}(e^{ius(E - p^*)}) = e^{-iusp^*}\mathbf{P}(e^{iusE}) = e^{-iusp^*}[1 + p(e^{ius} - 1)]$$

$$= (1 - p)e^{-iusp^*} + pe^{ius(1 - p^*)}$$

(here, and elsewhere, it is sufficient to apply the property relating to additive and multiplicative constants: $\mathbf{P}(e^{iu(cX + k)}) = e^{iuk}\mathbf{P}(e^{icuX})$, in other words, change $\phi(u)$ into $e^{iuk}\phi(cu)$).

In the particular case where $s = 2$, $p = p^* = \frac{1}{2}$, we have a fair bet at the game of Heads and Tails with gains ± 1: the above reduces to

$$(37) \qquad \phi(u) = \tfrac{1}{2}(e^{iu} + e^{-iu}) = \cos u,$$

whereas

$$(37') \qquad \mathbf{P}(e^{iuE}) = \tfrac{1}{2}(1 + e^{iu}).$$

In the case of n independent tosses, the gain $2Y - n$, and the number of successes $Y = E_1 + E_2 + \ldots + E_n$, have characteristic functions

(38) $\phi(u) = \cos^n u,$

and

(38') $\phi(u) = [\tfrac{1}{2}(1 + e^{iu})]^n,$

respectively (the sum of independent random quantities = convolution = product of characteristic functions; in particular, this becomes a power if the distributions are identical).

Similarly, if p is now taken to be general (and we continue to assume stochastic independence), the number of successes, Y, has the characteristic function

(38") $\phi(u) = [1 + p(e^{iu} - 1)]^n .$

This is the so-called Bernoulli distribution: the limit-case, obtained by letting n tend to ∞ with the prevision $np = a$ held constant, is called the Poisson distribution. This gives $p_h = e^{-a}a^h/h!$, and hence its characteristic function is given by

(39) $$\phi(u) = \lim\left[1 + \frac{a}{n}(e^{iu} - 1) \right]^n = e^{a(e^{iu} - 1)}.$$

In all cases where the possible values are non-negative integers (like the above, relating to Y, 'the number of successes') the characteristic function is a polynomial (or a power series) in $t = e^{iu}$ with coefficients $p_h = \mathbf{P}(Y = h)$:

$$\phi(u) = \sum_h p_h t^h = \sum_h p_h e^{iuh}.$$

Knowing this, we could have obtained (38'), (38") and (39) directly from the knowledge of the p_h; conversely, to find the latter from the characteristic function we expand in powers of e^{iu}.

Let us have another look at three distributions of this type (having integer values): we consider the *uniform, geometric* and *logarithmic*.

For the *uniform* distribution ($p_h = 1/n$; $1 \leqslant h \leqslant n$) one has

$$\phi(u) = 1/n \sum_{h=1}^{n} e^{iuh} = \frac{e^{iu(n+1)} - e^{iu}}{n(e^{iu} - 1)}$$

(40)

$$= \frac{1}{n} e^{iu\frac{1}{2}(n+1)} \frac{e^{\frac{1}{2}iun} - e^{-\frac{1}{2}iun}}{e^{\frac{1}{2}iu} - e^{-\frac{1}{2}iu}} = e^{iu\frac{1}{2}(n+1)} \frac{\sin \frac{1}{2}nu}{n \sin \frac{1}{2}u}.$$

For the *geometric* distribution ($p_h = Kq^h$, $0 < q < 1$, $K = 1 - q$; $0 \leqslant h < \infty$) one has

(41) $\phi(u) = (1 - q) \sum_{h=0}^{\infty} q^h e^{iuh} = K/(1 - qe^{iu}) = (1 - q)/(1 - qe^{iu}).$

For the *logarithmic* distribution $(p_h = Kq^h/h, \ 0 < q < 1, \ K = -\log(1-q); \ 1 \leqslant h < \infty)$ one has

(42) $\quad \phi(u) = K \sum_{h=1}^{\infty} q^h e^{iuh}/h = -K\log(1-qe^{iu}) = \log(1-qe^{iu})/\log(1-q)$.

6.11.3. Let us now turn our attention to some continuous distributions: we shall present the density functions $f(x)$ and the characteristic functions $\phi(u)$, always expressed in the most convenient standard form (since any transformation from X to $cX + k$ can be easily dealt with).

The *normal* distribution (sometimes known as the 'error' distribution) will be well-known to everyone, although we have not yet dealt with it explicitly. We shall give a more extensive treatment in Chapter 7 (Section 7.6).

The *standardized*, or normalized, distribution, with prevision $= 0$ and variance $= 1$, has density and characteristic function given by

(43) $$f(x) = Ke^{-\frac{1}{2}x^2} \quad \left(K = \frac{1}{\sqrt{2\pi}}\right),\dagger$$

(44) $$\phi(u) = e^{-\frac{1}{2}u^2}.$$

Direct calculation is straightforward (if we operate in the complex field, using the substitution $y = x - iu$; a little less straightforward if we proceed differently, or if we do not assume the form we want).

A convolution of normal distributions is also a normal distribution (in other words, the sum of independent random quantities having normal distributions also has a normal distribution). We express this fact by saying that the normal distribution is *stable*. In fact, one has $e^{-\frac{1}{2}(au)^2}e^{-\frac{1}{2}(bu)^2} = e^{-\frac{1}{2}(cu)^2}$; in other words

(45) $\quad \phi(au)\phi(bu) = \phi(cu), \quad \text{where} \quad c = \sqrt{(a^2 + b^2)}.$

The scale parameters (a, b, c) are in fact the standard deviations, so it follows that the composition should take place according to Pythagoras' theorem (as is always the case for a finite sum). Observe also that

(46) $$\phi^n(u) = \phi(\sqrt{n} \cdot u)$$

and that

(46') $$\phi^t(u) = \phi(\sqrt{t} \cdot u)$$

for any positive real t (and not only for integer n). The fact that $\phi^t(u)$ is always a characteristic function means that the distribution is *infinitely divisible*

---

† That this is the value of the normalization constant is well-known from analysis. We shall, in any case, prove this (Chapter 7, 7.6.7) at a more appropriate and meaningful time.

(e.g. into $(\phi^{1/n}(u))^n)$. We have already encountered another example of an infinitely divisible distribution (although we did not point it out at the time), the Poisson, whose characteristic function (39), contains an arbitrary constant as exponent (in (39) it was denoted by a). We shall soon come across other examples; the general form of infinitely divisible distributions, and the sub-class of the stable distributions, will be given in Chapter 8, along with some of the important properties.

The *uniform* distribution (taken over $[-1, +1]$) has

(47) $$f(x) = \tfrac{1}{2}(|x| \leqslant 1),$$

(48) $$\phi(u) = \frac{\sin u}{u}.$$

The calculation is straightforward (it can also be obtained from the discrete case, (40), by letting $n \to \infty$ with $nu =$ constant, along with obvious changes of origin and scale).

For the sum of two (independent) random quantities having this distribution we obtain

(49) $$f(x) = \tfrac{1}{2}(1 - \tfrac{1}{2}|x|)(|x| \leqslant 2),$$

(50) $$\phi(u) = (\sin u)^2/u^2;$$

which is called the 'triangular distribution', on account of the form of the graph of the density function (this could be deduced from the definition without any need for calculations: it is the orthogonal projection onto the diagonal of a square of a mass uniformly distributed on it; cf. Figure 6.8).

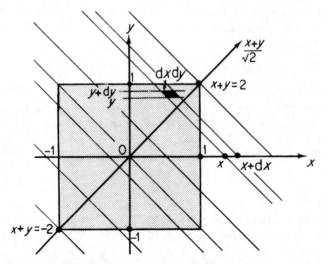

Figure 6.8   The convolution of uniform distributions

The characteristic function is positive: it follows immediately, therefore, that, conversely, there is a distribution (on $-\infty \leqslant x \leqslant +\infty$) with density and characteristic function given by

$$(51) \qquad f(x) = \frac{1}{\pi}\frac{(\sin x)^2}{x^2},$$

$$(52) \qquad \phi(u) = (1 - \tfrac{1}{2}|u|)(|u| \leqslant 2).$$

This distribution is not, in itself, very interesting. It is, however, of great importance in that one can immediately deduce from it conclusions of some generality. By means of mixtures of triangular distributions (on different ranges) we can obtain any distribution whose density has a polygonal graph (symmetric with respect to the origin, decreasing and concave upwards on either side of the origin). In the limit, we can obtain any curve with these kind of properties. By inversion, any function having such behaviour *is a characteristic function*: this is Pólya's criterion. The fact that in this way we can obtain characteristic functions which are zero outside a finite interval is also of some importance.

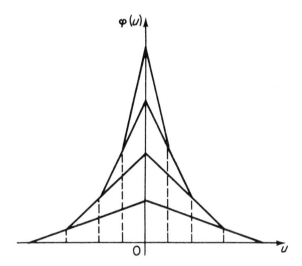

Figure 6.9 Characteristic functions constructed on the basis of Pólya's argument

In a similar way, we obtain $\phi(u) = (\sin u)^n / u^n$ as the characteristic function of the sum of n independent random quantities which are uniformly distributed in $[-1, +1]$; this corresponds to the density of the projection onto the diagonal of an n-dimensional cube of a mass uniformly distributed on it, and is represented by polynomials of degree $n - 1$ which vary on each of the n intervals of length 2 into which the interval $[-n, +n]$ is divided by the projections of the vertices of the cube. Think of the ordinary cube, $n = 3$ (the areas of the sections are first triangular, then hexagonal, then triangular again). The inversion (as for $n = 2$) can be made for any even n (since then the characteristic function has to be positive).

For the *exponential distribution*,

(53) $$f(x) = e^{-x} \ (x \geqslant 0)$$

and

(54) $$\phi(u) = 1/(1 - iu);$$

this is a special case ($t = 1$) of the *gamma* distribution, defined by

(55) $$f(x) = Kx^{t-1}e^{-x} \ (x \geqslant 0) \quad \text{with} \quad K = 1/\Gamma(t),$$

$$\Gamma(t) = \int_0^\infty x^{t-1}e^{-x}\,\mathrm{d}x = (t-1)! \quad \text{for integer } t, \ t > 0,$$

(56) $$\phi(u) = 1/(1 - iu)^t.$$

The fact that t appears as an exponent in $\phi(u)$ (or, more precisely, in $\phi'(u)$) implies that these distributions have the property of being infinitely divisible.

By symmetrization of the gamma distribution, we obtain distributions whose characteristic functions are given by

(57) $$\phi(u) = [1/(1 - iu)^t][1/(1 - iu)^t]^* = 1/(1 + u^2)^t$$

(and these are also infinitely divisible). In particular, for $t = 1$, we have the two-sided exponential distribution, with

(58) $$f(x) = \tfrac{1}{2}e^{-|x|},$$

(59) $$\phi(u) = 1/(1 + u^2).$$

By inversion, we obtain

(60) $$f(x) = 1/[\pi(1 + x^2)],$$

(61) $$\phi(u) = e^{-|u|};$$

the Cauchy distribution. This is infinitely divisible (for $t > 0$, $(e^{-|u|})^t = e^{-|tu|}$ is the characteristic function of $f(x) = K/(1 + (x/t)^2)$), and, since $f(x)$ remains invariant (apart from changes of scale), the distribution is also *stable* (like the normal). Its invariance is infinite, as can be seen directly (from $f(x)$ being of second-order smallness) or from the irregularity of $\phi(u)$ at the origin.

6.11.4. Knowing the characteristic functions of certain distributions enables us—using products, powers, conjugation, linear combinations, limits, etc.—to obtain innumerable others, as required for various applications, and corresponding to distributions whose densities cannot in many cases be expressed explicitly.

Let us examine some of the more interesting examples of mixtures; those given by the sum of N independent, identically distributed random quantities X_h when N itself is also random. If at each step there is a probability p of stopping and $q = 1 - p$ of continuing, the N has a geometric distribution; that is,

$$p_n = \mathbf{P}(N = n) = Kq^n \quad (K = (1 - q)).$$

If it turned out that $N = n$, the characteristic function of the sum would be $\chi^n(u)$, where $\chi(u)$ denotes the characteristic function of each X_h; the characteristic function of the unconditional sum is hence given by the mixture

$$(62) \qquad \phi(u) = \sum_{n=0}^{\infty} Kq^n\chi^n(u) = K/[1 - q\chi(u)] = (1 - q)/[1 - q\chi(u)].$$

Formally, it is sufficient to substitute in (41), replacing the characteristic function e^{iu} of each of the summands '1' by the characteristic function $\chi(u)$ of X_h. Following the same rule in the general case, one obtains.

$$(63) \qquad\qquad\qquad \phi(u) = \sum_n p_n\chi^n(u),$$

and, in the particular cases of N having the Bernoulli or Poisson distribution, we have (cf. (38″) and (39))

$$(64) \qquad\qquad\qquad \phi(u) = [1 + p(\chi(u) - 1)]^m$$

and

$$(65) \qquad\qquad\qquad \phi(u) = e^{a[\chi(u) - 1]},$$

respectively. In (64) we used m in the exponent rather than n (which is now used to denote particular values of N); the interpretation (for example in the case of a game) is as follows: an individual has the right to m trials, each having probability p of success; he then has n successes ($0 \leqslant n \leqslant m$), and receives a random prize X_h for each success. The Poisson case can, for the present, be regarded as a limit-case (but will be seen to have a much more interesting interpretation when viewed as a 'random process'; see Chapter 8).

When a characteristic function $\chi(u)$ is infinitely divisible, that is, $\chi^t(u)$ is a characteristic function for any $t > 0$ (not only for t integer), one need not limit oneself to mixtures involving integer powers (61), but can also consider sums of the form

$$(66) \qquad\qquad \phi(u) = \sum_n p_n\chi^{t_n}(u), \quad \text{for any } t_n > 0,$$

or even

(67) $\phi(u) = \int_0^\infty p(t)\chi^t(u)\,dt$ (with $p(t) \geqslant 0$, $\int_0^\infty p(t)\,dt = 1$).

6.11.5. If we take a random quantity X and add on a random quantity Δ, which is small and has appropriate regularity properties, then $X + \Delta$ will differ only slightly from X (it is as though we intentionally measure X with a small error), but will enjoy the regularity properties possessed by Δ (and perhaps some others as well). As we shall see, this can turn out to be very useful.

For example, suppose Δ is chosen to have a uniform distribution between $\pm \delta$, with density $1/2\delta$. In this case, $X + \Delta$ will always have a density $\leqslant 1/2\delta$, whatever the distribution of X (cf. 6.9.8). If we take a triangular distribution for Δ ($f(x) = K(1 - |x|/\delta): K = 1/\delta$), $X + \Delta$ will have a density which is $\leqslant 1/\delta$ everywhere, and the derivatives of the density will also be $\leqslant 1/\delta^2$ (in absolute value). Similar bounds obtain when Δ is taken to be normal ($m = 0$, $\sigma = \delta$).

In terms of characteristic functions, this results in $\phi(u)$, the characteristic function of X, being multiplied by the characteristic function of Δ; in the cases mentioned above, we consider

$$\phi(u)(\sin \delta u)/\delta u, \quad \phi(u)(\sin \delta u)^2/(\delta u)^2, \quad \phi(u)e^{-\frac{1}{2}(\delta u)^2}.$$

This device often enables us to reduce problems posed in terms of general distributions to a framework in which suitable regularity conditions are obeyed.

In particular, we observe that if Δ is assumed to have the first form mentioned above (uniform over $\pm \delta$) then $f_\delta(x)$, the density of $X + \Delta$, is precisely the *average density* of X in the interval $x \pm \delta$; in other words,

$$f_\delta(x) = [F(x + \delta) - F(x - \delta)]/2\delta.\dagger$$

Informally, this formula says the following: the probability of $X + \Delta$ lying between $x \pm \frac{1}{2}dx$ is the probability (of the necessary condition) that X lies inside $x \pm \delta$, since, conditional on $X = x_0$ (x_0 any point in $x \pm \delta$) the density of $X + \Delta$ at x is always the same, $1/2\delta$. More formally, considering the convolution for $X + \Delta$ (see 6.9.6), we have

$$f_\delta(z) = \int f(x).(1/2\delta)(|z - x| \leqslant \delta)\,dx$$

$$= (1/2\delta)\int_{z-\delta}^{z+\delta} f(x)\,dx = 1/2\delta[F(z + \delta) - F(z - \delta)]$$

---

† The fact that $f_\delta(x)$ is discontinuous and undefined at those points (at most a countable number) at distance δ (to the left or right) from discontinuity points of $F(x)$ (points with concentrated mass for X) is irrelevant.

(clearly, we could have considered the general case straightaway, by writing $dF(x)$ instead of $f(x)\,dx$). This formula may be used to obtain $F(x'') - F(x')$ for a preassigned interval (x', x''). In fact, it suffices to put $z = \frac{1}{2}(x' + x'')$, $\delta = \frac{1}{2}(x'' - x')$. In particular, to obtain $F(x) - F(0)$, it is enough to put $z = \delta = \frac{1}{2}x$. We have, therefore,

$$F(x'') - F(x') = (x'' - x')f_{\frac{1}{2}(x'' - x')}(\tfrac{1}{2}(x' + x'')), \quad F(x) - F(0) = xf_{\frac{1}{2}x}(\tfrac{1}{2}x).$$

The characteristic function of $f_\delta(x)$ is given by $\phi(u)(\sin \delta u)/\delta u$, so that we obtain the following inversion formula for passing from the characteristic function $\phi(u)$ to the distribution function $F(x)$:

$$(68) \quad F(x) - F(0) = \frac{x}{2\pi} \int_{-\infty}^{*+\infty} e^{-\frac{1}{2}iux}\phi(u) \frac{\sin \frac{1}{2}ux}{\frac{1}{2}ux}\,du = \int_{-\infty}^{*+\infty} \phi(u) \frac{e^{iux} - 1}{iu}\,du.$$

This (or one of the alternative forms) is the standard result, usually proved on the basis of the Dirichlet integral; this is a more laborious method, and, in the words of Feller (II, p. 484), 'detracts from the logical structure of the theory'.

6.11.6. A more intuitive and expressive way of interpretating and explaining this is as follows: we think of the characteristic function—and let us take the simplest case, $\phi(u) = \sum p_h e^{iux_h}$—as a mixture of oscillations of various frequencies x_h and intensities p_h (the variable u being thought of as time). The formula for determining the components of the mixture given $\phi(u)$ (or, if we think in terms of light, for separating it into its monochromatic components), corresponds to a device capable of filtering out lines or bands. In order to discover whether a component of frequency x_0 exists, and in this case to isolate it and determine its intensity p_0, we must have a monochromatic filter. This is precisely what is achieved by the operation of computing the mean value (over a long period) of $\phi(u)$ multiplied by e^{-iux_0}; in more precise terms, the operation of computing

$$(69) \quad \lim_{a \to \infty} \frac{1}{2a} \int_{-a}^{a} \phi(u)e^{-iux_0}\,du = \lim_{a \to \infty} \sum_h p_h \frac{1}{2a} \int e^{iu(x_h - x_0)}\,du.$$

We see immediately, however, that if $x_0 \neq x_h$ the mean value (over any period, and hence asymptotically, on any very long interval $[-a, a]$) is zero. Only if x_0 coincides with one of the x_h does the integrand reduce to $e^{iu0} \equiv 1$, the mean value to 1, and the result to p_0 (that is, the p_h for which x_h is our x_0).

The other operations can be regarded as band filters, used to obtain the sum of the p_h corresponding to frequencies x_h contained in some given interval $[x', x'']$, etc.

6.12   SOME REMARKS CONCERNING THE DIVISIBILITY OF DISTRIBUTIONS

A distribution obtained by the convolution of others is said to be divisible into the latter (its factors); G is a factor of F if we can write $F = G * H$ (for some suitable H). In terms of characteristic functions, this means that $\phi(u)$ can be expressed as a product of functions $\phi_h(u)$, each of which is also a characteristic function.

We have already seen the example of infinitely divisible distributions that can be defined (in the simplest, but also the most meaningful way) as those for which $\phi(u) = [\phi(u)^{1/n}]^n$ for any n (that is, $\phi(u)^{1/n}$ is a characteristic function for every n). Although we shall have no reason to give a systematic treatment of this topic, we shall, from time to time, come across problems where divisibility enters in. For this reason, it is appropriate at this stage to briefly mention it, for the sole purpose of warning against the errors that can arise if one proceeds by analogy with factorization as it occurs in arithmetic or algebra. Anyone interested in pursuing the subject more deeply should read P. Lévy (1937), pp. 190–195, and the references cited there, or the recent survey by M. Fisz, in *Ann. Math. Stat.* (1962), 68–84; the latter contains a bibliography.

There exist distributions which are not divisible: for example, it is clear that a distribution with only 4 possible values, $0, a, b$ and c (in increasing order) cannot be divisible unless $c = a + b$ (in which case we must have $Z = X + Y$, with 0 and a the possible values for X, 0 and b the possible values for Y); given this fact, we see that the 4 probabilities p_0, p_a, p_b and p_c cannot be chosen arbitrarily (subject only to their sum $= 1$), because they must be of the form $p_0 = (1 - \alpha)(1 - \beta)$, $p_a = \alpha(1 - \beta)$, $p_b = (1 - \alpha)\beta$ and $p_c = \alpha\beta$ (where $\alpha = \mathbf{P}(X = a)$ and $\beta = \mathbf{P}(X = b)$; an extra condition must hold, leaving two degrees of freedom instead of three).

In general, there are no uniqueness type properties for factorizations; a distribution always admits a decomposition into an infinitely divisible distribution and indivisible ones (Khintchin's theorem); there may be infinitely many of the latter, or none; or it may be that the former is not present. We can also have, in general, various, different factorizations, combining different factors, without even a sharp dividing line between the factor which is infinitely divisible and the others. In fact, it can happen that an infinitely divisible distribution turns out to be a product of indivisible factors when factorized in a particular way.

There are, however, important cases in which the factorization is unique, and, in fact, reduces to the *trivial* factorization—the decomposition into factors $[\phi(u)]^{t_h}$ (with $t_h > 0$, $\sum t_h = 1$) with $\phi(u)$ infinitely divisible. This is the case for the normal distribution (so that, if $X + Y = Z$ has a normal distribution and X and Y are independent, than X and Y both have normal distributions; Cramèr's theorem), and also for the Poisson distribution (same result; Raikov's theorem).

Finally, if we turn to the question of factorizations of infinitely divisible distributions *which remain in the ambit of infinitely divisible distributions* (i.e. we require that the factors also be such), we can say straightaway that in this case the answer is straightforward and complete. We shall deal with this in Chapter 8, 8.4 (at the present time we do not have at our disposal the concepts required for taking this any further).

It is instructive to point out the following rather surprising fact: given a factorization $\phi(u) = \phi_1(u)\phi_2(u)$, this does not imply that if one factor is kept fixed the other is uniquely determined (in other words, we can also have $\phi(u) = \phi_1(u)\phi_3(u)$, with $\phi_3 \neq \phi_2$). Clearly, we can only have $\phi_3(u) \neq \phi_2(u)$ when $\phi_1(u) = 0$; but we have already seen that a characteristic function can be zero (like the triangular case, $1 - \frac{1}{2}|u|$; cf. 6.11.3) outside an interval (in this example, for $|u| \geq 2$). In fact, the counter example given by Khintchin consists precisely in taking ϕ_1 to be such a triangular function; for ϕ_2 and ϕ_3, one can take, for example, concave polygonal functions (cf. Pólya's theorem) which are the same in $(-2 \leq u \leq 2)$ but differ outside.

Index

Reference is made to the important ideas and terms, and to the authors cited. A detailed bibliography is *not* provided; the interested reader is referred to those given in the following texts:

W. Feller, *An Introduction to Probability Theory and its Applications*, Vol. 1, 1958, and Vol. 2, 1966, Wiley, New York;

B. de Finetti, *Probability, Induction and Statistics*, 1972, Wiley, New York;

H. E. Kyburg and H. E. Smokler, *Studies in Subjective Probability*, 1964, Wiley, New York.